Digital Resources for Students

This new textbook includes six (6) months of access to the Clarke Publishing ***Learning Management System (LMS)***, which provides a range of digital learning resources. These may include video lectures from trusted sources, step-by-step tutorials on digital security, programming concepts, and online safety, as well as guidance on identifying digital threats and other essential computer skills.

Students are provided with a safe and secure learning environment where they can engage with peers and instructors and connect with learners around the world. Through our LMS, they can do the following:

1. Complete assigned quizzes and hands on projects
2. Access lab files and images
3. View Grades
4. Participate in course discussions
5. Communicate with Instructors, and access a wide range of additional learning tools

How to Use Access Students Online Resources?

Now that you have purchased this textbook, please request your **Access Code** from your instructor. Once you receive it, follow the steps below to register and gain access to our LMS eLearning Platform.

1. Go to ***https://www.clarkepublish.com***
2. Click Learning Platform (on the top right of the page)
3. Click Register (if your instructor has not already done so)
4. Enter your Access Code (**the Access Code can only be used once**)
5. Follow the on-screen instructions to create a login name and password

Use the login username and password you created during registration to start using the LMS online resource that accompany your textbook

This subscription provides access for four (4) months during your course and an additional two (2) months after course completion upon activation. Subscriptions are non-transferable. If your access code does not function, you may purchase a subscription directly from the LMS login page to access the course's online resources.

For technical support, go to ***https://www.clarkepublish.com/contact***

Modern Computer Applications for Colleges and Universities

1st Edition

Emmanuel Clarke, PhD

Oludele awodele, PhD

Seun Ebiesuwa, PhD

United States of America

Executives

Emmanuel Clarke, PhD., Chairman & CEO

Calvin Harris, CFO

Wante Mekey Clarke, COO

Regina Emma Gaye, VP Marketing

Nigeria

Executives

Dogbahgen Alphonso Yarseah, COO

Liberia

Executives

Stanley Gaye, COO

France Addy Williams, Senior Sales Executive

Emmanuel Jerry Jr., Business & Partner Development Executive

Copyright Statement

Modern Computer Applications for Colleges and Universities

COPYRIGHT © 2027 Clarke Publishing Communications & BPO, Inc.

PCN: 011-900-381

ALL RIGHTS RESERVED. No part of this work covered by the copyright herein may be reproduced or distributed in any form or by any means, except as permitted by U.S. copyright law, Nigeria copyright law, and Liberia copyright law, without the prior written permission of the copyright owner.

Unless otherwise noted, all content is Copyright © Clarke Publishing Communications & BPO, Inc.

Microsoft and the Office logo are either registered trademarks or trademarks of Microsoft Corporation in the United States and/or other countries. This product is an independent publication and is neither affiliated with, nor authorized, sponsored, or approved by, Microsoft Corporation.

Some of the product names and company names used in this book have been used for identification purposes only and may be trademarks or registered trademarks of Microsoft Corporation in the United States and/or other countries.

For product information and technology assistance, contact us at Clarke Publishing Customer & Sales Support, +336-602-1769 or support@clarkepublish.com.

For permission to use material from this text or product, submit all requests online at https://www.clarkepublish.com/contact.

Student Edition ISBN: 979-8-9953639-1-0

Looseleaf is available as part of a digital bundle per request from institutions

This book was printed in the United States of America on acid-free paper.

USA

8025 North Point Boulevard

Winston Salem, NC 27104 Winston Salem, NC 27104

USA
Website: https://www.clarkepublish.com

Email: info@clarkepublish.com

Locations

Nigeria

5 Tafawa Balewa Square Opposite Ministry of Defence, Suite 16A East Pavilion

Moloney Lagos, 100001

Nigeria, West Africa

Website: https://www.clarkepublish.com

Email: info@clarkepublish.com

Clarke Publishing Communications & BPO, Inc. is a leading publisher of trade books, and provider of customized learning solutions with employees currently residing in three countries and sales in 11 countries around the world. If you have question you can find your local customer & sales support representative at support@clarkepublish.com.

Online Learning Platform

To access Clarke Publishing Learning Platforms and gain access to textbook resources, including textbook contents, files and photos, hands on labs, test banks (instructors only) and other learning solutions, visit https://learning.clarkepublish.com to register and access your online learning solutions. If the link does not work if clicked, copy and paste it.

For Teachers & Students

With the Learning Platform, instructors have access to teaching resources including test banks, and lab solutions files etc. On the platform, instructors can create chapter tests, lab, grade tests and labs. Students can take tests, access lab files and textbooks resources.

Notice to the Reader

Publisher does not warrant or guarantee any of the products described herein or perform any independent analysis in connection with any of the product information contained herein. Publisher does not assume, and expressly disclaims, any obligation to obtain and include information other than that provided to it by the manufacturer. The reader is expressly warned to consider and adopt all safety precautions that might be indicated by the activities described herein and to avoid all potential hazards. By following the instructions contained herein, the reader willingly assumes all risks in connection with such instructions. The publisher makes no representations or warranties of any kind, including but not limited to, the warranties of fitness for particular purpose or merchantability, nor are any such representations implied with respect to the material set forth herein, and the publisher takes no responsibility with respect to such material. The publisher shall not be liable for any special, consequential, or exemplary damages resulting, in whole or part, from the readers' use of, or reliance upon, this material.

***** Please note that the names of the files, folders, images, tables in the figures within this lab manual/textbook may be different from the actual files or folders in the actual lab. The reason for the slack difference is due to textbook updates.**

About the Authors

EMMANUEL CLARKE, PhD, is a distinguished professor with over two decades of extensive teaching experience across various higher education institutions. His academic career includes positions at Mercer County Community College in Trenton, New Jersey; Burlington County College in Burlington, New Jersey; United Methodist University in Liberia; the University of Liberia, where he serves as a lecturer; and Forsyth Technical College in Winston-Salem, North Carolina. Dr. Clarke's diverse teaching portfolio reflects his commitment to fostering educational excellence in both local and international contexts.

Dr. Clarke has contributed significantly to educational literature, authoring and co-authoring numerous influential textbooks. His publications encompass a broad spectrum of topics, primarily focusing on technology and computer education, including Computer Concepts for Liberian Colleges and Universities, Computer Concepts for Liberian Schools, Jr. & Sr. High Edition, Computer Concepts for Liberian Schools, Sr. High Lab Manual, Computer Concepts for Liberian Schools, Jr. High Lab Manual, Growing Up with Technology, Primary First Edition, Elementary Computer Lab Manual, Project Management for a Modern Liberia.

His work has been instrumental in shaping computer literacy and technology education in Liberia and beyond. Dr. Clarke's expertise extends beyond academia into the corporate and publishing sectors, where he has held roles such as editor, author, and Chief Executive Officer.

He holds a Doctorate in Computer Science from The New Jersey Institute of Technology. He also holds professional certificates in Project Management, CompTIA Security+, SSCP, Oracle Certified Professional, Certified Instructional Designer, to name a few.

Dr. Clarke is married and has four children. He is also passionate about music, engaging in songwriting and scriptwriting. Currently, he resides in Winston-Salem, North Carolina.

AWODELE, Oludele, PhD, is a Professor of Computer Science and Artificial Intelligence in the School of Computing, Babcock University, Ilisan-Remo. He was Head of the Department of Computer Science (2009-2016), Dean, School of Computing and Engineering Sciences (2016-2020), and currently the Director of Academic Planning, Babcock University, Nigeria. His current research is on Artificial Intelligence, Blockchain Technology, Data Communications and Computer Security.

About the Authors

He has successfully supervised 22 PhDs and currently mentoring 5 PhD. Scholars in different areas of Artificial Intelligence, Computer Science and Block chain Technology. He is a Fellow, Nigeria Computer Society (NCS), he has published well over 150 papers both in International and Local Journals and attended several academic conferences across the globe.

He was the Editor-in-Chief of the following Nigeria Computer Society Journals: International Journal of Information Security Privacy & Digital Forensics and the Journal of Computer Science and Its Applications (2017-2021). Senior Member of the Institute of Electrical and Electronics Engineers, Member Governing Council, Association of Applied Information, Management Professionals, Member, National Executive Committee of the Nigeria Computer Society (2017- 2025), Member Governing Council of the Computer Professional Registration Council of Nigeria (2019- 2025) and National President of Practical Artificial Intelligence Development Foundation (PAIDeF) also known as Nigeria Foundation for Artificial Intelligence, an interest group of the Nigeria Computer Society (2024-2025).

In recognition of his hard work, diligence, and immeasurable contribution to the body of knowledge and societal development, he has been conferred with numerous distinguished awards.

He is happily married and blessed with lovely children.

EBIESUWA Seun (Ph.D.) is a Senior Lecturer and the current Head of the Department of Computer Science in the School of Computing, Babcock University, Ilishan-Remo, Ogun State, Nigeria. He served as the Director for Students Industrial Work Experience Scheme (SIWES) at Babcock University from 2021 to 2024. His research efforts revolve around Data Science, Machine Learning, and Information Systems.

Dr. Seun Ebiesuwa is a multiple award winner of several academic gold medals, he has successfully supervised a number of Postgraduate students in the Computer Science sub-fields of Data Science, Machine Learning, and Information Systems. He is a member of the following Professional bodies: Nigeria Computer Society (NCS), Computer Professionals Registration Council of Nigeria (CPN), Nigeria Mathematical Society (NMS), and the Institute of Electrical and Electronics Engineers (IEEE).

Dr. Ebiesuwa has published over fifty internationally peer-reviewed scholarly articles in the field of Computer Science.

He is happily married and blessed with adorable children.

New to Edition and Organization of Text

In This Edition

Clarke Publishing Communications and BPO, Inc. is proud to bring you this all-new edition of Microsoft Office Suite. This edition was designed to provide a robust learning experience that is not dependent upon a specific version of Office.

Microsoft supports several versions of Office.

Office 365: A cloud-based subscription service that delivers Microsoft's most up-to-date, feature. **Office 2021 and 2024:** Microsoft's "on-premises" version of the Office apps, available for both PCs and Macs. **Office Online:** A free, simplified version of Office web applications (Word, Excel, PowerPoint, and Access).

Organization of the Text

This manual provides a comprehensive introduction to the core functionalities of Microsoft Office applications, tailored specifically for beginners. It serves as a practical guide, offering step-by-step instructions to facilitate effective learning and skill development in using productivity software such as Word, Excel, PowerPoint, and Access.

Organization and Structure

The content is systematically organized to enhance understanding and retention. Key features include:

- **Color-Coded Tasks and Concepts:** To aid navigation and comprehension, tasks and concepts are distinguished by specific colors:
- Green: Introduces new concepts, providing foundational knowledge necessary for subsequent activities.
- Orange: Highlights new activities, encouraging practical application for learning.
- Blue: Contains explicit instructions that guide students through each task, ensuring clarity and ease of execution.

The activities within this manual are designed to be straightforward and can typically be completed within a single session. The sequential nature of the tasks is intentional, with each activity building upon the skills acquired in the previous one. This progressive approach ensures that students develop a solid understanding of each application's functionalities before advancing to more complex tasks. The main instructions are in **Bold** letters.

Students are encouraged to complete each activity thoroughly before moving on to the next. This methodical progression fosters confidence and competence in using Microsoft Office tools effectively, laying a strong foundation for more advanced learning in the future.

Acknowledgments

Dr. Emmanuel Clarke

To my wife and children, thanks for making the ultimate sacrifice for time not spent with you all during the writing of this book. Thanks for the support. I love you all.

Dr. Oludele AWODELE

This is the Lord's doing and It's marvelous in our eyes! I thank all my professional colleagues and students for the opportunity of knowledge exchange. Thanks to my beloved wife and God given Children for the shower of love and support all the way.

Dr. Seun Ebiesuwa

To God Almighty be the glory for providing the inspiration for this book. Special appreciation to my wife and children for their support throughout this endeavor. Thanks to my friends and well wishers for the moral support given.

The Clarke Publishing Communications Team

We extend our sincere appreciation to our editors for their invaluable expertise in the development of this publication. Our gratitude also goes to the dedicated team at Clarke Publishing, including Ivor Augustus Mitchell (Mr. Graphics), Emmaree Thomas, Emmanuella Luah Clarke, Albert Kazou, Eukey Smallwood, and Joseph Pambu, whose unwavering support was instrumental in bringing this project to fruition.

Lest we forget, we would like to thank the Technology Research Team for their contributions to the Technology Timeline and other emerging technology.

Clarke Publishing Communications & BPO, Inc is also located in Monrovia Liberia. Please contact us at info@clarkepublish.com for our more information. If you need technical support, please contact us at support@clarkepublish.com.

Table of Contents

CHAPTER 1

CHAPTER 1 HIGHLIGHT

Microsoft Office Suite Overview: Getting Started with Computer and its Applications

After completing this chapter, you will be able to:

1. Define Computer and Computer Literacy
2. Define and Understand the various Office Applications
3. Define and Identify Microsoft Windows Operating System and it various Version
4. Differentiate Computers by the Types
5. Work with Computer Application, Navigation the Screen with Mouse
6. Set Up a Computer for Use by Creating a User Account
7. Manage Files, Folders and Documents
8. Differentiate Networks, Internet Vs. Intranet.

CHAPTER 1 | Getting Started With Office

Computer is an electronic device, operating under the control of instructions stored in its own memory, that can accept data, process that data according to specified rules, produce results, and store the results for future use. Computer uses application software to perform all types of tasks for users. You will learn about computer applications in this lab manual. The computer is currently impacting the world in so many ways whether you like it or not. It does not matter which country you live in across the continent of Africa. Computer is becoming ubiquitous across Africa, and around the world. You see, computer knowledge is very critical in acquiring so many things such as, an education, starting a business, acquiring a good job upon graduation, contributing to the development of your country or Africa, being a great contributor to global communications, and effectively participating in the ever expending and interdependent international community.

It doesn't matter if you live in Liberia, Nigeria, Ghana, Kenya, or Sierra Leone, throughout a typical day, you might use a computer to complete a school assignment, compare prices on a new cellphone or a used vehicle, exchange text messages with friends, and listen to music as you walk to class or while sitting in a Keke (tricycle taxicab) while on your way home. As you further your college education and gain computer skills and knowledge, you will enhance these experiences and prepare yourself for new ones. Learning and applying computer knowledge makes you computer literate, an essential requirement for a career in today's economy.

Computer Literacy Fundamentals

For you as an African student to achieve success in your career and personal life, it is essential to be comfortable and proficient in using a computer, regardless of your background or location. There a millions of people across Africa and around the world that are afraid to use a computer. Many of them think if they push a key on the keyboard too hard, they would break something inside of the computer. But this is really not true. While we may have given a technical definition of a computer, by a simple definition and at the most basic level, a computer is an electronic device that receives data (**Input**), processes and stores the data, and then produces a result (**Output**). **Data** is a raw fact, such as text or numbers that has not been processed.

The modern computer which are now becoming common across Africa is over 80 years old. Prior to discussing Microsoft Productivity Applications/Apps and the many social influence of computers, it is pertinent to examine the four major applications/apps that are widely used around the world. These applications have become the benchmark for measuring a person's computer literacy. The computer and its applications that we see nowadays, have come a long way. This lab manual with its many hands on activities will take you from being a novice to being a confident user of computer and its applications.

CHAPTER 1 | Getting Started With Office

In the contemporary academic environment, proficiency in modern computer applications such as Microsoft Word, Access, Excel, and PowerPoint as seen in **Figure 1-1** has become essential for college students. These tools are not only fundamental for completing coursework efficiently but also serve as vital skills for future professional endeavors. Understanding the importance of these applications can significantly enhance a student's academic performance and career readiness. Currently, many companies in Liberia, Nigeria, Ghana, Sierra Leon, and in other countries across Africa are requiring employees and prospective employees to have a fundamental computer and application knowledge prior to employment.

Figure 1-1 Microsoft Applications

Microsoft Word is a widely used word processing application that enables students and professionals to create, edit, and format documents with ease. Mastery of Word allows students to produce well-structured essays, reports, and research papers. It also provides skills in using templates, styles, and referencing tools, which are crucial for maintaining academic integrity and professionalism. Additionally, Word's collaboration features facilitate group projects and peer reviews, fostering teamwork and communication skills.

Microsoft Excel is a powerful spreadsheet application used for data analysis, visualization, and financial calculations. Proficiency in Excel enables students and professionals to handle complex data sets, perform statistical analysis, and create charts and graphs for presentations. These capabilities are essential for coursework in mathematics, economics, engineering, and sciences. Moreover, Excel skills are often prerequisites for internships and job roles in data analysis, accounting, and project management.

Microsoft PowerPoint is a presentation software that allows students and professionals to communicate ideas effectively through visual aids. Mastery of PowerPoint helps students develop skills in designing engaging presentations, which are crucial for class projects, seminars, and conferences. The ability to convey information clearly and persuasively enhances a student's confidence and professional image. Additionally, PowerPoint skills are transferable to various industries, where presentation and communication are key components of success.

Microsoft Access is a database management system that helps students, working professionals, to organize, store, and analyse large amounts of data. Learning Access is particularly beneficial for students involved in research, data analysis, or projects requiring data

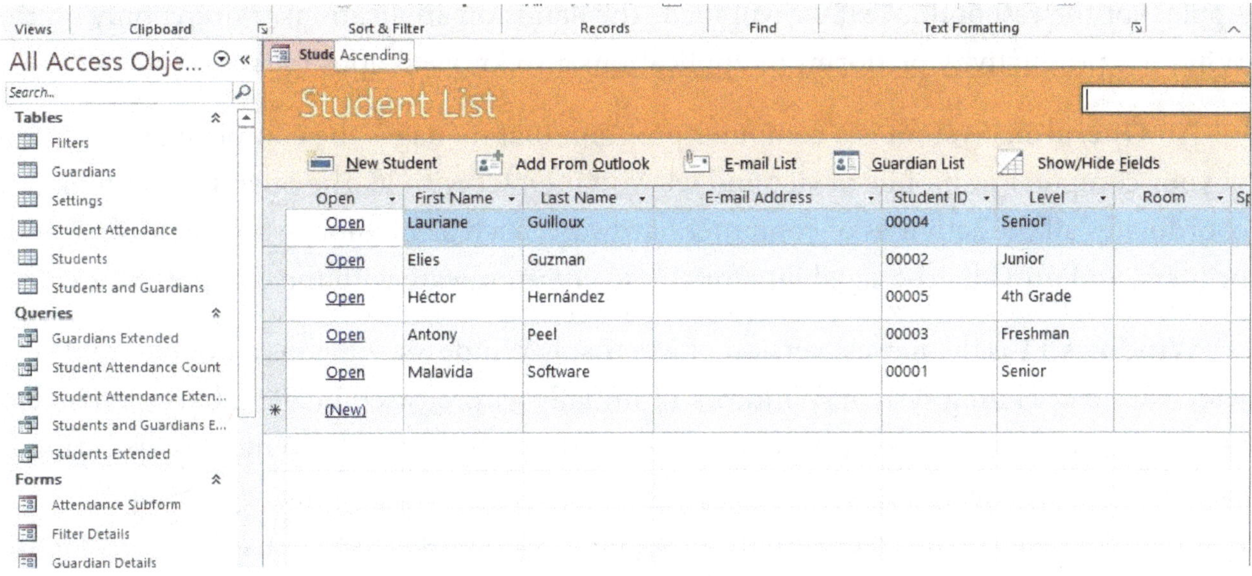

Figure 1-2 Microsoft Access

management. It enhances logical thinking and problem-solving abilities by teaching students how to design databases, create queries, and generate reports. These skills are highly valued in fields such as business, information technology, and social sciences. **Figure 1-2** above shows a Student List screen of a Microsoft Access database.

It doesn't matter whether you live in Abuja or Monrovia, acquiring proficiency in Microsoft Office Suite and other specialized computer applications equips college students and working professionals with essential digital literacy skills. These tools streamline academic and professional tasks, improve productivity, and foster critical thinking. Furthermore, they prepare students and other professionals for the technological demands of the modern workplace, making them more competitive in the job market. As technology continues to evolve, staying updated with these applications ensures that students and professionals will remain adaptable and capable of leveraging digital tools for success in their academic and professional lives.

CHAPTER 1 | Getting Started With Windows 11

What Is Windows 11?

While this textbook is dedicated to Microsoft Office Application/App, we would like to do a quick overview of **Microsoft Windows Operating System** which is often referred to as **Windows**. Currently in the world of computer, there are two types of operating systems, Microsoft Windows Operating System and MacIntosh or Mac Operating from Apple. For the rest of this text we will focus the hands on applications training only on the Windows productivity platforms or applications also know as apps for short.

An **Operating System** is a computer program that manages the complete operation of your computer or mobile device and lets you interact with it. The operating system coordinates all the activities of computer hardware, such as memory, storage devices, and printers, and provides the capability for you to communicate with the computer.

Windows 11 is the newest version of Microsoft Windows, which is a popular and widely used operating system. Windows 11 initially was released in 2021, but Microsoft releases updates to the operating system several times per year. The Windows operating system simplifies the process of working with documents and apps by organizing the manner in which you interact with the computer. Windows is used to start apps. An **App**

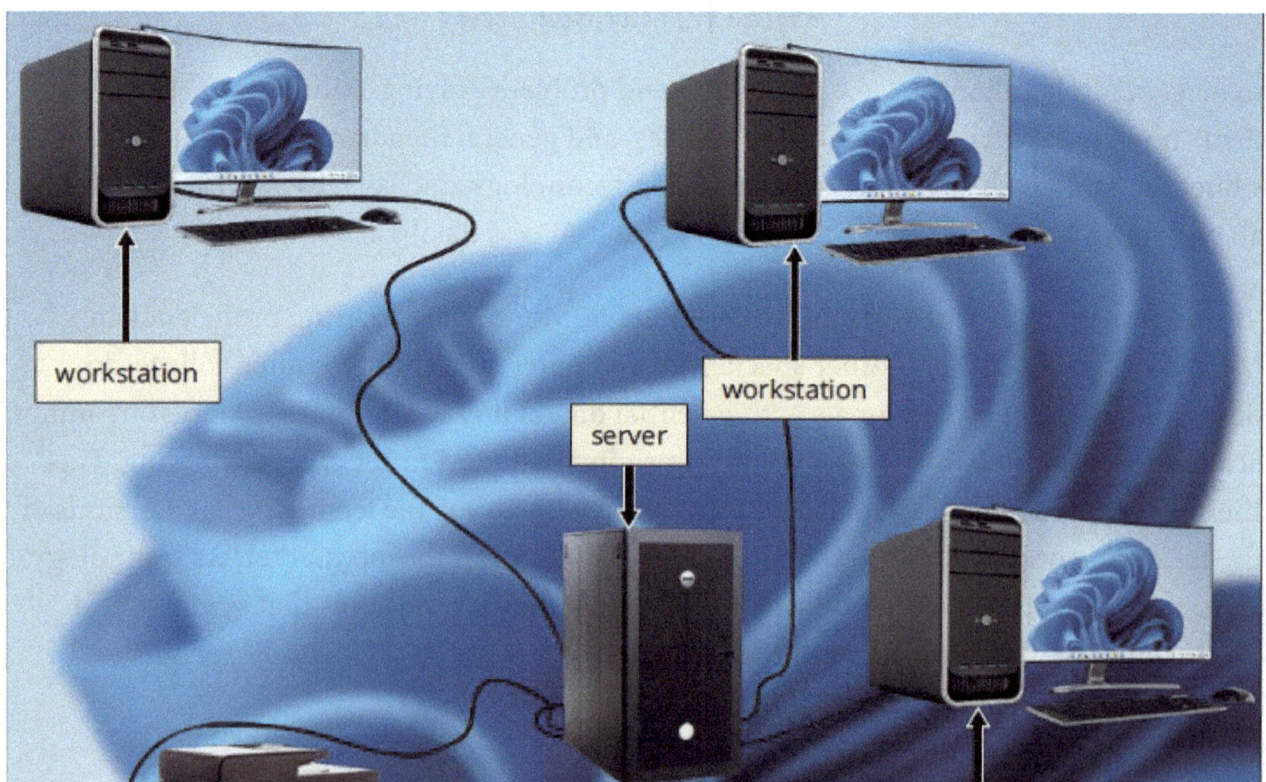

Figure 1-3 Workstations and Servers

(short for application), also called a program, is a computer program that performs specific tasks. Apps are designed to make users more productive and/or assist them with personal tasks, such as word processing or browsing the web.

Windows commonly is used on desktops, laptops and other mobile devices, and workstations. A **Workstation** is a computer connected to a server. A **Server** on the other hand is a powerful, high-capacity computer you access using the Internet or other network; it stores files and "serves" them, that is, makes the files available to, users; usually grouped at a location called a data center. The server controls access to the hardware and software on a network and provides a centralized storage area for programs, data, and information. **Figure 1–3** on the previous page illustrates a simple computer network consisting of a server, three workstations, and a printer connected to the server.

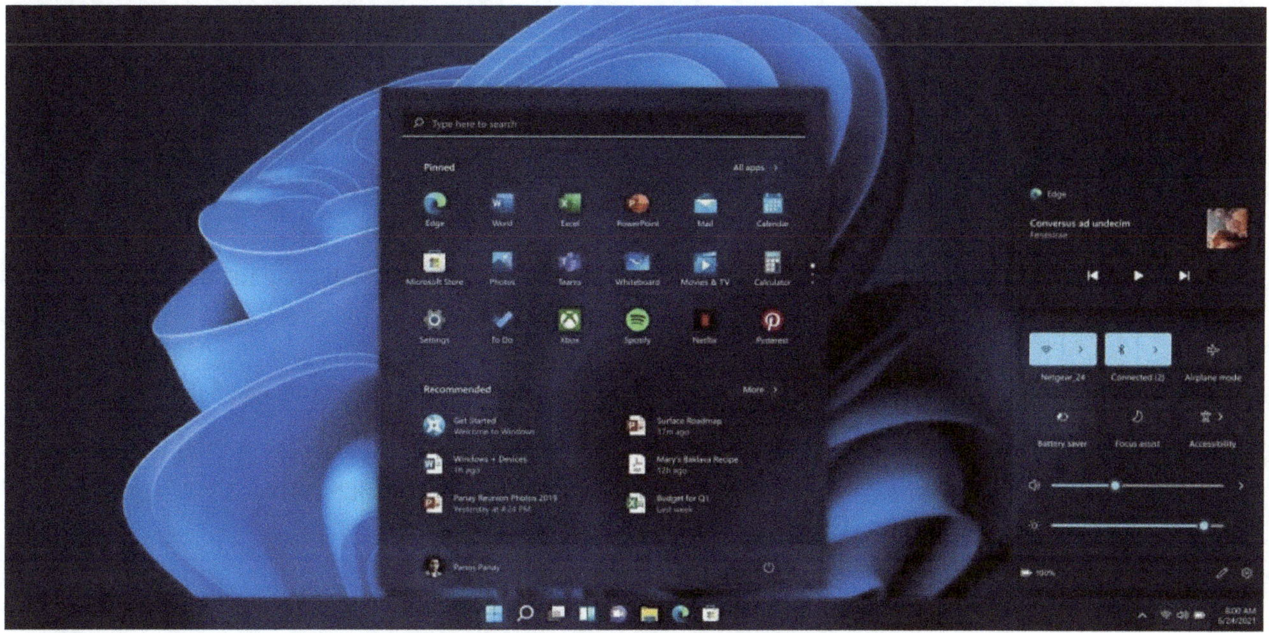

Figure 1-4 Windows 11 Home Screen

Windows is easy to use and can be customized to fit individual needs. **Figure 1-4** shows a typical Windows 11 Home screen. The operating system simplifies working with documents and programs, transferring data between documents, interacting with the different components of the computer, and using the computer to access information on the Internet or an intranet. As you may have already read in the accompanying textbook for this hands on textbook, the **Internet** is a global collection of millions of computers linked together to share information, and gives users the ability to use this information, send messages, and obtain products and services. An **intranet** is an internal network site used by a group of people who work together that uses Internet technologies.

Windows 11 is designed to provide a similar user interface across multiple devices such as desktops, laptops, and tablets. The Windows 11 interface combines some of the most successful features of previous versions of Windows, such as the Start menu, an optimal interface for touch input, and enhanced search functionality. Several other improvements over previous versions of Windows make Windows 11 a suitable choice for all users.

This textbook demonstrates how to use Windows 11 to control the computer and communicate with other computers, both on a network and the Internet. In this chapter, you will learn about Windows and how to use some of its basic features and application.

Multiple Editions of Windows 11

Windows 11 is available in multiple editions. The first, **Windows 11 Home**, is simplified and designed primarily for home and small office users. **Windows 11 Pro** is designed for businesses and technical professionals. **Windows 11 Enterprise** has the same features as Windows 11 Pro but is designed for large enterprises where IT professionals need to manage and secure computers and mobile devices easily. Windows 11 Education contains many of the same features as Windows 11 Enterprise but is designed for faculty, staff, and students.

For a computer, minimum system requirements specify that the processor is 1 GHz or faster on a 64-bit architecture, random access memory (RAM) is at least 4 GB, the hard drive has at least 64 GB available space, and the video card supports DirectX 12 graphics with WDDM (Windows Display Driver Model) 2.0 or higher driver.

Windows can be customized using a Microsoft account. When you add a Microsoft account, you can sign in to the account and then sync (synchronize) your information with all of your Windows devices. This allows you to set your desktop background and color settings, for example, and then sync those settings with your other devices. When you sign in to your Microsoft account with another Windows device, your settings will appear the same as they do on your other Windows 11 devices.

Navigating Using Touch or a Mouse

Windows 11 provides touch support. With touch, you can use your fingers to control how Windows functions. For example, you can swipe your finger from the right to display the Action Center. (The Action Center is discussed in greater detail later in this book.)

Touch also allows Windows 11 to more easily work on touch-enabled devices, such as laptops and tablets.

Using a Touch Screen

Windows users who have computers or devices with touch screen capability can interact with the screen using gestures. A **Gesture** is a motion you make on a touch screen with the tip of one or more fingers or your hand. Touch screens are convenient because they do not require a separate device for input. **Table 1–1** presents common ways to interact with a touch screen.

Motion	Description	Common Uses	Equivalent Mouse Operation
Tap	Quickly touch and release one finger one time.	Activate a link (built-in connection). Press a button. Start a program or an app.	Click
Double-tap	Quickly touch and release one finger two times.	Start a program or an app. Zoom in (show a smaller area on the screen so that contents appear larger) at the location of the double-tap.	Double-click
Press and hold	Press and hold one finger to cause an action to occur or until an action occurs.	Display a shortcut menu (immediate access to allowable actions). Activate a mode enabling you to move an item with one finger to a new location.	Right-click
Swipe	Press and hold one finger and then move the finger horizontally or vertically on the screen.	Select an object. Swipe from edge to display the Action Center.	Drag

Stretch	Move two fingers apart.	Zoom in (show a smaller area on the screen so that contents appear larger).	None
Pinch	move two fingers together.	Zoom out (show a larger area on the screen so that contents appear smaller).	None

Figure 1-1 Touch Screen Gestures

USING AN ON-SCREEN KEYBOARD

When using touch, you can access an on-screen keyboard that allows you to enter data using your fingers. To display the on-screen keyboard, click the Touch keyboard button on the taskbar. You tap a key on the keyboard to enter data or manipulate what you see on the screen. Figure 1–2 displays the on-screen keyboard.

Figure 1-5 On-Screen Keyboard

USING A MOUSE

Windows users who do not have touch screen capabilities typically work with a mouse that has at least two buttons. For a right-handed user, the left button usually is the primary mouse button, and the right mouse button is the secondary mouse button. Left-handed people, however, can reverse the function of these buttons.

Table 1–2 explains how to perform a variety of mouse operations. Some apps also use keys in combination with the mouse to perform certain actions. For example, when you hold down ctrl while rolling the mouse wheel, text on the screen may become larger or smaller based on the direction you roll the wheel. The function of the mouse buttons and the wheel varies depending on the app.

Operation	**Mouse Action**	**Example**	**Equivalent Touch Gesture**
Point	Move the mouse until the pointer on the desktop is positioned on the item of choice.	Position the pointer on the screen.	None
Click	Press and release the primary mouse button, which usually is the left mouse button.	Select or deselect items on the screen or start an app or app feature.	Tap
Right-click	Press and release the secondary mouse button, which usually is the right mouse button.	Display a shortcut menu.	Press and hold
Double-click	Quickly press and release the primary mouse button twice without moving the mouse.	Start an app or app feature.	Double-tap
Triple-click	Quickly press and release the primary mouse button three times without moving the mouse.	Select a paragraph.	Triple-tap

Drag	Point to an item, hold down the primary mouse button, move the item to the desired location on the screen, and then release the mouse button.	Move an object from one location to another or draw pictures.	Drag or slide
Right-drag	Point to an item, hold down the right mouse button, move the item to the desired location on the screen, and then release the right mouse button.	Display a shortcut menu after moving an object from one location to another.	Press and hold, then drag
Rotate wheel	Roll the wheel forward or backward.	Scroll vertically (up and down).	Swipe
Free-spin wheel	Whirl the wheel forward or backward so that it spins freely on its own.	Scroll through many pages in seconds.	Swipe
Press wheel	Press the wheel button while moving the mouse.	Scroll continuously.	None
Tilt wheel	Press the wheel toward the right or left.	Scroll horizontally (left and right).	None
Press thumb button	Press the button on the side of the mouse with your thumb.	Move forward or backward through webpages and/or control media, games, etc.	None

Figure 1-2 Mouse Operations

Scrolling

A **Scroll Bar** is a bar on the bottom edge (horizontal scroll bar) or right edge (vertical scroll bar) of a document window that lets you view a document that is too large to fit on the screen at once (**Figure 1-6**). A scroll bar contains scroll arrows and a scroll box. **Scroll Arrows** are small triangular up and down arrows at each end of a scroll bar that you use to adjust your window view in small increments. A **Scroll Box** is a box in a scroll bar that you can drag, or click above and below, to display different parts of a window. Clicking the up and down scroll arrows moves the screen content up or down one line. You also can click above or below the scroll box to move up or down a section, or drag the scroll box up or down to move to a specific location.

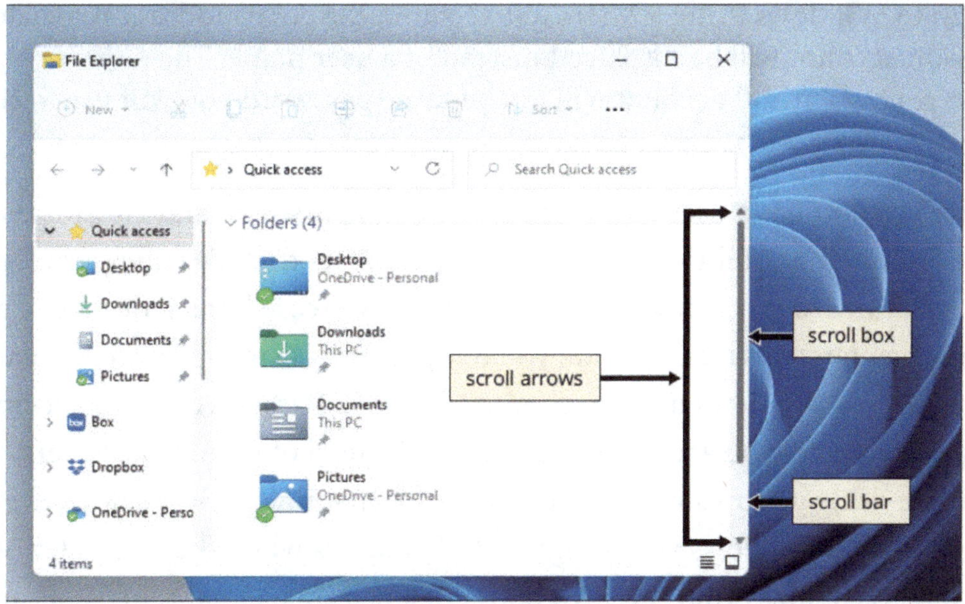

Figure 1-6 Scroll Arrows and Scroll Bar

Using Keyboard Shortcuts

In many cases, you can use the keyboard instead of the mouse to accomplish a task. To perform tasks using the keyboard, you can use a **Keyboard Shortcut**, which is a key or combination of keys that you press to access a feature or perform a command, instead of using a mouse or touch gestures. Some keyboard shortcuts consist of a single key, such as f3. Other keyboard shortcuts consist of multiple keys, in which case a plus sign separates the key names, such as ctrl+esc. This notation means to press and hold down the first key listed, press one or more additional keys, and then release all keys. For example, another way to display the Start menu is by pressing **ctrl+esc;** that is, hold down **ctrl**, press **esc**, and then release both keys.

STARTING WINDOWS 11

It is not unusual for multiple people to use the same computer in a work, educational, recreational, or home setting. Windows enables each user to establish a **User Account**, which identifies to Windows the resources, such as apps and storage locations, a user can access when working with the computer.

Smart Tips

Microsoft Accounts

If you sign in to Windows using a Microsoft account, the email address associated with your Microsoft account will be displayed instead of a user name. The password you use to sign in to your Microsoft account will be the same password you will use to sign in to Windows.

Each user account has a user name and may have a password and an icon, as well. A **User Name** is a unique combination of letters or numbers that identifies a specific user to Windows. A **Password** is a string of uppercase and lowercase letters, numbers, and symbols that, when entered correctly, allows you to open a password-protected database or to obtain access to a Windows user's account. A **Picture Password** is a replacement for a text password that gives you access to your device; instead of typing characters, you swipe across a picture using gestures you previously set for that image. Some devices support **Windows Hello**, which allows you to sign in to Windows when the computer or mobile device's camera recognizes your face. An **icon** is a small image that represents an object; thus, a **User Icon** is a picture associated with a user name.

Smart Tips

PINs

If In addition to passwords and picture passwords, you also can sign in to Windows using a PIN (personal identification number). If desired, Windows 11 also allows you to include letters and symbols in your PIN, making it more secure than using just numbers.

When you turn on a computer, Windows starts and displays a lock screen. A **lock screen**, which appears before you sign into Windows or after your computer has been idle for a certain amount of time, displays the date and time and other information, such as network and battery status (shown in **Figure 1-8**). After tapping, sliding, or clicking anywhere on the lock screen, or pressing ctrl+alt+delete, depending on your computer's settings, Windows may or may not display a sign-in screen that shows the user names and user icons for users who have accounts on the computer as shown in (**Figure 1-7**). This **Sign-In Screen**, sometimes called a **Log-In Screen**, enables you to sign in to your user account and makes the computer available for use. Clicking the user icon begins the process of signing in to, also called **Logging On** to, your user account.

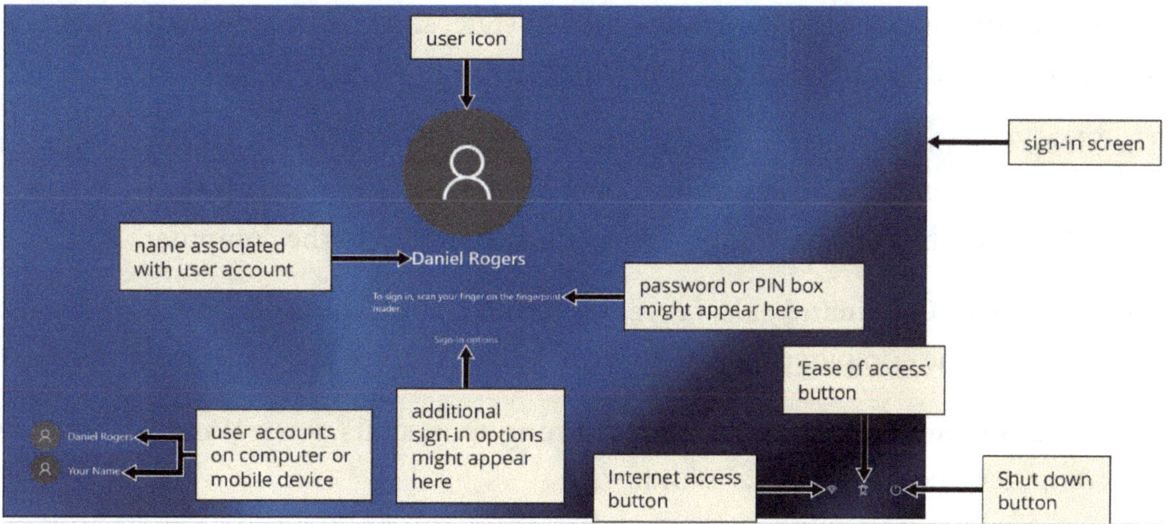

Figure 1-7 Windows Sign-In Screen

Smart Tips

Strong Passwords

You should consider using a strong password with your Windows user account. A strong password is more secure because it is more difficult to guess. Strong passwords have at least eight characters and contain a combination of uppercase and lowercase letters, numbers, and special characters.

At the bottom of the sign-in screen are the Internet access button, the 'Ease of access' button, and a Shut down button. The Internet access button shows the current status of the network connection. Your Internet access button may look different, depending on the type of network connection you are using (wired or wireless). Clicking the 'Ease of access' button displays the Ease of access menu, which provides tools to optimize a computer to accommodate the needs of users with mobility, hearing, and vision impairments.

Smart Tips

Shut Down Options

If you are walking away from your computer for only a brief period, you should put the computer in sleep mode instead of turning it off completely. Keeping the computer in sleep mode for this short period often uses less power than powering on the computer. Do not ever leave your computer screen opened and walkaway from it.

Clicking the Shut down button displays a menu containing commands related to restarting the computer, putting it in a low-power state, and shutting down the computer. The commands available on your computer may differ.

- The **Sleep Command** saves your work, turns off the computer fans and hard drive, and places the computer in a lower-power state. To wake the computer from sleep mode, press the power button or lift a laptop's cover, and sign in to the computer.
- The **Shut down Command** exits currently running apps, shuts down Windows, and then turns off the computer.
- The **Restart Command** exits currently running apps, shuts down Windows, restarts the computer, and then restarts Windows.

To Sign In to an Account

The following steps, which use the user account for a made up character we called Daniel Rogers, sign in to an account based on a typical Windows installation. Why? After starting Windows, you might be required to sign in to access the computer's resources. If you don't already know, you may need to ask your instructor how to sign in to your account.

To Sign in to your Windows Account do this:

Figure 1-8 Windows' Locked Screen

(1). **Click** the lock screen seen in (**Figure 1–8**) to display a sign-in screen.

(2). If necessary, **Click** the desired user icon on the sign-in screen, which, depending on settings, either will display a second sign-in screen that contains a sign-in method, such as a password text box (**Figure 1–9**), or will display the Windows desktop.

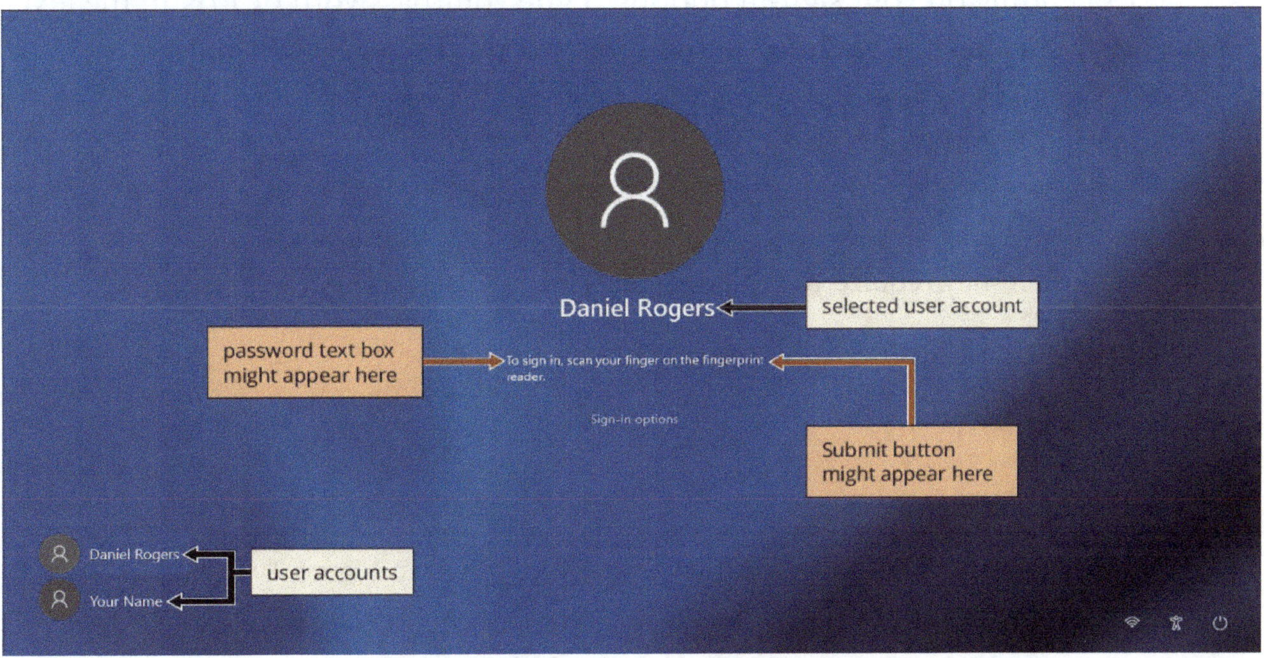

Figure 1-9 Sign-In Screen With Options

Figure 1-9 above shows the Windows sign-in screen. The screen background is a standard user screen with a dark blue background. At the bottom left, icons indicate

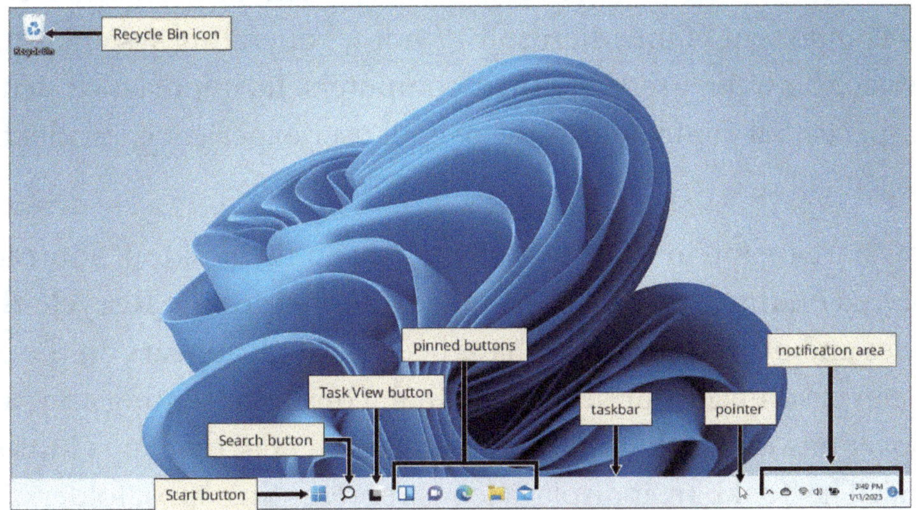

Figure 1-10 Windows Desktop

that two user accounts are on the computer or mobile device: Daniel Rogers (selected) and Your Name. The selected user account, Daniel Rogers, appears below the user icon in the middle of the screen. A password text box might appear at the space below the user name. The Submit button might appear to the right of the password text box or below it.

(3). Depending on your sign-in options, **Type** your password or PIN in the text box and then click the **Submit** button or **Scan** your finger on the fingerprint reader to sign in to your account and display the Windows desktop as shown in (**Figure 1–7**) on the next page.

Quick Review

1. Your computer may require you to type a user name instead of clicking an icon.
2. How can I get past the lock screen if I do not have a mouse?
3. What is a text box?
4. What is a text box?
5. Why does my desktop look different from the one shown in Figure 1-7?

The Windows 11 Desktop

Think of the Windows desktop as an electronic version of the top of your desk. You can perform tasks such as placing objects on the desktop, moving the objects around the desktop, and removing items from the desktop. Windows Desktop is a fundamental component of the Microsoft Windows operating system, serving as the primary interface between users and their computers. Its importance stems from several key aspects that contribute to the overall user experience, productivity, and system management.

When you start an app in Windows, it appears on the desktop. Some icons also may be displayed on the desktop. For instance, the icon for the **Recycle Bin**, the location of files and other objects that have been deleted, appears on the desktop by default. You can customize your desktop so that icons representing apps and files you use often appear on your desktop. When you start an app, that app's button appears on the taskbar. The **Taskbar**; it displays icons representing apps, folders, and/or files on the left, and the notification area, containing the date and time and special program messages, on the right. Buttons for **Microsoft Edge** (Microsoft's browser) and the Widgets, Chat, File Explorer, and Mail apps are pinned to the taskbar. Pinned app buttons always are displayed on the taskbar, regardless of whether the app is running or not. The right side of the taskbar contains the notification area, date, and time. The **Notification Area** is an area on the right side of the Windows 11 taskbar that displays the current time, as well as icons representing selected information; the Notifications button displays pop-up messages and, when selected, the Action Center. For example, the notification area can tell you if your virus protection is out of date, how much battery life you have remaining (if you are using a mobile device), and whether you are connected to a network. The taskbar also displays the **Search Button** to the left of the app buttons, which allows you to type a term to locate files or folders containing that term in the location you are searching.

Below are the significance of the Windows 11 Desktop:

Smart Tips

Pinned Buttons

If you use an app frequently, you should consider pinning that app button to the taskbar so that you can access it easily. Pinning and unpinning app buttons is discussed later in this module.

- **User Interface and Accessibility:** The Windows Desktop provides a visual environment where users can access applications, files, and system settings through icons, menus, and taskbars. Its intuitive design allows users of varying technical skills to navigate and operate their computers efficiently. The desktop environment supports customization, enabling users to personalize their workspace with wallpapers, widgets, and shortcuts, thereby enhancing comfort and productivity.

Figure 1-11 Windows Desktop

- **Central Hub for System Operations:** The desktop acts as a central hub where users can launch programs, manage files, and access system tools. The Start menu, taskbar, and desktop icons facilitate quick access to frequently used applications and documents, streamlining workflow and reducing time spent searching for resources.
- **Multitasking and Workflow Management:** Windows Desktop supports multitasking by allowing users to open multiple windows and applications simultaneously. Features such as virtual desktops, snap assist, and task view improve workflow management, enabling users to organize their tasks efficiently and switch between activities seamlessly.
- **Security and System Control:** The desktop environment integrates security features such as user account controls, notifications, and system alerts. It provides access to system settings and control panels, empowering users to configure security options, update software, and troubleshoot issues effectively.
- **Compatibility and Software Ecosystem:** Windows Desktop ensures compatibility with a vast array of software applications, hardware devices, and peripherals. This extensive ecosystem makes Windows a versatile platform suitable for personal, educational, and professional use.

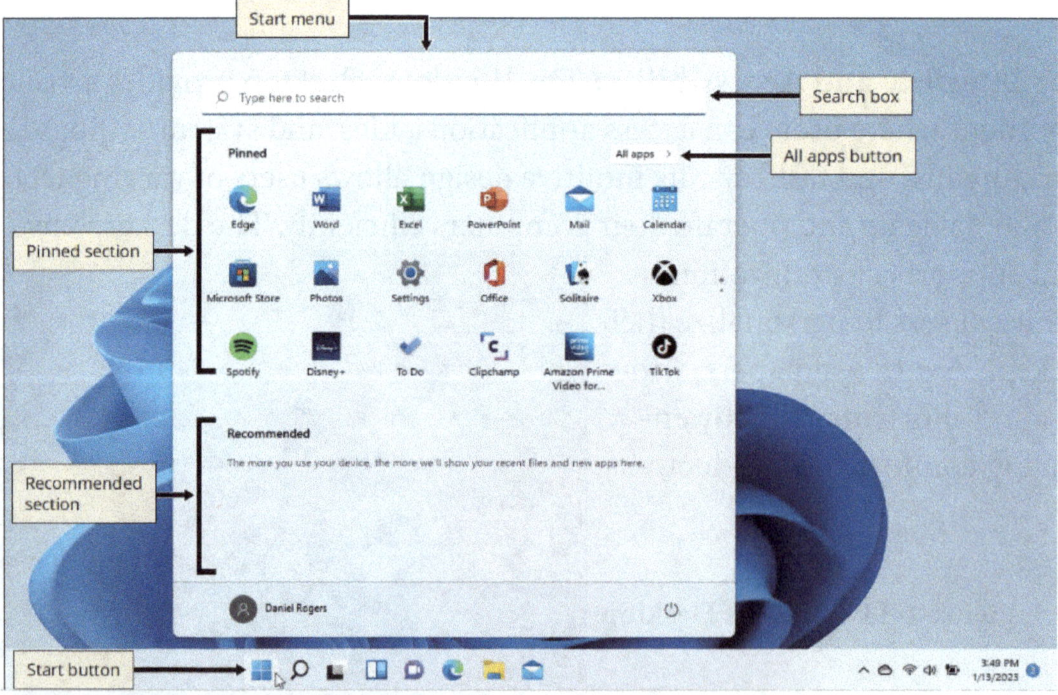

Figure 1-12 Windows Desktop Buttons

GETTING STARTED WITH FILE MANAGEMENT

Working on a Windows PC typically starts with opening an application/app or a file. You use applications such as Microsoft Word to open and edit files, also called documents. Understanding how to save and organize files means you can easily find and open them when necessary, a practice called **File Management**. Knowing how to save, locate, and organize computer files makes you more productive when working with a computer.

In addition to storing documents locally on desktops, laptops, or mobile devices, the new default option for saving documents in Microsoft 365 uses the cloud for file storage. But due to the high cost of connecting to the Internet here in Liberia, or Nigeria and many parts of Africa, you will be saving your lab projects locally on a USB flash drive. If you can afford the cost of Internet connection, you are free to use Microsoft cloud storage. OneDrive is the name given to Microsoft's cloud storage services. You may be aware that Google has Google Drive and Apple has iCloud. Microsoft provides OneDrive for you to store and exchange data online.

MANAGING YOUR FILES

Depending on the make, model of computer, and the version of the operating system as well as the application program, a computer can store folders and files in different locations, including removable media such as USB flash drives or external hard drives, an internal hard drive permanently housed in a computer, and a cloud location such as OneDrive. The internal hard drive is assigned as drive C, also called the root directory. The remaining drives can have any other letters but are usually assigned in the order that the drives were installed on the computer—so your USB drive might be drive D, drive E, or drive F.

File Explorer is the file manager in Microsoft Windows Operating System. To open

Figure 1-3 File Explorer on Taskbar

File Explorer, click the File Explorer icon on the taskbar as shown in **Figure 1-3** above. The File Explorer window shown in **Figure 1-4** displays the Quick Access area, which by default is the first section of the navigation pane on the left. The Quick Access area shows your most recently accessed folder, making it easy to find the files you work with often. OneDrive, Microsoft's cloud storage service, has a dedicated folder below the Quick Access area. Other folders on your computer are available in a tree view below OneDrive.

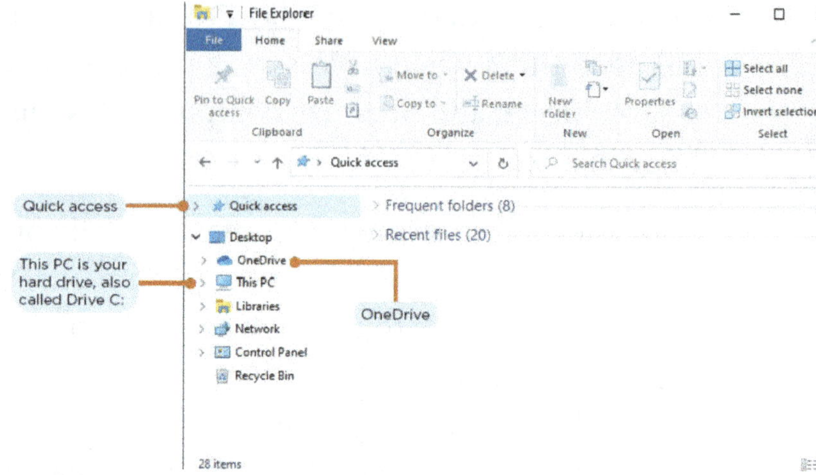

Figure 1-4 File Explorer Window

Smart Tips

Organizing Files and Folders

- File management is the system of organization to keep track of folders and files on a computer, external drive, or cloud service.
- File Explorer is a Windows file manager app that you use to create folders and move files from one folder to another.

tions called panes. The left pane is the navigation pane, which contains icons and links to locations on your computer. The right pane displays the contents of the location selected in the navigation pane. If the navigation pane showed all the contents on your computer at once, it could be a very long list. So Windows provides a folder structure that includes the Desktop, Documents, Downloads, Music, Pictures, and Videos locations. The Documents folder is designed to store your files—your MLA or APA papers, resume, spreadsheets, presentations, and other files that you create, edit, and manipulate in an app. File Explorer displays the file path in the address bar so you can keep track of your current location as you navigate drives and folders. The Downloads folder is the default folder that stores all your downloaded files from the Internet such as things that you download from your favorite web browser.

CREATING A FOLDER

After you use an app to create a file, save it in a folder so that you can find it easily and work efficiently. The best folder structure is one that mimics the organization of your classes. For example, consider creating a new folder in the Documents folder for each class every semester or quarter of the school year. A top-level folder named for each class makes it easy to find your work. **Figure 1-5** shows how you could organize files on your computer if you were taking a full semester of college classes. To duplicate this organization, you would open the main folder for your documents, such as Documents, create four subfolders—one each for the Computer Concepts, English, History, and Math courses—and then store the files by class, such as the writing assignments you complete in the English subfolder.

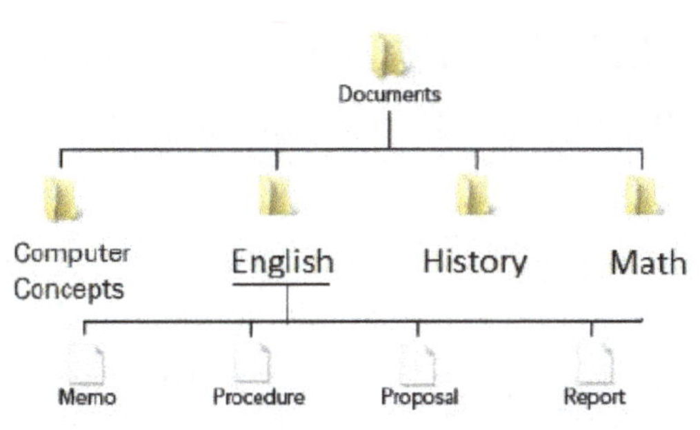

Figure 1-5 Folders and Subfolders

Even if you store most of your files in a cloud service such as OneDrive, Google Drive or on removable media such as USB drives, you still need to organize those files into folders and subfolders.

You create folders in File Explorer using one of three methods: using the New folder button in the New group on the Home tab, using the New folder button on the Quick Access Toolbar, or by right-clicking to display a shortcut menu that includes the New command. Keep the following guidelines in mind as you create folders:

» Keep folder names short yet descriptive of the folder's contents. Long folder names can be more difficult to display in their entirety in folder windows, so use names that are short but clear. Choose names that will be meaningful later, such as project names or course numbers.

» Create subfolders to organize files. If a file list in File Explorer is so long that you must scroll the window, you should probably organize those files into subfolders.

» Develop standards for naming folders. Use a consistent naming scheme that is clear to you, such as one that assigns a project name to the main folder and includes step numbers in each subfolder name (for example, 1-Outline, 2-First Draft, 3-Final Draft, and so on).

To create a folder named Computer Concepts:

(1). Start File Explorer, and then **Click** the Home tab on the ribbon.

(2). In the New group, **Click** the New folder button. A folder icon with the label "New folder" appears in the right pane of the File Explorer window.

(3). Type Computer Concepts and then press **ENTER**. The new folder is named Computer Concepts and is the selected item in the right pane, as shown in **Figure 1-6** on the previous page.

RENAMING A FOLDER

After creating and naming a folder or a file, you might realize that a different name would be more meaningful or descriptive. You can rename a folder or a file using the Rename command on the file's shortcut menu.

To rename the Computer Concepts folder:

(1). In the right pane of the File Explorer window, **Right-Click** the Computer Concepts folder and then **Click Rename** on the shortcut menu. The filename is highlighted, and a box appears around it.

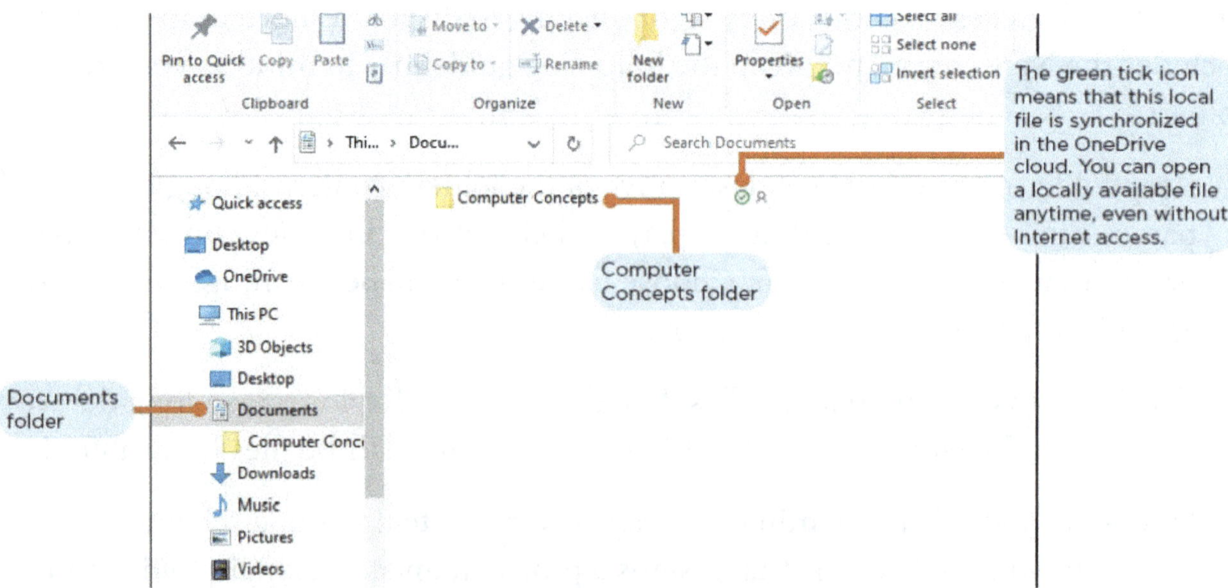

Figure 1-6 Creating a New Folder

(2). Type Information Technology and then press ENTER. The file now appears with the new name, as shown in **Figure 1-7**below.

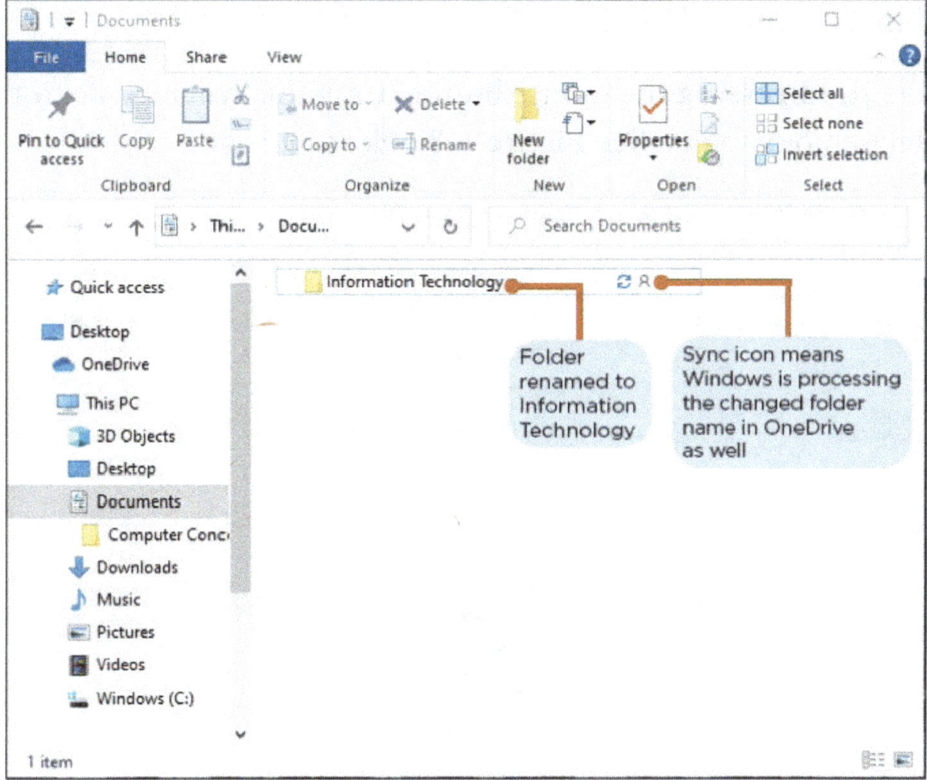

Figure 1-7 Renaming a Folder

Moving and Copying Files and Folders

After creating and organizing your folder structure, you can either move or copy a file from its current location to a new location. Moving a file removes it from its current location and places it in a new location that you specify. Copying a file places a duplicate version of the file in a new location that you specify while leaving the original file intact in its current location. You can also move and copy folders. When you do, you move or copy all the files contained in the folder.

In File Explorer, you can move and copy files by using the Move to or Copy to buttons in the Organize group on the Home tab, the Copy and Cut commands on a file's shortcut menu, or keyboard shortcuts. When you copy or move files using these methods, you are using the Clipboard, a temporary storage area for files and information that you copy or move from one place to another. You can bypass the Clipboard by dragging files or folders to move or copy them.

To move a file named Syllabus by dragging it:

(1). In File Explorer, **Point** to the Syllabus file in the right pane, and then press and hold the left mouse button.

(2). While still **Pressing** the mouse button, drag the Syllabus file to the Information Technology folder. See **Figure 1-8** below.

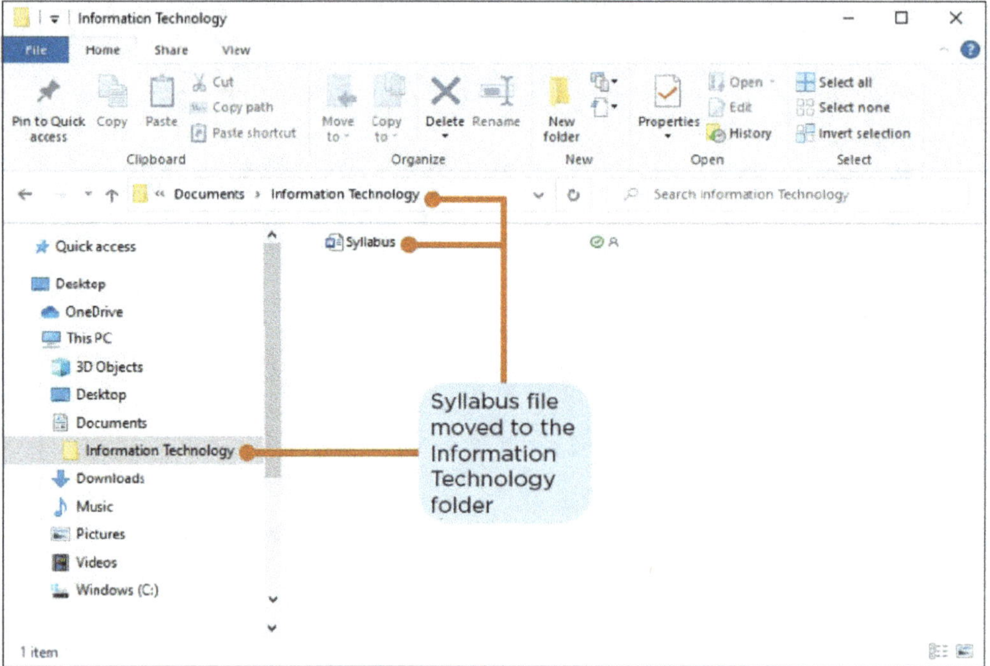

Figure 1-8 Dragging a File

FEW IMPORTANCE OF FILE MANAGEMENT IN MICROSOFT WINDOWS

File management in Microsoft Windows is a fundamental aspect of computer usage that significantly impacts productivity, data security, and system organization. Proper file management ensures that users can efficiently locate, access, and organize their data, which is essential for both individual and organizational tasks. A few key aspects are:

- Organization and Structure
- Data Accessibility
- Data Security and Backup
- System Performance
- Collaboration and Sharing
- Time Efficiency
- Data Integrity
- Disaster Recovery

CHAPTER 2 HIGHLIGHT

Microsoft Word: Creating and Editing a Document

After completing this chapter, you will be able to:

1. Create and save a document
2. Enter text and correct errors as you type
3. Use AutoComplete and Auto-Correct
4. Select text and move the insertion point
5. Undo and redo actions
6. Adjust paragraph spacing, line spacing, and margins
7. Preview and print a document
8. Create an envelope
9. Open an existing document
10. Use the Editor pane
11. Change page orientation, font, font color, and font size
12. Apply text effects and align text
13. Copy formatting with the Format Painter
14. Insert a paragraph border and shading
15. Delete, insert, and edit a photo
16. Add a page border
17. Create bulleted and numbered lists
18. Use Microsoft Word Help

The overview image below shows a blank Word document window with the following elements labeled and defined: Quick Access Toolbar at the top left of window's title bar (collection of buttons for common commands such as Save, Undo Repeat); tab on the ribbon (includes command related to particular activities or tasks); title bar (displays the name of the open file and program); ribbon bar of buttons above the document window (main set of buttons and tools to complete tasks, organized into tabs and groups); insertion point in the document (shows where characters will appear when typing); margins (dark gray areas on the ruler,

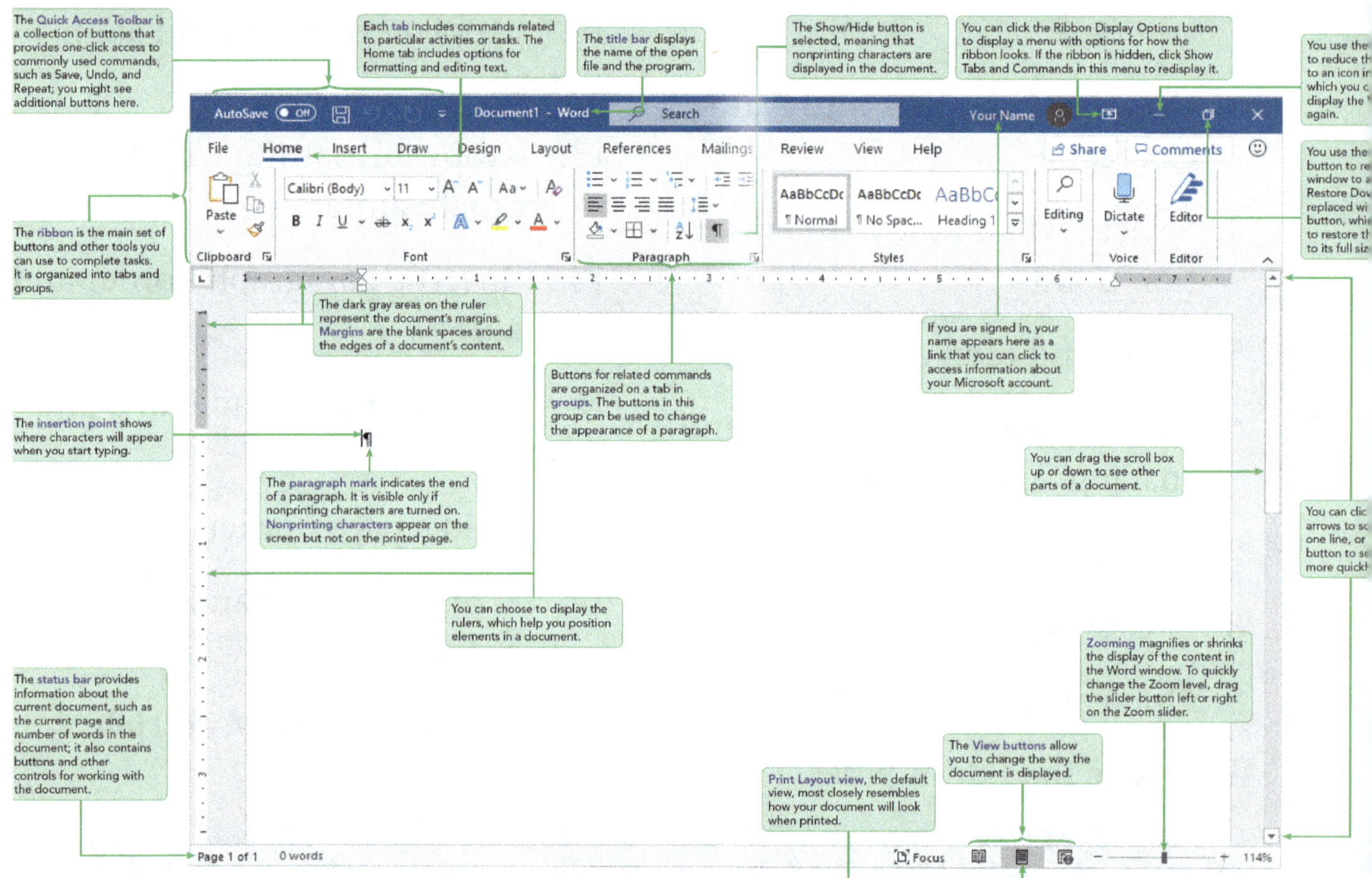

Figure 2-1 Word Overview

representing blank spaces around the edges of a document's content); paragraph mark in the document (indicates the end of a paragraph); nonprinting characters (appear on the screen but not on the printed page); groups on the ribbon (buttons for related commands on a tab are organized in groups); rulers below and to the left of the ribbon (help position elements in a document); status bar at the bottom of the window (provides information about the current document, such as current page and number of words, and contains buttons and controls for working with the document); Show/Hide button on the ribbon (when selected, nonprinting characters display); Ribbon Display Options button on the title bar (when clicked, displays a menu with options for how the ribbon looks); sign in information on the title bar (if you are signed in to your Microsoft account, your name appears on the title bar, and you can click it to access information about your account); Minimize button on the title bar (used to reduce the Word window to an icon on the taskbar, which you can click later to display the window again); Restore Down button on the title bar (reduces with window to a smaller size; the button then is replaced with the Maximize button, which you can click to restore the window to its full size); scroll arrows on the right of the document window (click the arrows to scroll up or down a line, or hold the mouse button to scroll up or down more quickly); scroll box on the right of the document window (drag up or down to see other parts of a document); zooming (a slider on the status bar that magnifies or shrinks the display of the content of the window by dragging the slider), View buttons on the status bar (change the way the document is displayed); Print Layout view (default view, most closely resembles how the document will look when printed).

Launching Word

Microsoft Word is a word processing application.

Purpose: You can use Word to create letters, flyers, resumés, research papers, and many other kinds of documents.

Tools: Word includes helpful tools for formatting text, inserting basic graphics, and checking spelling and grammar.

How you access the Word application will vary depending on what kind of device and installation you have. For this course, you will need the installed, desktop version of Office Suite or Office. If you are student and you have not yet installed Office, your school might provide steps for you to install Word, or you can get detailed instructions from the Microsoft website link below:

https://support.microsoft.com/en-us/office/download-install-or-reinstall-microsoft-365-or-office-2024-on-a-pc-or-mac-4414eaaf-0478-48be-9c42-23adc4716658

If the above link is no longer working, please use your favorite web browser to search for Microsoft Office 360 2024 download for student.

When you install Word, you will need a Microsoft Office license. Microsoft Office is an office suite of applications that includes the Word application. Office offers three licensing options for you to use their applications:

Office Online, also called Office for the web, includes web app versions of Word, Excel, and PowerPoint. Office Online is free to use. It works on any computer or mobile device with a browser. You can also install free Office apps on many mobile devices, including iPhones, iPads, and Android tablets and smartphones. Remember, though, that the free Office Online web apps and mobile apps are not as powerful as the desktop applications on a laptop or desktop computer.

Office 2024 includes Word, Excel, and PowerPoint on Mac computers, and it adds Access on Windows computers. Depending on the requirements for your collage course, you may or may not need Access. Office 2024 requires a one-time payment up front to use the applications on one computer, and you never have to pay for that license again. However, this price does not include upgrades to the applications.

Microsoft 365 includes Word, Excel, and PowerPoint on Mac computers, and it adds Access on Windows computers. Microsoft 365 requires a regular subscription payment and includes automatic upgrades. You can choose from various subscription plans. Some schools provide students with Microsoft 365 subscriptions. Check with your school for specific instructions on how to access that benefit if it is offered.

In most cases, you can use either Office 2019 for this course or Microsoft 365. Both give you the installed, desktop version of Word. Your school might offer a license as part of your student resources.

If you are using a Windows computer, complete the activity "To Open Word in Windows." If you are using a macOS computer, you may have consult your instructor to help you "To Open Word in macOS."

To Open Word in Windows

When the Word application is installed on your computer, you must open the application to work with documents in Word. Why? The installed version of Word offers many more tools than does Word Online in a browser. The following steps open the Word application in Windows.

CHAPTER 2 | MICROSOFT WORD

STARTING WORD

With Word, you can quickly create polished, professional documents. You can type a document, adjust margins and spacing, create columns and tables, add graphics, and then easily make revisions and corrections. In this session, you will create one of the most common types of documents—a block-style business letter.

To begin creating the letter, you first need to start Microsoft Word and then set up the Word window.

To Start Word:

(1). On the Windows Taskbar **Click** the **Start** button in the lower-left corner of your screen to open the Start menu as shown in (**Figure 2-2**).

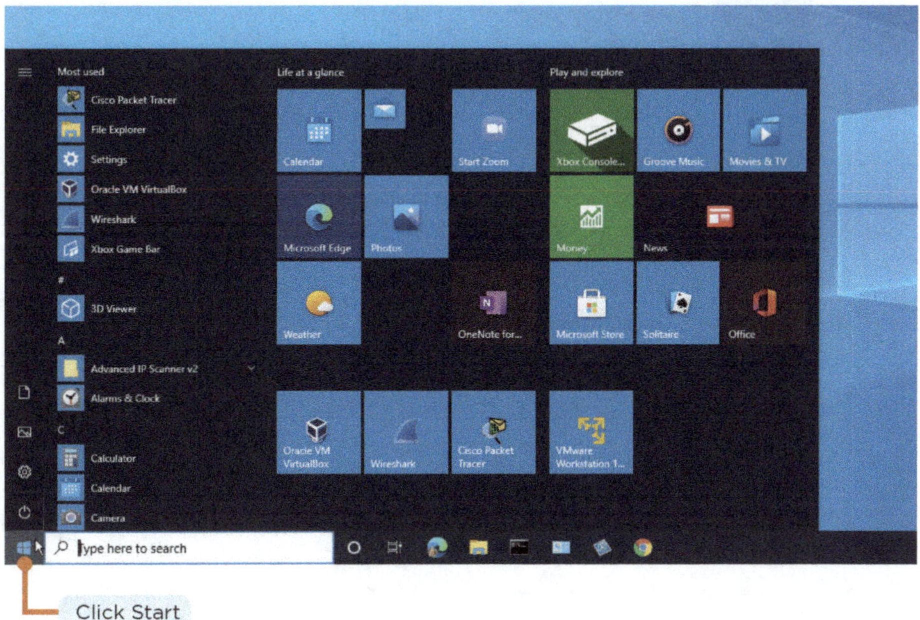

Figure 2-2 Word Start Menu

(2). Type On the **Start Menu**, scroll the list of apps, and then **click Word**. Word starts and displays the Recent screen in Backstage view. **Backstage View** provides access to various screens with commands that allow you to manage files and Word options. **Figure 2-3** on the next page shows the Backstage View.

(3). **Click** the **Blank Document** icon. The Word window opens, with the ribbon displayed.

Are You Having Trouble? If you don't see the ribbon, **Click** the **Ribbon Display Options** button, as shown in Word Overview, and then click Show Tabs and Commands.

Don't be concerned if your Word window doesn't match the Figure 2-1 in Word Overview exactly. You'll have a chance to adjust its appearance shortly.

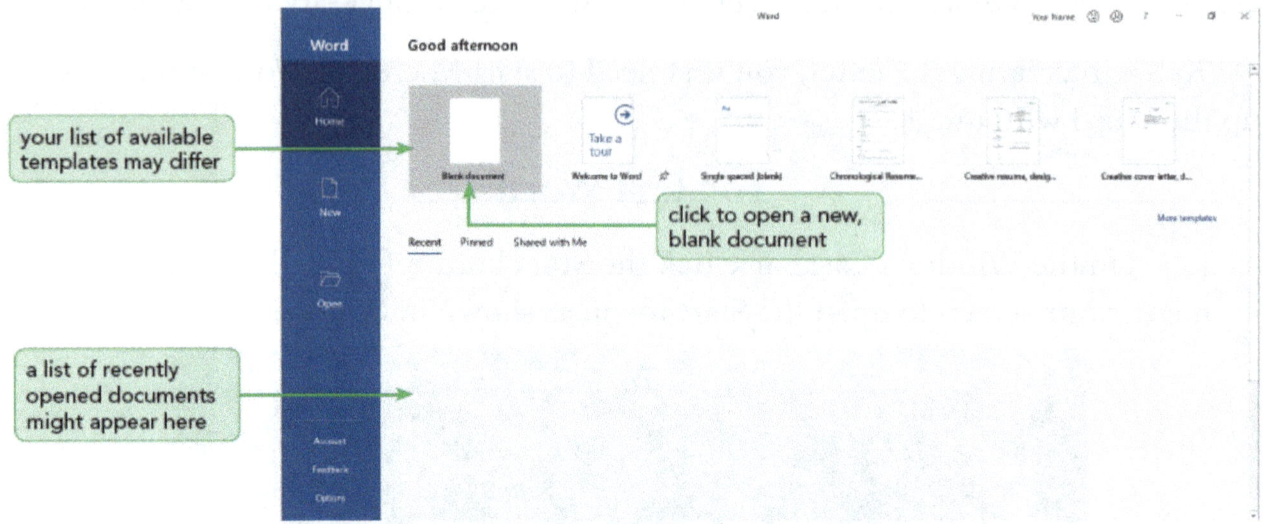

Figure 2-3 screen in Backstage view

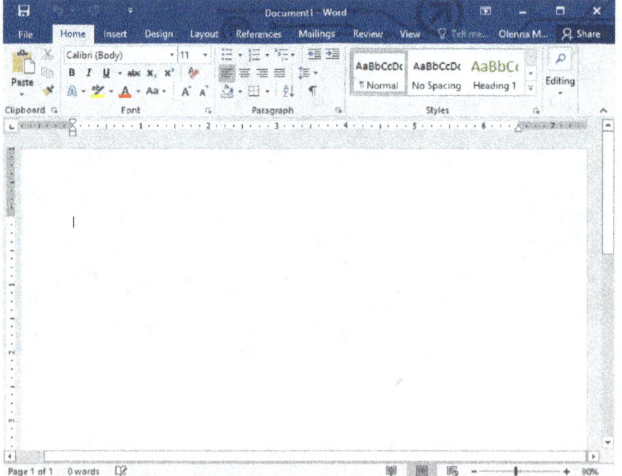

Figure 2-4 A Blank Opened Word Window

WORKING IN TOUCH MODE

You can interact with the Word screen using a mouse, or, if you have a touch-screen, you can work in **Touch Mode**, using a finger instead of the pointer. In Touch Mode, extra space around the buttons on the ribbon makes it easier to tap the specific button you need. The figures in this text show the screen with Mouse Mode on, but it's helpful to learn how to switch back and forth between Touch Mode and Mouse Mode.

Caution: The steps in this section assume that you are using a mouse. If you are instead using a touch device, please read these steps but don't complete them so that you remain working in Touch Mode.

To switch between Touch and Mouse Mode:

(1). On the Quick Access Toolbar, **Click** the **Customize Quick Access Toolbar** button to open the menu. The Touch/Mouse Mode command near the bottom of the menu does not have a checkmark next to it, indicating that it is currently not selected.

Are You Having Trouble? If the Touch/Mouse Mode command has a checkmark next to it, **Press ESC** to close the menu, and then skip to **Step 3**.

(2). On the menu, **Click Touch/Mouse Mode**. The menu closes, and the Touch/Mouse Mode button appears on the Quick Access Toolbar.

(3). On the Quick Access Toolbar, **Click** the **Touch/Mouse Mode** button . A menu opens with two options—Mouse and Touch. The icon next to Mouse is shaded gray to indicate it is selected.

Are You Having Trouble? If the icon next to Touch is shaded gray, press ESC to close the menu and skip to **Step 5**.

(4). On the menu, **Click Touch**. The menu closes, and the ribbon increases in height so that there is more space around each button on the ribbon. See **Figure 1-5** on the next page.

Are You Having Trouble? If you are working with a touchscreen and want to use Touch Mode, skip **Steps 5 and 6.**

(5). On the Quick Access Toolbar, **Click the Touch/Mouse Mode** button , and then **Click Mouse**. The ribbon changes back to its Mouse Mode appearance, as shown in the Word Overview.

(6). On the Quick Access Toolbar, **Click** the **Customize Quick Access Toolbar** button , and then click **Touch/Mouse Mode** to deselect it. The Touch/Mouse Mode button is removed from the Quick Access Toolbar.

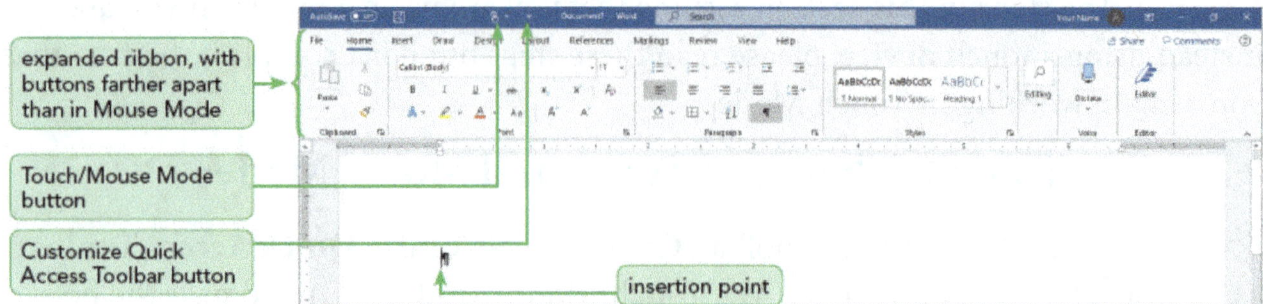

Figure 2-5 Word Window Touch Mode

RIBBONS AND TOOLBARS

Near the top of the Word window is a series of tabs as shown in **Figure 2-6**. If you have used one of the Google web apps, such as Google Docs, some of these tabs might seem familiar. The ribbons are almost all alike in Microsoft Office applications. The difference in the Tabs and Ribbons will be based on functionality and feature.

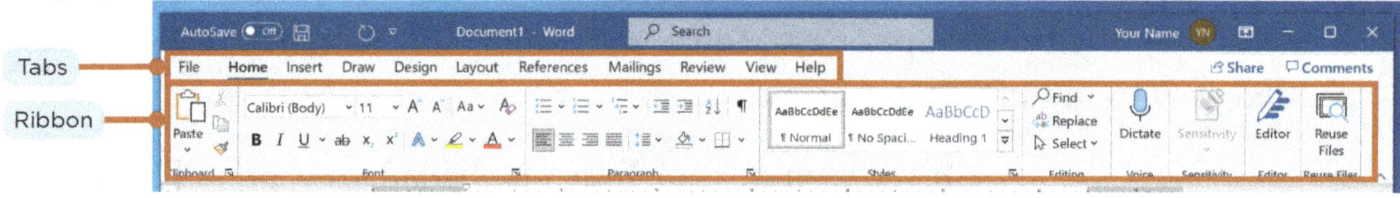

Figure 2-6 Toolbars on the Ribbons

Click each of these tabs to select different ribbons. Each ribbon contains a collection of tools. For example, the Home ribbon offers basic text formatting tools. The Insert ribbon offers many tools for inserting—or adding—tables, pictures, shapes, videos, and more.

It is easier to find the tool you need when you think about which of these ribbons is the best fit for that tool. For example, when you are ready to review and edit your document, you will need some of the tools on the Review ribbon. If you want to change layout options for your document, such as the space around the text or the size of the document, you will need tools found on the Layout ribbon. Explore the various ribbons to find the kinds of tools each ribbon contains.

Unless the default settings have been changed, the Quick Access Toolbar is located along the top of the Word window as shown in (**Figure 2-7**) on the next page.

- **Undo:** The backward arrow is the Undo button. If you make a mistake on your document, click the Undo button to undo the action. For example, if you accidentally delete a paragraph, clicking Undo will bring the paragraph back.

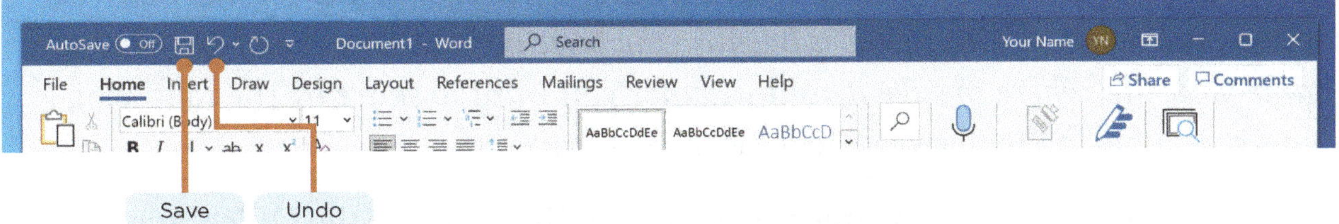

Figure 2-7 Quick Access Toolbar

- **Save:** The Save button is also located on the Quick Access Toolbar. When you are working on a document in Word, make sure you click the Save icon often so you do not lose your work. You might also see an option for AutoSave. You will learn more about AutoSave later in this module.

Smart Tips

Save Icon

The Save icon looks like a floppy disk. Before flash drives and cloud storage, files could be saved on a floppy disk like the one shown in **Figure 2-8**. This disk could be inserted into a computer to save a file, and then removed from the computer and inserted into a different computer to transfer the file to the other computer. While floppy disks are no longer used often today, the design of the Save icon continues to look like a floppy disk.

Figure 2-8 A Floppy Disk

TO SAVE A FILE

When you click the Save icon to save a new Word document, you have several options to consider. ***Why?*** *This is because the Save options let you choose what to name your file, where to save it, and whether you want to share it.* The following steps save a file to OneDrive or to a local hard drive.

(1). With your Word document still open, **Click Save** on the Quick Access toolbar to open the Save this file dialog box as shown in (**Figure 2-8**).

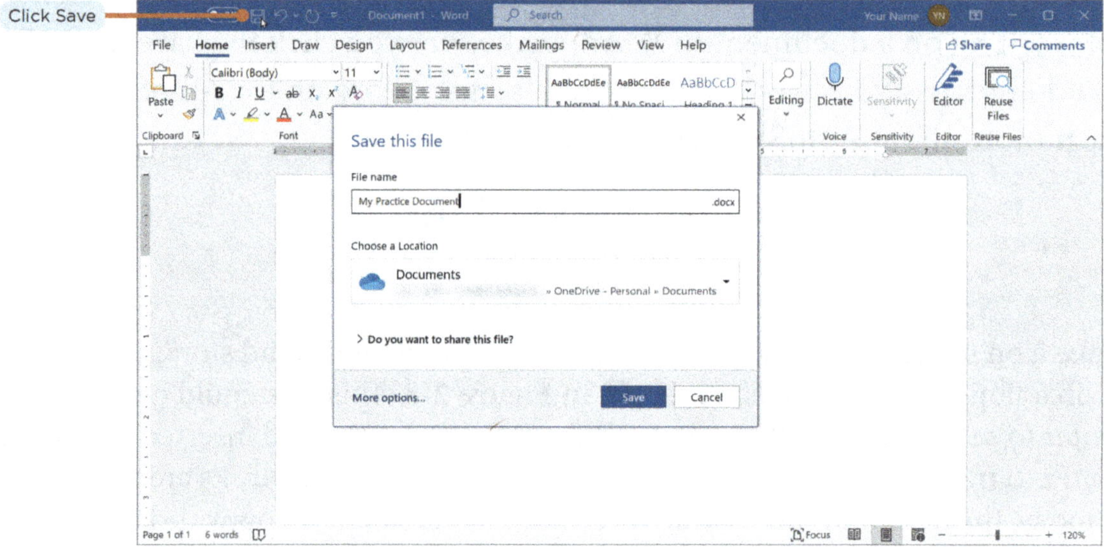

Figure 2-8 Save This File Dialog Box

Smart Tips

Naming Files

Giving your document a meaningful name will make it easier later to remember what the file contains. You can use any letter or number, and you can use some symbols. Many symbols, such as the slash /, are not allowed in file names because these symbols are used in other ways to track files on your computer.

(2). **Type** a meaningful name for your file, such as **My Practice Document** or something similar.

(3). By default, Word will probably offer to save your file in the Documents folder in your OneDrive. **Click More Options…** to open the Save As window. Here, you can choose from folders you have used recently.

(4). **Click Browse** to find a different folder on your computer or in your One-Drive. It is helpful to organize your files into folders. This makes it easier to find the file you want later.

(5). **Click Save** to save your file to the location you have selected.

INSTALLED WORD VS. WORD ONLINE

Knowing the differences between the installed, desktop version of Word and Word Online will help you make better decisions about which option will serve you best.

TOOLS

The tools available for the installed, desktop version of Word are very different than those for the free Word Online you access through your browser. **Figure 2-9** shows the Home ribbon for the installed version of Word in Windows. **Figure 2-10** shows the Home ribbon for Word in a browser.

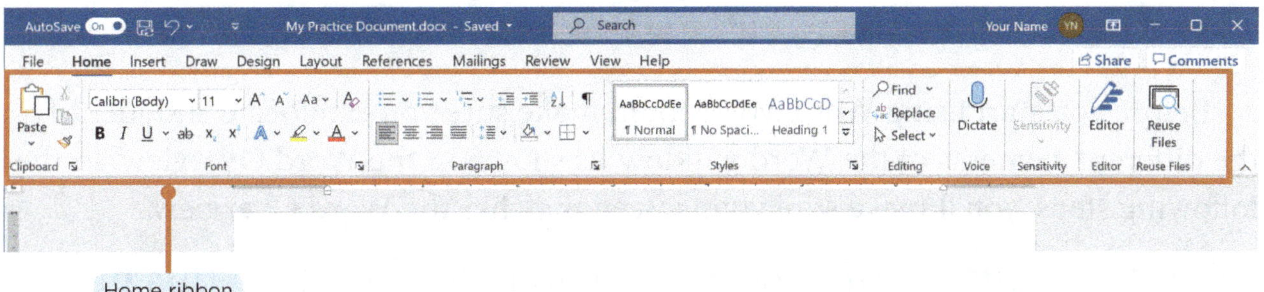

Figure 2-9 Home Ribbon in the Installed Version of Word

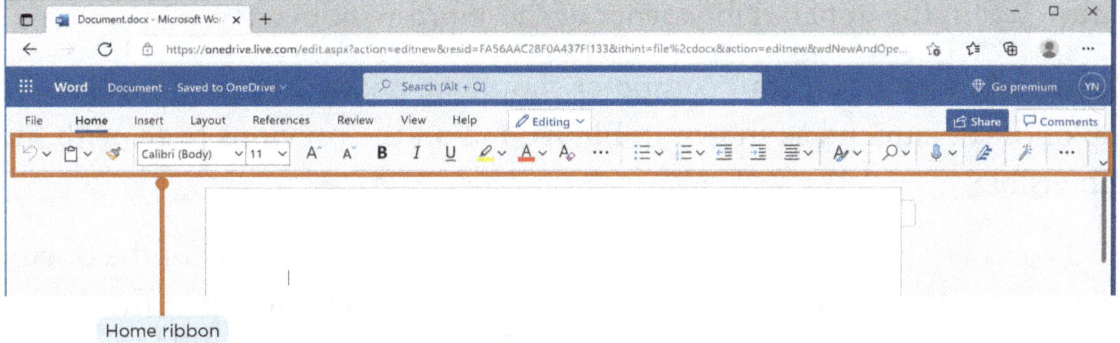

Figure 2-10 Home Ribbon in the Online Version of Word

Notice that Word Online has fewer tabs than desktop Word does. Word Online also has fewer tools on each ribbon. This course teaches you how to use some of the tools available only in the installed, desktop version of Word. As you can see in the figures, this means you will learn more about Word by practicing with the installed version.

Save Locations

Another difference between Word Online and the installed, desktop version of Word is how you save your files. Both versions of Word can save files to your OneDrive that is linked to your Microsoft account. If OneDrive is installed on your computer and you choose OneDrive as you save your file, you are saving your file in the cloud. You can access that file from any computer with Internet access, either through the OneDrive app or through the OneDrive website in a browser.

Word Online automatically saves changes to your document—notice there is no Save icon in Word Online. If you need to save a copy of your file on your local hard drive, **Click File** and then **Click Save as**. **Click Download** a **Copy** and choose a location on your hard drive.

Setting up the Word Window

Before you start using Word, you should make sure you can locate and identify the different elements of the Word window, as shown in the Word Overview. In the following steps, you'll make sure your screen matches the Word Overview.

To set up your Word window to match the figures in this book:

(1). If the Word window does not fill the entire screen, **Click** the **Maximize** button in the upper-right corner of the Word window.

The insertion point on your computer should be positioned about an inch from the top of the document, as shown in **Figure 1-2** on the previous page, with the top margin visible.

Are You Having Trouble? If the insertion point appears at the top of the document, with no white space above it, position the pointer between the top of the document and the horizontal ruler, until it changes to , double-click, and then scroll up to top of the document.

(2). On the ribbon, **Click** the **View** tab. The ribbon changes to show options for changing the appearance of the Word window.

(3). In the Show group, **Click** the **Ruler** check box to insert a checkmark, if necessary. If the rulers were not displayed, they are displayed now.

Next, you'll change the Zoom level to a setting that ensures that your Word window will match the figures in this book. To increase or decrease the screen's magnification, you could drag the slider button on the Zoom slider in the lower-right corner of the Word window. But to choose a specific Zoom level, it's easier to use the Zoom dialog box.

(4). In the Zoom group, **Click** the **Zoom** button to open the Zoom dialog box. Double-click the current value in the **Percent** box to select it, type **120**, and then **Click OK** to close the Zoom dialog box.

(5). On the status bar, click the **Print Layout** button to select it, if necessary. As shown in the Word Overview, the Print Layout button is the middle of the three View buttons located on the right side of the status bar. The Print Layout button in the Views group on the View tab is also now selected.

Caution: Before typing a document, you should make sure nonprinting characters are displayed. Nonprinting characters provide a visual representation of details you might otherwise miss. For example, the (¶) character marks the end of a paragraph, and the (•) character marks the space between words.

To verify that nonprinting characters are displayed:

(1). On the ribbon, click the **Home** tab.

(2). In the blank Word document, look for the paragraph mark (¶) in the first line of the document, just to the right of the blinking insertion point.

Are You Having Trouble? If you don't see the paragraph mark, click the Show/ Hide ¶ button ¶ in the Paragraph group.

In the Paragraph group, the Show/Hide ¶ button should be highlighted in gray, indicating that it is selected, and the paragraph mark (¶) should appear in the first line of the document, just to the right of the insertion point.

SAVING A DOCUMENT

Before you begin working on a document, you should save it with a new name.

When you click the File tab and then click Save to save a document for the first time, Word displays the Save As screen in Backstage view. On the Save As screen, you can select the location where you want to store your document. After that, you can click the Save button on the Quick Access Toolbar to save your document to the same location you specified earlier and with the same name.

To save the document:

(1). **Click** the **File Tab** on the ribbon, and then **Click Save**. Word switches to the Save As screen in Backstage view, as shown in **Figure 2-11**.

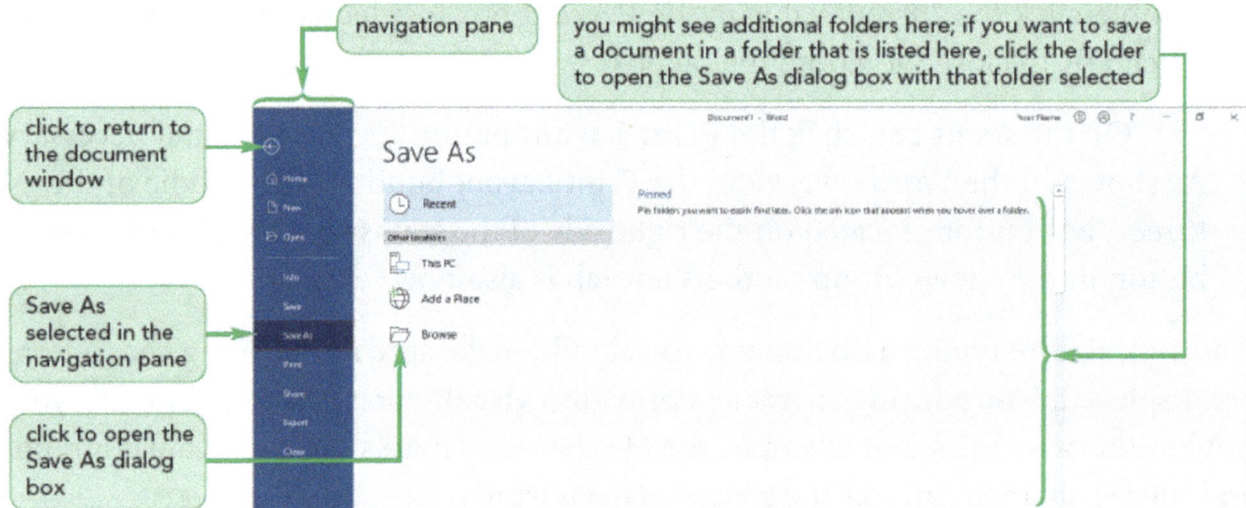

Figure 2-11 Save As screen in Backstage view

Because a document is now open, more commands are available in Backstage view than when you started Word. The **Navigation Pane** on the left contains commands for working with the open document and for changing settings that control how Word works.

(2). **Click** the **Browse** button. The **Save As** dialog box opens.

Are You Having Trouble? If your instructor wants you to save your files to your OneDrive account, click OneDrive, and then log in to your account.

(3). Navigate to the location specified by your instructor. The default file name, “Doc1,” appears in the File name box. You will change that to something more descriptive. See **Figure 2-12** on the next page.

***** Please note that the names of the file in your folder might be different from thata which is in this lab manual/textbook. It is due to textbook update.**

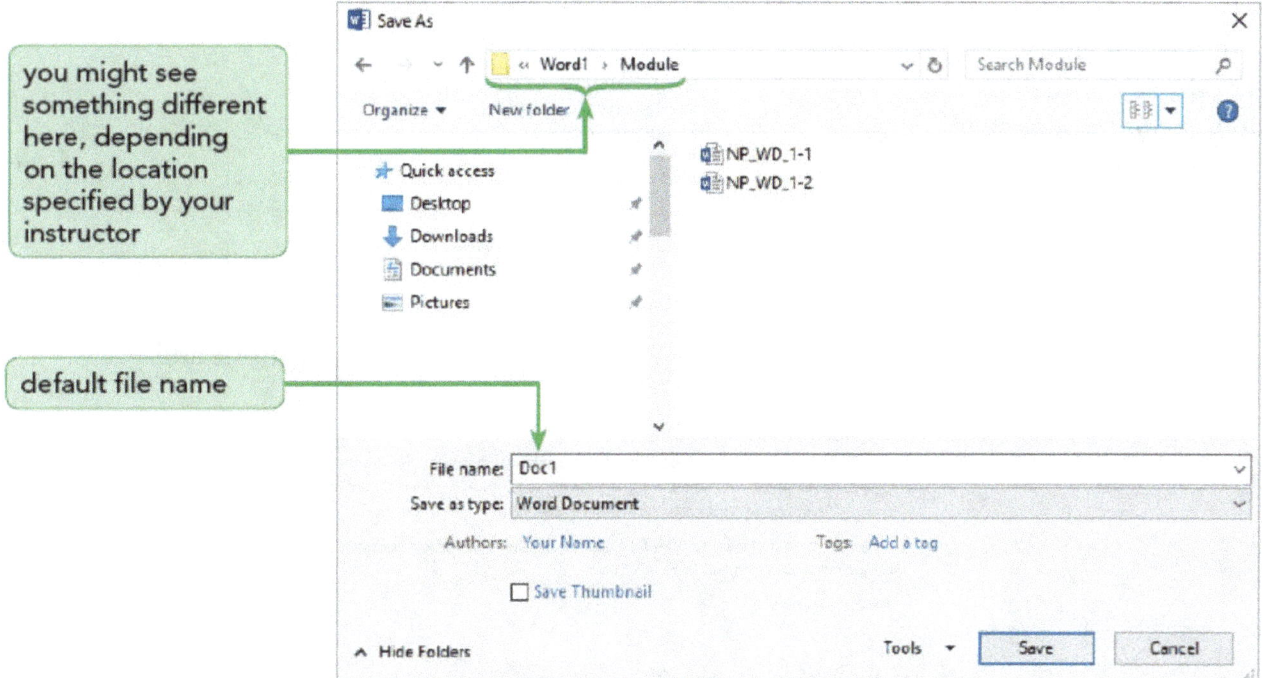

Figure 2-12 Save As dialog box

(4). **Click** the **File Name** box, and then **Type Chapter 2 Word Letter 1**. The text you type replaces the selected text in the File name box.

(5). **Click Save**. The file is saved, the dialog box and Backstage view close, and the document window appears again, with the new file name in the title bar.

Now that you have saved the document, you can begin typing the letter. You boss David Kollie has asked you to type a block-style letter to accompany some water conservation brochures that will be sent to Flomo Tunde. **Figure 2-13** on the next page shows the block-style letter you will create in this chapter.

SMARTSKILLS Written Communication: Creating a Business Letter

Several styles are considered acceptable for business letters. The main differences among the styles have to do with how parts of the letter are indented from the left margin. In the block style, which you will use in this module, each line of text starts at the left margin. In other words, nothing is indented. Another style is to indent the first line of each paragraph. The choice of style is largely a matter of personal preference, or it can be determined by the standards used in a particular business or organization. To further enhance your skills in writing business correspondence, you should consult an authoritative book on business writing that provides guidelines for creating a variety of business documents, such as Business Communication: Process & Product by Mary Ellen Guffey and Dana Loewy.

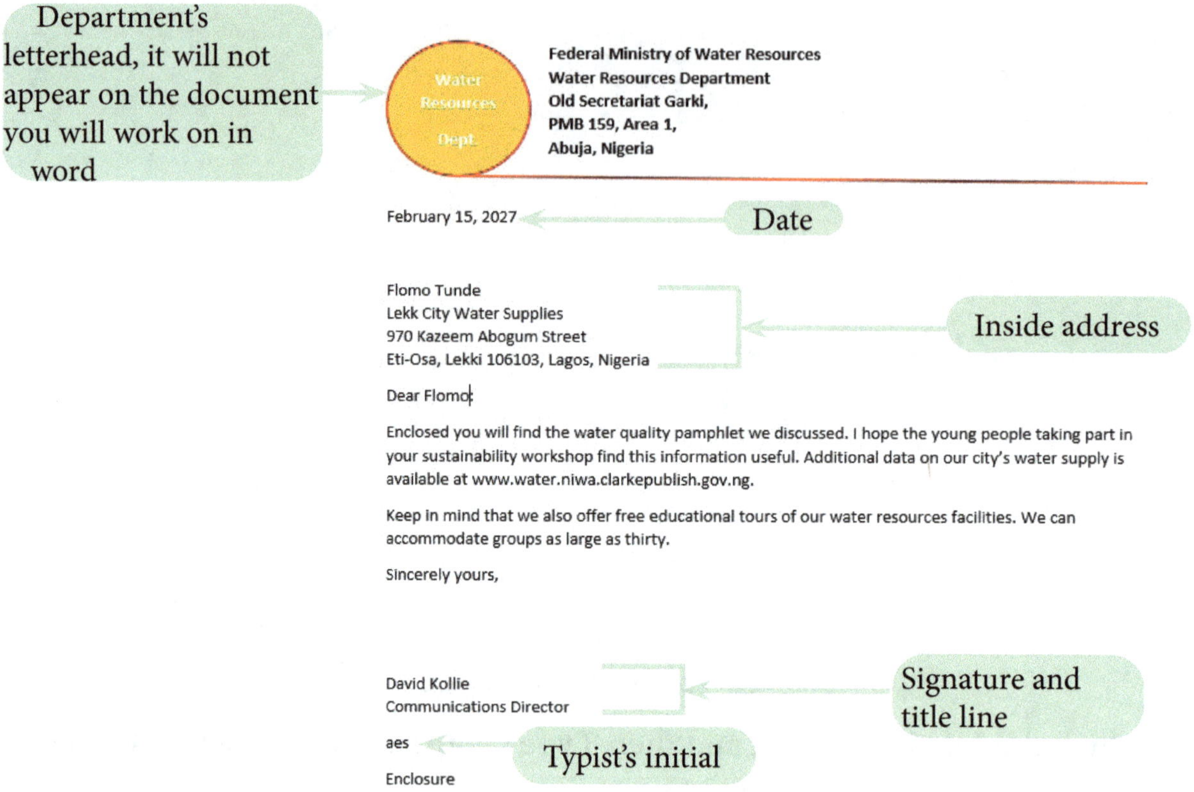

Department's letterhead, it will not appear on the document you will work on in word

Water Resources Dept.

Federal Ministry of Water Resources
Water Resources Department
Old Secretariat Garki,
PMB 159, Area 1,
Abuja, Nigeria

February 15, 2027 ← Date

Flomo Tunde
Lekk City Water Supplies
970 Kazeem Abogum Street
Eti-Osa, Lekki 106103, Lagos, Nigeria ← Inside address

Dear Flomo

Enclosed you will find the water quality pamphlet we discussed. I hope the young people taking part in your sustainability workshop find this information useful. Additional data on our city's water supply is available at www.water.niwa.clarkepublish.gov.ng.

Keep in mind that we also offer free educational tours of our water resources facilities. We can accommodate groups as large as thirty.

Sincerely yours,

David Kollie
Communications Director ← Signature and title line

aes ← Typist's initial

Enclosure

Figure 2-13 Completed block-style letter

ENTERING TEXT

The letters you type in a Word document appear at the current location of the blinking insertion point.

INSERTING A DATE WITH AUTOCOMPLETE

The first item in a block-style business letter is the date. David plans to send the letter to Flomo on February 15, so you need to insert that date into the document. To do so, you can take advantage of AutoComplete, a Word feature that automatically suggests dates and other regularly used items for you to insert. In this case, you can type the first few characters of the month and let Word insert the rest.

To insert the date:

(1). **Type Febr** (the first four letters of "February"). A ScreenTip appears above the letters, as shown in **Figure 2-14** on the next page, suggesting "February" as the complete word.

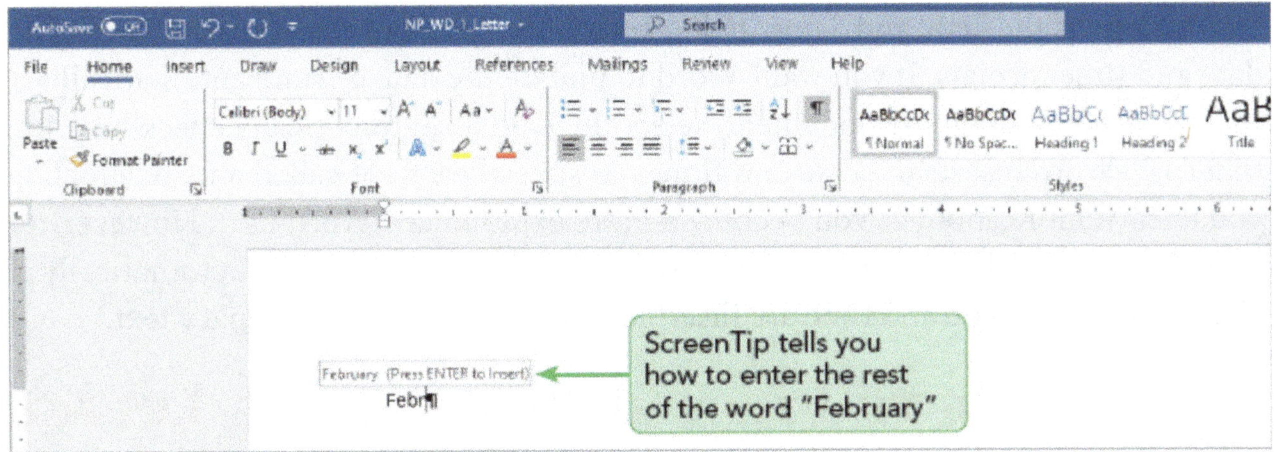

Figure 2-14 AutoComplete Suggestion

A **ScreenTip** is a label with descriptive text or an explanation about what to do next.

On the other hand, if you wanted to type something other than "February," you could continue typing to complete the word. In this case, you want to accept the AutoComplete suggestion.

(2). **Press ENTER**. The rest of the word "February" is inserted in the document. Note that AutoComplete works for long month names like February but not shorter ones like May, because "Ma" could be the beginning of many words besides "May."

(3). **Press SPACEBAR, Type 15, 2027** and then **Press ENTER Twice**, leaving a blank paragraph between the date and the line where you will begin typing the inside address, which contains the recipient's name and address. Notice the nonprinting character (•) after the word "February" and before the number "15," which indicates a space. Word inserts this nonprinting character every time you press **SPACEBAR**.

Are You Having Trouble? If February happens to be the current month, you will see a second AutoComplete suggestion displaying the current date after you press SPACEBAR. To ignore that AutoComplete suggestion, continue typing the rest of the date, as instructed in **Step 3**.

Note that you can also insert the current date (as well as the current time) by using the Insert Date and Time button in the Text group on the Insert tab.

This opens the Date and Time dialog box, where you can select from a variety of date and time formats. If you want Word to update the date or time automatically each time you reopen the document, select the Update automatically check box. In that case, Word inserts the date and time as a special element called a field, which you'll learn more about as you become a more experienced Word user. However, for typical correspondence, it makes more sense to deselect the Update automatically check box so the date and time are inserted in the document as ordinary text.

Continuing to Type the Block-Style Letter

In a block-style business letter, the inside address appears below the date, with one blank paragraph in between. Some style guides recommend including even more space between the date and the inside address. But in the short letter you are typing, more space would make the document look out of balance.

To insert the inside address:

(1). Type the following information, pressing **ENTER** after each item:

Flomo Tunde

Lekki City Water Supply

Water Resources Department

Lekki City Water Supply

970 Kazeem Abogum Street

Eti-Osa, Lekki 106103

Lagos, Nigeria

Remember to **Press ENTER** after you type the area code. Your screen should look like **Figure 2-15**. Don't be concerned if the lines of the inside address seem too far apart. You'll use the default spacing for now, and then adjust it after you finish typing the letter.

Are You Having Trouble? If you make a mistake while typing, press BACKSPACE to delete the incorrect character, and then type the correct character.

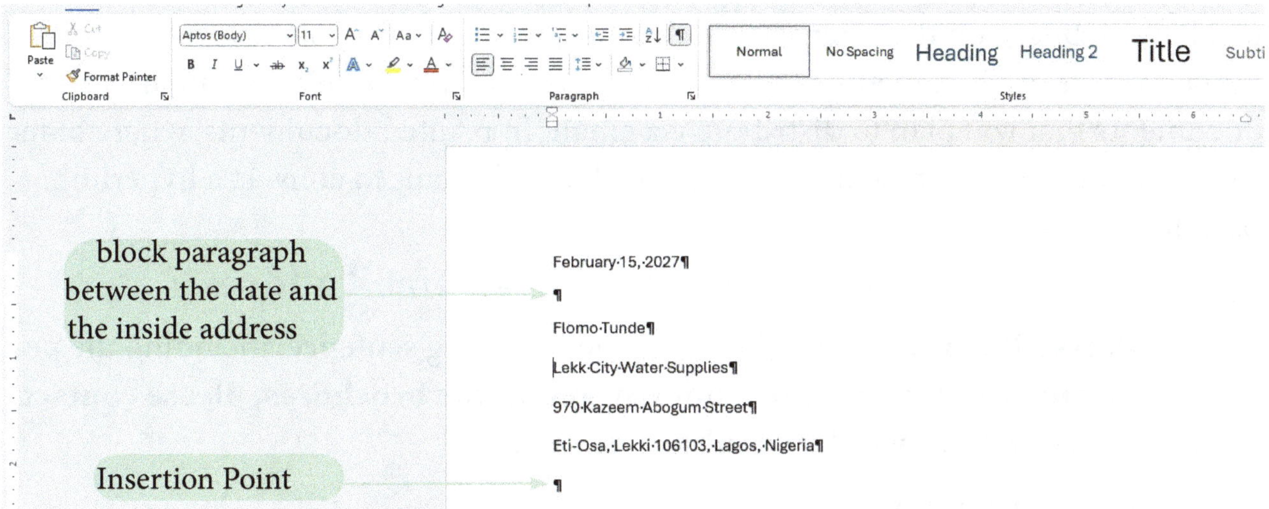

Figure 2-15 Letter With Inside Address

Now you can move on to the salutation and the body of the letter. As you type the body of the letter, notice that Word automatically moves the insertion point to a new line when the current line is full.

To type the salutation and the body of the letter:

(1). **Type Dear Flomo**: and then **Press ENTER** to start a new paragraph for the body of the letter.

(2). Type the following sentence, including the period: **Enclosed you will find the sustainability brochures we discussed**

(3). **Press SPACEBAR**. Note that you should only include one space between sentences.

(4). Type the following sentence, including the period: **Thank you for agreeing to distribute them at next month's National Water Quality Conference.**

(5). On the Quick Access Toolbar, **Click** the **Save** button. Word saves the document as **Chapter 2 Word Letter 1** to the same location you specified in this chapter earlier.

TYPING A HYPERLINK

When you type an email address and then press the SPACEBAR or ENTER, Word converts it to a hyperlink, with blue font and an underline. A **Hyperlink** is a specially formatted word, phrase, or graphic that, when clicked or tapped, lets you display a webpage on the Internet, another file, an email, or another location within the same file; it is sometimes called hypertext or a link. Hyperlinks are useful in documents that you plan to distribute via email. In printed documents, where blue font and underlines can be distracting, you'll usually want to convert a hyperlink back to regular text.

To add a sentence containing an email address:

(1). **Press SPACEBAR**, and then type the following sentence, including the period: **If you have any questions not covered in the brochures, please contact me at www.water.niwa.clarkepublish.gov.ng.**

(2). **Press ENTER**. Word converts the email address to a hyperlink, with blue font and an underline. The same thing would happen if you pressed SPACEBAR instead of ENTER.

(3). Position the pointer over the hyperlink. A ScreenTip appears, indicating that you could press and hold **CTRL** and then click the link to follow it—that is, to open an email message addressed to the Water Resources Department.

(4). With the pointer positioned over the hyperlink, **Right-Click**—that is, press the right mouse button. A shortcut menu opens with commands related to working with hyperlinks.

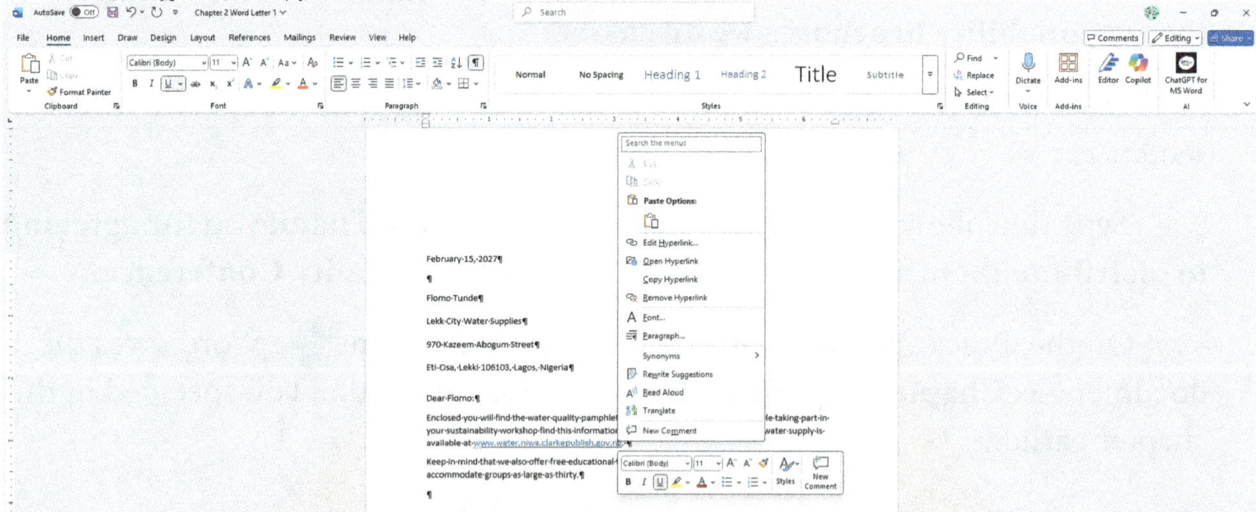

Figure 2-16 Shortcut Menu

You can right-click many items in the Word window to display a **Shortcut Menu** with commands related to the item you right-clicked. The **Mini Toolbar** also appears when you **Right-Click** or **Select Text**, giving you easy access to the buttons and settings most often used when formatting text. See **Figure 2-16** on the previous page.

(5). **Click Remove Hyperlink** in the shortcut menu. The shortcut menu and the Mini toolbar are no longer visible. The email address is now formatted in black, like the rest of the document text..

(6). On the Quick Access Toolbar, **Click** the **Save** button .

USING THE UNDO AND REDO BUTTONS

When you first open Word, you see the Undo button and the Repeat button in the Quick Access Toolbar. To undo (or reverse) the last thing you did in a document, you can click the Undo button on the Quick Access Toolbar. Once you click the Undo button, the Repeat button is replaced with the Redo button. To restore your original change, click the Redo button, which reverses the action of the Undo button (or redoes the undo). To undo more than your last action, you can continue to click the Undo button, or you can click the Undo arrow on the Quick Access Toolbar to open a list of your most recent actions. When you click an action in the list, Word undoes every action in the list up to and including the action you clicked.

David asks you to change "sustainability" to "water conservation" in the first sentence you typed. You'll make the change now. If David decides he doesn't like it after all, you can always undo it. To delete a character, space, or blank paragraph to the right of the insertion point, you press DEL, or to delete an entire word, you can press CTRL+DEL. To delete a character, space, or blank paragraph to the left of the insertion point, you press BACKSPACE, or to delete an entire word, you can press CTRL+BACKSPACE.

To change the word "sustainability":

(1). **Press** the ↑ key twice and then press the ← key as necessary to move the insertion point to the left of the first "s" in the word "sustainability."

(2). **Press** and hold **CTRL**, and then press **DEL** to delete the word "sustainability."

(3). Type **water conservation** as a replacement, and then press **SPACEBAR**. After reviewing the sentence, David decides he prefers the original wording, so you'll undo the change.

(4). On the Quick Access Toolbar, **click** the **Undo** button . The phrase "water conservation" is removed from the sentence.

(5). **Click** the **Undo** button again to restore the word "sustainability."

David decides that he does want to use "water conservation" after all. Instead of retyping it, you'll redo the undo.

(6). On the Quick Access Toolbar, **Click** the **Redo** button twice. The phrase "water conservation" replaces "sustainability" in the document, so that the phrase reads "… the water conservation brochures we discussed."

(7). **Press** and hold **CTRL**, and then **Press END** to move the insertion point to the blank paragraph at the end of the document.

Are You Having Trouble? If you are working on a small keyboard, you might need to press and hold a key labeled "Function" or "FN" before pressing END.

(8). On the Quick Access Toolbar, **Click** the **SAVE** button . Word saves your letter with the same name and to the same location you specified earlier.

In the previous steps, you used the arrow keys and a key combination to move the insertion point to specific locations in the document. For your reference, **Figure 2-17** summarises the most common keystrokes for moving the insertion point in a document.

To Move the Insertion Point	Press
Left or right one character at a time	← or →
Up or down one line at a time	↑ or ↓
Left or right one word at a time	CTRL+ ← or CTRL+ →
Up or down one paragraph at a time	CTRL+ ↑ or CTRL+ ↓
To the beginning or to the end of the current line	HOME or END
To the beginning or to the end of the document	CTRL+HOME or CTRL+END
To the previous screen or to the next screen	PAGE UP or PAGE DOWN
To the top or to the bottom of the document window	ALT+CTRL+PAGE UP or ALT+CTRL+PAGE DOWN

Figure 2-17 Keystrokes for moving the insertion point

CORRECTING ERRORS AS YOU TYPE

As you have seen, you can press BACKSPACE or DEL to remove an error, and then type a correction. In many cases, however, the AutoCorrect feature will do the work for you. Among other things, **AutoCorrect** automatically detects and corrects common typing errors, such as typing "adn" instead of "and." For example, you might have noticed AutoCorrect at work if you forgot to capitalize the first letter in a sentence as you typed the letter. After you type this kind of error, AutoCorrect automatically corrects it when you press SPACEBAR, TAB, or ENTER.

Word draws your attention to other potential errors by marking them with **Red Squiggles** or **Underlines**. If you type a word that doesn't match the correct spelling in the Word dictionary, or if a word is not in the dictionary at all, a wavy red line appears beneath it. A wavy red underline also appears if you mistakenly type the same word twice in a row. Misused words (for example, "you're" instead of "your") are underlined with a double blue line, as are problems with punctuation and potential grammar errors, such as a singular verb used with a plural subject. Possible wordiness is marked with a dotted purple underline, although keep in mind that this feature does not produce consistent results. Word might mark a phrase as wordy in one document, but then not mark the same phrase in a different document. This feature can be a helpful guide, but ultimately you'll need to make your own decisions about whether a phrase could be more concise.

You'll see how this works as you continue typing the letter and make some intentional typing errors.

To learn more about correcting errors as you type:

(1). **Type** the following sentence, including the errors: **In you're phone message, you asked me me to include the maps of the city of Lekki reservoirres. They are not back from the printer yet, but I will send them as soon as they actually ready.**

As you type, AutoCorrect changes the lowercase "i" at the beginning of the sentence to uppercase. It also changes "mesage" to "message" and "teh" to "the." Also, the incorrectly used word "you're" is marked with a double blue underline. The second "me" and the spelling error "reservoirres" are marked with wavy red underlines.

(2). **Press ENTER**. One additional error is now visible—the phrase “actually ready” is marked with a dotted purple underline, indicating a lack of conciseness. See **Figure 2-18**.

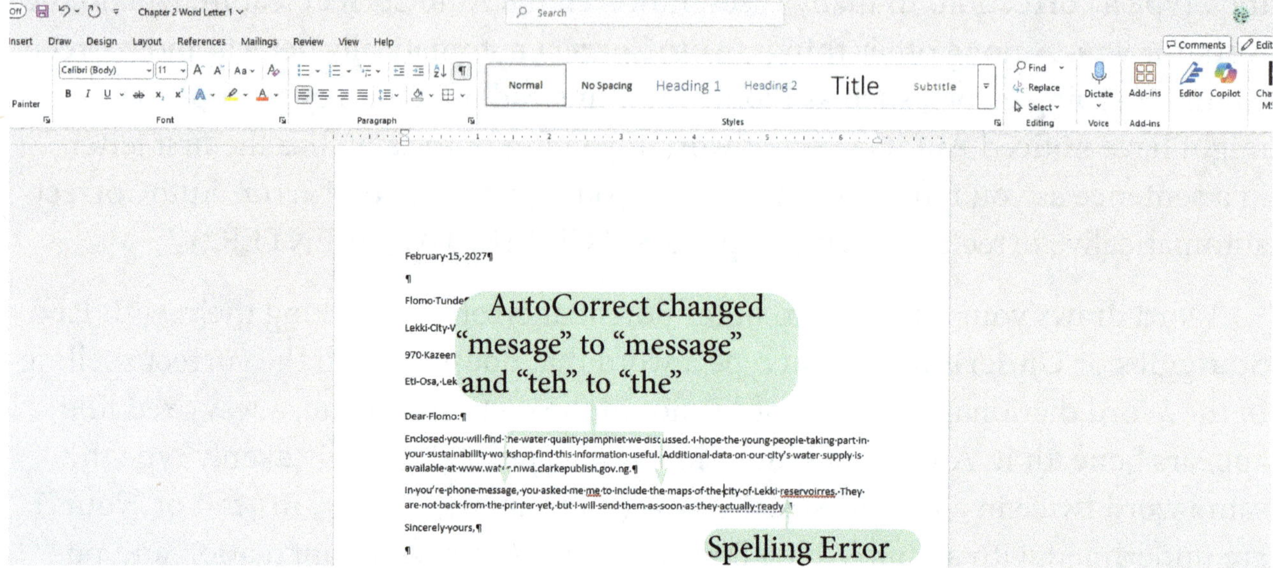

Figure 2-18 Errors marked in the document

To correct an error marked with an underline, you can right-click the error and then click a replacement in the shortcut menu. If you don’t see the correct word in the shortcut menu, click anywhere in the document to close the menu, and then type the correction yourself. You can also bypass the shortcut menu entirely and simply delete the error and type a correction.

To correct the spelling, grammar, and wordiness errors:

(1). **Right-Click** you’re to display the shortcut menu shown in **Figure 2-19**.

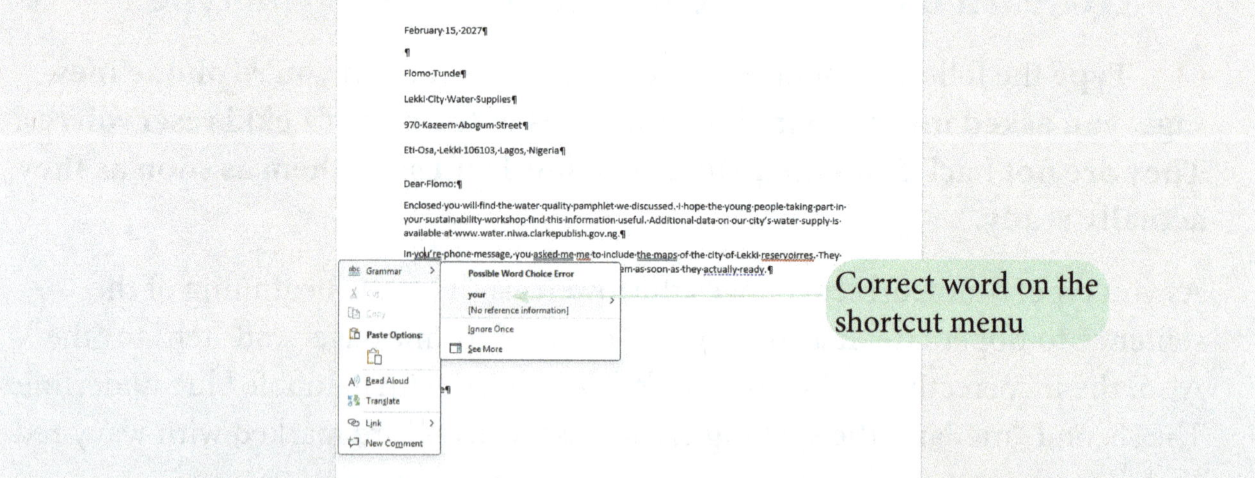

Figure 2-19 Shortcut Menu with Suggested Spelling

Are You Having Trouble? If you see a shortcut menu other than the one shown in **Figure 1-11**, you didn't right-click exactly on the word "you're." Press ESC to close the menu, and then repeat **Step 1**.

(2). On the shortcut menu, **Click your**. The correct word is inserted into the sentence, and the shortcut menu closes.

(3). Use a shortcut menu to replace the spelling error "reservoirres" with the correct word "reservoirs."

You could use a shortcut menu to remove the second instance of "me," but in the next step you'll try a different method—selecting the word and deleting it.

(4). **Double-click** anywhere in the underlined word **me**. The word and the space following it are highlighted in gray, indicating that they are selected. The Mini toolbar is also visible, but you can ignore it.

Are You Having Trouble? If the entire paragraph is selected, you triple-clicked the word by mistake. Click anywhere in the document to deselect it, and then repeat **Step 4**.

(5). **Press DEL**. The second instance of "me" and the space following it are deleted from the sentence. Finally, you need to correct the error related to concise language.

(6). **Right-Click** the phrase **actually ready** and use the shortcut menu to choose the more concise option, **ready**.

(7). On the Quick Access Toolbar, click the **Save** button.

You can see how quick and easy it is to correct common typing errors with AutoCorrect and the multicolored underlines, especially in a short document that you are typing yourself. If you are working on a longer document or a document typed by someone else, you'll also want to have Word check the entire document for errors. You'll learn how to do this soon in this lab manual.

Next, you'll finish typing the letter.

To finish typing the letter:

(1). **Press CTRL+END**. The insertion point moves to the end of the document.

(2). **Type Sincerely yours,** (including the comma).

(3). **Press ENTER** three times to leave space for the signature.

(4). **Type David Kollie** and then **press ENTER**. Because David's last name is not in the Word dictionary, a red squiggle appears below it. You can ignore this for now.

(5). Type your first, middle, and last initials in lowercase, and then press **ENTER**. AutoCorrect wrongly assumes your first initial is the first letter of a new sentence and changes it to uppercase. If your initials do not form a word, a red squiggle appears beneath them. You can ignore this for now.

(6). On the Quick Access Toolbar, **Click** the **Undo** button . Word reverses the change, replacing the uppercase initial with a lowercase one.

(7). **Type Enclosure** so your screen looks like **Figure 2-20**.

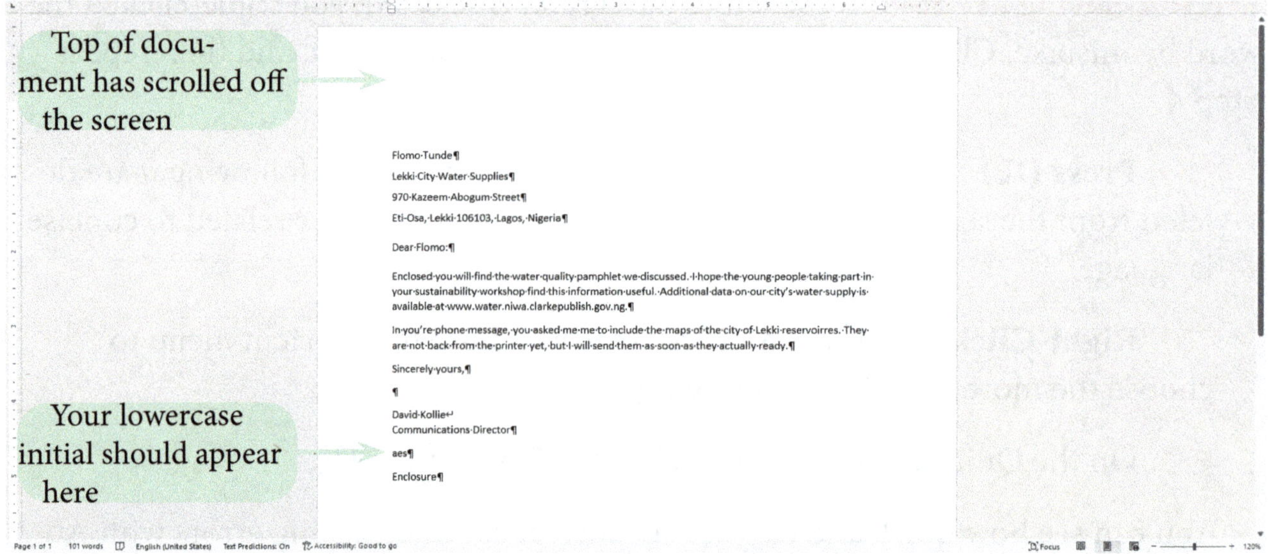

Figure 2-20 Letter to Flomo Tunde

Notice that as you continue to add lines to the letter, the top part of the letter scrolls off the screen. For example, in **Figure 2-20**, you can no longer see the date. Don't be concerned if more or less of the document has scrolled off the screen on your computer.

(8). Save the document.

PROOFREADING A DOCUMENT

After you finish typing a document, you need to proofread it carefully from start to finish. Part of proofreading a document in Word is removing all wavy underlines, either by correcting the text or by telling Word to ignore the underlined text because it isn't really an error. For example, David's last name is marked as an error, when in fact it is spelled correctly. You need to tell Word to ignore "Kollie" wherever it occurs in the letter. You need to do the same for your initials.

To proofread and correct the remaining marked errors in the letter:

(1). **Right-Click Kollie**. A shortcut menu opens.

(2). On the shortcut menu, **Click Ignore All** to indicate that Word should ignore the word "Kollie" each time it occurs in this document. (The Ignore All option can be particularly helpful in a longer document.) The wavy red underline disappears from below David's last name.

(3). If you see a wavy red underline below your initials, right-click your initials. On the shortcut menu, **Click Ignore All** to remove the red wavy underline. To choose to ignore something just once in a document, you could click See More in the shortcut menu, and then click Ignore Once in the Editor pane. You'll learn how to use the Editor pane in this lab textbook

(4). Read the entire letter to proofread it for typing errors. Correct any errors using the techniques you have just learned.

Are You Having Trouble? If you have not already developed the habit of proofreading your text messages, emails, essays, letters, resume, reports, and important documents, you should start doing so right now. What you write as a person represents you. What you write will somehow open or close doors onto you.

(5). **Scroll Up**, if necessary, so you can see the complete inside address, which you'll work on next, and then save the document.

The text of the letter is finished. Now you need to think about its appearance—that is, you need to think about the document's **formatting**. First, you need to adjust the spacing in the inside address.

ADJUSTING PARAGRAPH AND LINE SPACING

When typing a letter, you might need to adjust two types of spacing—paragraph spacing and line spacing. **Paragraph Spacing** is the space that appears directly above and below a paragraph. In Word, any text that ends with a paragraph mark symbol (¶) is a paragraph. So, a **Paragraph** can be a group of words that is many lines long, a single word, or even a blank line, in which case you see a paragraph mark alone on a single line. A paragraph can also contain a picture instead of text. Paragraph spacing is measured in points; a **Point** is **1/72** of an inch. The default setting for paragraph spacing in Word is 0 points before each paragraph and 8 points after each paragraph. When laying out a complicated document, resist the temptation to simply press ENTER to insert extra space between paragraphs. Changing the paragraph spacing gives you much more control over the final result.

Line Spacing is the space between lines of text within a paragraph. Word offers a number of preset line spacing options. The **1.0** setting, which is often called **Single-Spacing**, allows the least amount of space between lines. All other line spacing options are measured as multiples of 1.0 spacing. For example, **2.0** spacing (sometimes called **Double-Spacing**) allows for twice the space of single-spacing. The default line spacing setting is 1.08, which allows a little more space between lines than 1.0 spacing.

Now consider the line and paragraph spacing in the letter. The four lines of the inside address are too far apart. That's because each line of the inside address is actually a separate paragraph. Word inserted the default 8 points of paragraph spacing after each of these separate paragraphs. See **Figure 2-21** on the next page shows line and paragraph spacing.

To follow the conventions of a block-style business letter, the four paragraphs that make up the inside address should have the same spacing as the lines of text within a single paragraph—that is, they need to be closer together. You can accomplish this by removing the 8 points of paragraph spacing after the first two paragraphs in the inside address. To conform to the block-style business letter format, you also need to close up the spacing between your initials and the word "Enclosure" at the end of the letter.

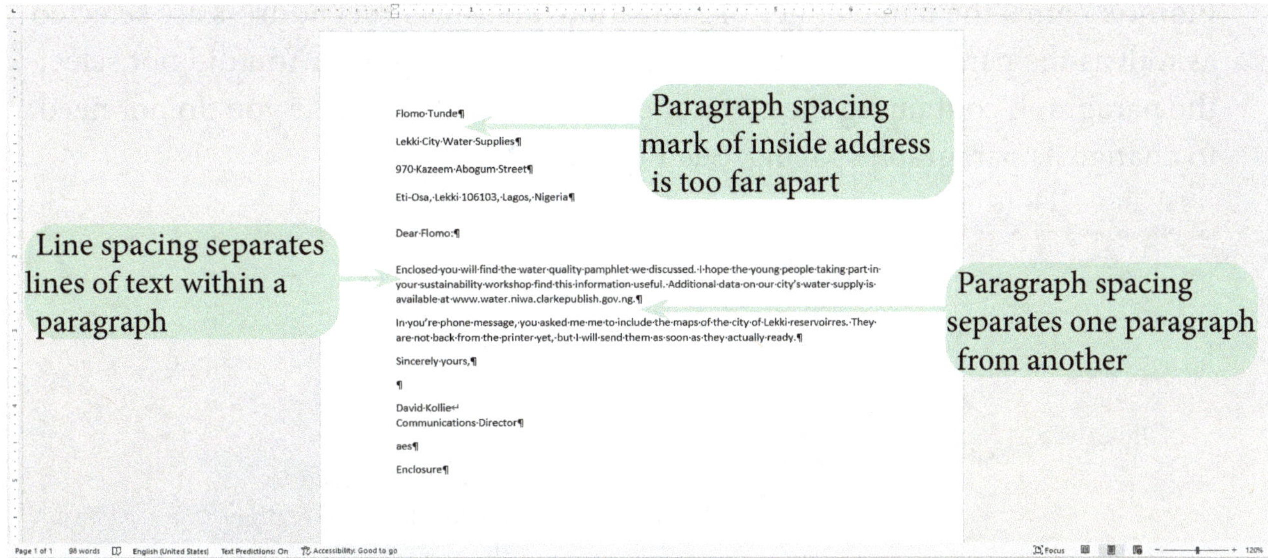

Figure 2-21 Line and paragraph spacing in the Letter to Flomo Tunde

To adjust paragraph and line spacing in Word, you use the Line and Paragraph Spacing button in the Paragraph group on the Home tab. Clicking this button displays a menu of preset line spacing options (1.0, 1.15, 2.0, and so on). The menu also includes two paragraph spacing options that allow you to add 12 points before a paragraph or remove the default 8 points of space after a paragraph.

Next you'll adjust the paragraph spacing in the inside address and after your initials. In the process, you'll also learn some techniques for selecting text in a document.

To adjust the paragraph spacing in the inside address and after your initials:

(1). **Move** the **pointer** to the white space just to the left of "Flomo Tunde" until it changes to a right-pointing arrow .

(2). **Click** the **Mouse button**. The entire name, including the paragraph symbol after it, is selected.

Are You Having Trouble? If the Mini toolbar obscures your view of David's name, move the pointer away from the address to close the Mini toolbar.

(3). **Press** and **Hold** the **Mouse button**, drag the pointer down to select the next two paragraphs of the inside address as well, and then release the mouse button.

Flomo's name, the name of her organization, and the street address are selected as well as the paragraph marks at the end of each paragraph. You did not select the paragraph containing the city, state, and postcode because you do not need to change its paragraph spacing. See **Figure 2-22**.

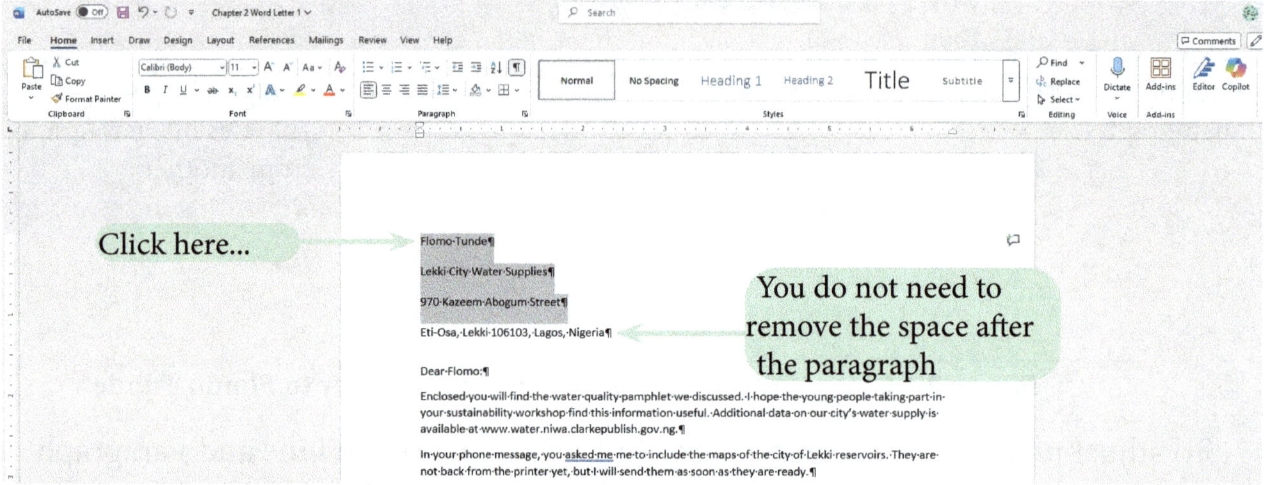

Figure 2-22 Inside address selected

(4). Make sure the Home tab is selected on the ribbon.

(5). In the Paragraph group on the Home tab, **Click** the **Line** and **Paragraph Spacing** button . A menu of line spacing options appears, with two paragraph spacing options at the bottom. See **Figure 2-23**.

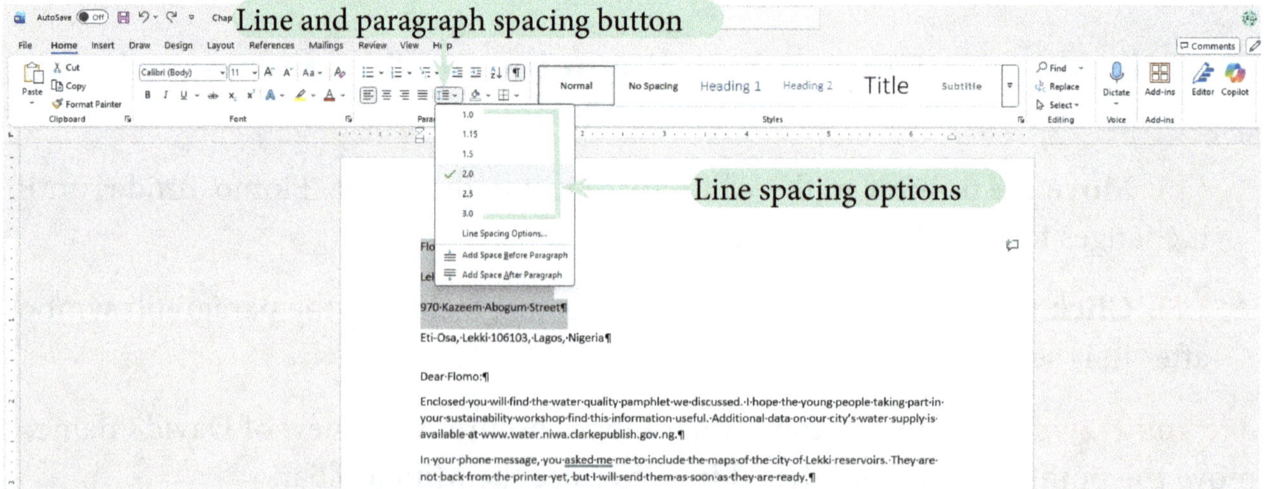

Figure 2-23 Line and paragraph spacing options

At the moment, you are interested only in the paragraph spacing options. Your goal is to remove the default 8 points of space after the first two paragraphs in the inside address.

(6). **Click Remove Space After Paragraph**. The menu closes, and the paragraphs are now closer together.

(7). **Double-Click** your initials to select them and the paragraph symbol after them.

(8). In the Paragraph group, **Click** the **Line and Paragraph Spacing** button, **Click Remove Space After Paragraph**, and then click anywhere in the document to deselect your initials.

Another way to compress lines of text is to press SHIFT+ENTER at the end of a line. This inserts a **Manual Line Break**, also called a **Soft Return**, which moves the insertion point to a new line without starting a new paragraph. You will use this technique now as you add David's title below his name in the signature line.

To use a manual line break to move the insertion point to a new line without starting a new paragraph:

(1). Click to the right of the "e" in "Kollie."

(2). **Press SHIFT+ENTER**. Word inserts a small arrow symbol, indicating a manual line break, and the insertion point moves to the line below David's name.

(3). **Type Communications Director**. David's title now appears directly below his name with no intervening paragraph spacing, just like the lines of the inside address.

(4). **Save** the document.

SMARTSKILLS Understanding Spacing between Paragraphs

When discussing the correct format for letters, many business style guides talk about single-spacing and double-spacing between paragraphs. In these style guides, to single-space between paragraphs means to press ENTER once after each paragraph. Likewise, to double-space between paragraphs means to press ENTER twice after each paragraph. With the default paragraph spacing in Word, however, you need to press ENTER only once after a paragraph. The space Word adds after a paragraph is not quite the equivalent of double-spacing, but it is enough to make it easy to see where one paragraph ends and another begins. Keep this in mind if you're accustomed to pressing ENTER twice; otherwise, you could end up with more space than you want between paragraphs.

As you corrected line and paragraph spacing in the previous set of steps, you used the mouse to select text. Word provides multiple ways to select, or highlight, text as you work. Figure 1-16 summarises these methods and explains when to use them most effectively. Note that there are multiple ways to select each element in a document. Three especially useful options are:

1. selecting an entire paragraph by triple-clicking it;
2. selecting nonadjacent text by pressing and holding CTRL, and then dragging the mouse pointer to select multiple blocks of text; and
3. selecting an entire document by pressing CTRL+A.

Figure 2-24 Selecting Texts

To Select	Mouse	Keyboard	Mouse and Keyboard
A word	Double-click the word	Move the insertion point to the beginning of the word, press and hold CTRL+SHIFT, and then press →	
A line	Click in the white space to the left of the line	Move the insertion point to the beginning of the line, press and hold SHIFT, and then press ↓	
A sentence	Click at the beginning of the sentence, then drag the pointer until the sentence is selected		Press and hold CTRL, then click any location within the sentence

Multiple lines	Click and drag in the white space to the left of the lines	Move the insertion point to the beginning of the first line, press and hold SHIFT, and then press ↓ until all the lines are selected	
A paragraph	Double-click in the white space to the left of the paragraph, or triple-click at any location within the paragraph	Move the insertion point to the beginning of the paragraph, press and hold CTRL+SHIFT, and then press ↓	
Multiple paragraphs	Click in the white space to the left of the first paragraph you want to select, and then drag to select the remaining paragraphs	Move the insertion point to the beginning of the first paragraph, press and hold CTRL+SHIFT, and then press ↓ until all the paragraphs are selected	
An entire document	Triple-click in the white space to the left of the document text	Press CTRL+A	Press and hold CTRL, and click in the white space to the left of the document text
A block of text	Click at the beginning of the block, then drag the pointer until the entire block is selected		Click at the beginning of the block, press and hold SHIFT, and then click at the end of the bloc
Nonadjacent blocks of text			Press and hold CTRL, then drag the mouse pointer to select multiple blocks of nonadjacent text

ADJUSTING THE MARGINS

Another important aspect of document formatting is the amount of margin space between the document text and the edge of the page. You can check the document's margins by changing the Zoom level to display the entire page.

To change the Zoom level to display the entire page:

(1). On the ribbon, **Click** the **View Tab**.

(2). In the Zoom group, **Click** the **One Page** button. The entire document is now visible in the Word window. See **Figure 2-25**.

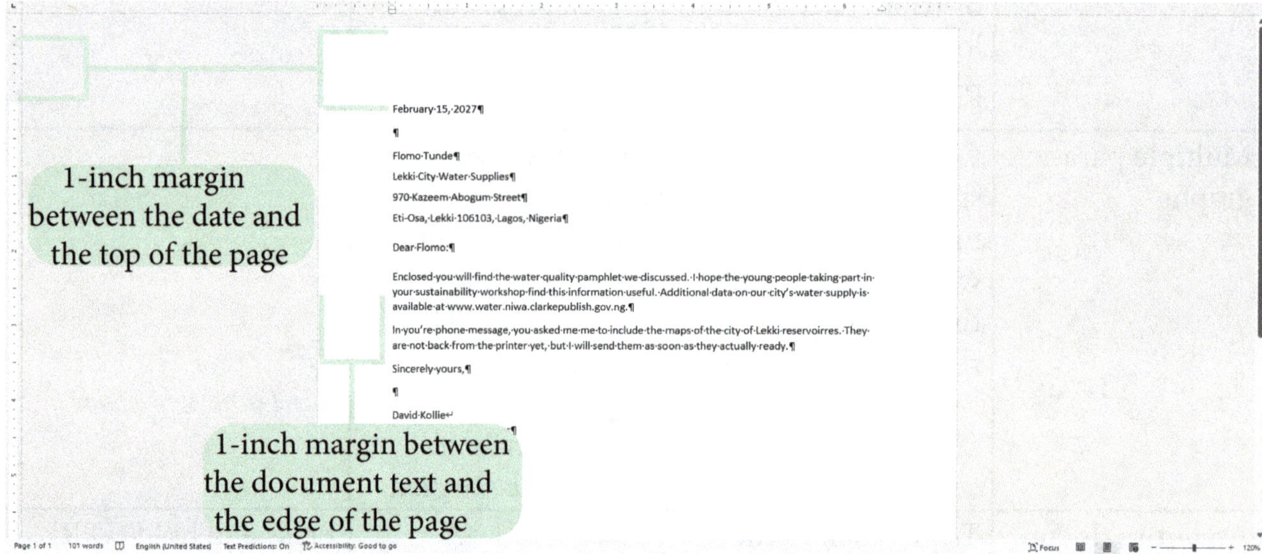

Figure 2-25 Document zoomed to show entire page

On the rulers, the margins appear dark gray. By default, Word documents include 1-inch margins on all sides of the document. By looking at the vertical ruler, you can see that the date in the letter, the first line in the document, is located 1 inch from the top of the page. Likewise, the horizontal ruler indicates the document text begins 1 inch from the left edge of the page.

Reading the measurements on the rulers can be tricky at first. On the horizontal ruler, the 0-inch mark is like the origin on a number line. You measure from the 0-inch mark to the left or to the right. On the vertical ruler, you measure up or down from the 0-inch mark.

David plans to print the letter on the Water Resources Department letterhead, which includes a graphic and the department's address. To allow more blank space for the letterhead, and to move the text down so that it doesn't look so crowded at the top of the page, you need to increase the top margin. The settings for changing the page margins are located on the Layout tab on the ribbon.

To change the page margins:

(1). On the ribbon, **Click** the **Layout Tab**. The Layout tab displays options for adjusting the layout of your document.

(2). In the Page Setup group, **Click** the **Margins** button. The Margins gallery opens, as shown in **Figure 2-26**.

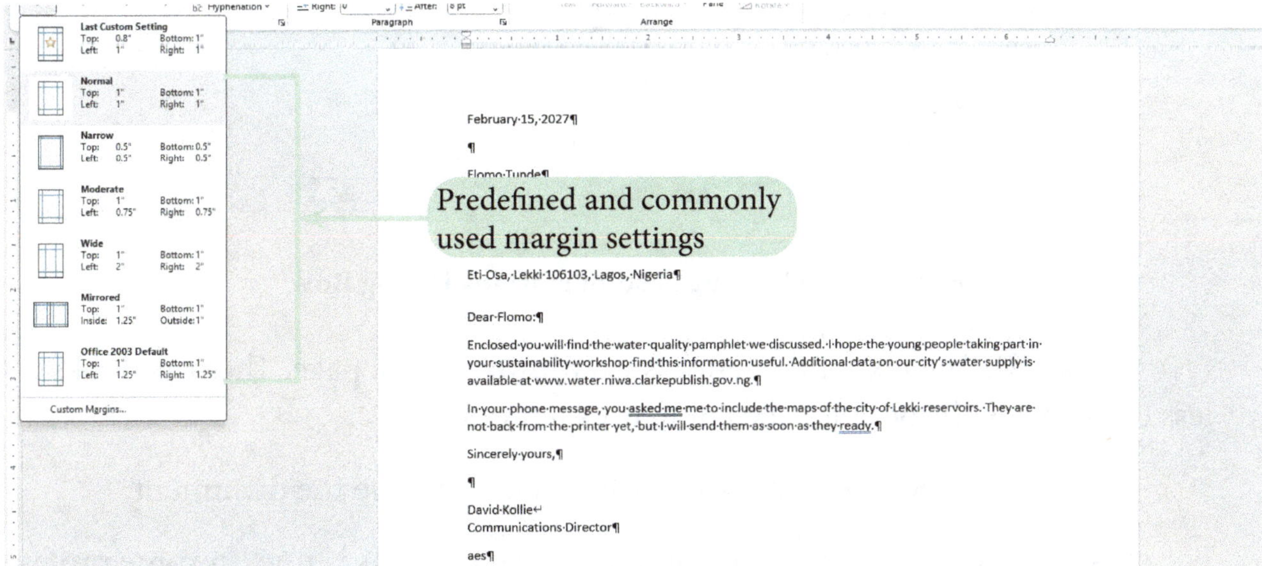

Figure 2-26 Margins gallery

In the Margins gallery, you can choose from a number of predefined margin options, or you can click the Custom Margins command to select your own settings. After you create custom margin settings, the most recent set appears as an option at the top of the menu. For the current document, you will create custom margins.

(3). **Click Custom Margins**. The Page Setup dialog box opens with the Margins tab displayed. The default margin settings are displayed in the boxes at the top of the Margins tab. The top margin of **1"** is already selected, ready for you to type a new margin setting.

(4). In the Top box in the Margins section, **Type 2.5**. You do not need to type an inch mark ("). See **Figure 2-27**.

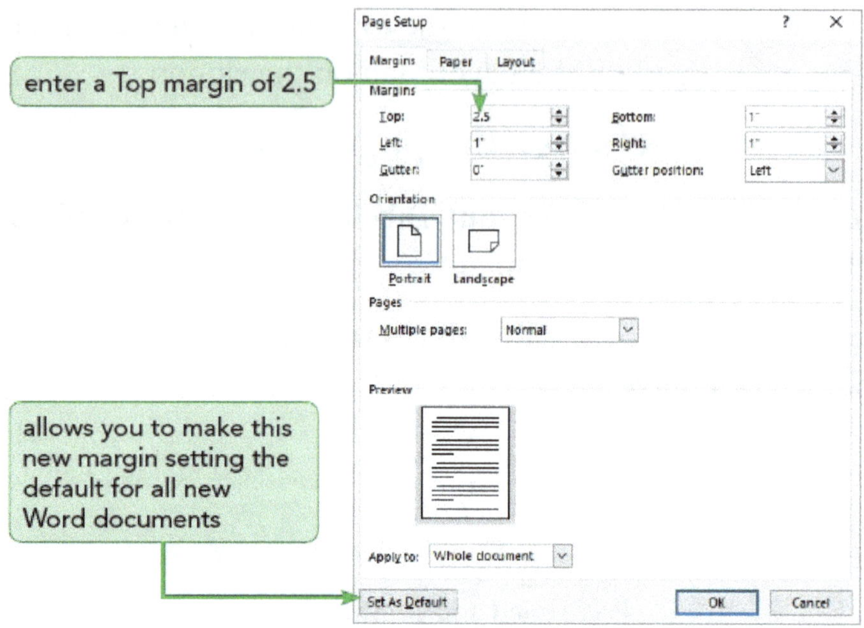

Figure 2-27 Creating custom margins Dialog Box

(5). **Click OK**. The text of the letter is now lower on the page. The page looks less crowded, with room for the company's letterhead.

(6). Change the Zoom level back to **120%**, and then save the document.

For most documents, the Word default of 1-inch margins is fine. In some professional settings, however, you might need to use a particular custom margin setting for all your documents. In that case, define the custom margins using the Margins tab in the Page Setup dialog box, and then click the Set As Default button to make your settings the default for all new documents. Keep in mind that most printers can't print to the edge of the page; if you select custom margins that are too narrow for your printer's specifications, Word alerts you to change your margin settings.

PREVIEWING AND PRINTING A DOCUMENT

To make sure the document is ready to print, and to avoid wasting paper and time, you should first review it in Backstage view to make sure it will look right when printed. Like the One Page zoom setting you used earlier, the Print option in

Backstage view displays a full-page preview of the document, allowing you to see how it will fit on the printed page. However, you cannot actually edit this preview. It simply provides one last opportunity to look at the document before printing.

To preview the document:

(1). Proofread the document one last time and correct any remaining errors.

(2). **Click** the File tab to display Backstage view.

(3). In the navigation pane, **Click Print**.

The Print screen displays a full-page version of your document, showing how the letter will fit on the printed page. The Print settings to the left of the preview allow you to control a variety of print options. For example, you can change the number of copies from the default setting of "1." The 1 Page Per Sheet button opens a menu where you can choose to print multiple pages on a single sheet of paper or to scale the printed page to a particular paper size. You can also use the navigation controls at the bottom of the screen to display other pages in a document. See **Figure 2-28**.

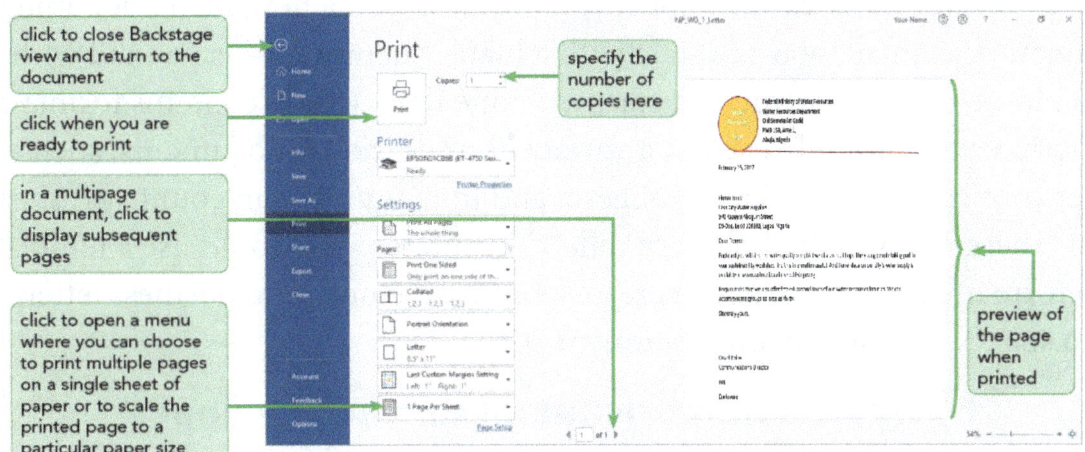

Figure 2-28 Print settings in Backstage view

(4). Review your document and make sure its overall layout matches that of the document in Figure **2-28**. If you notice a problem with paragraph breaks or spacing, **Click** the **Back** button at the top of the navigation pane to return to the document, make any necessary changes, and then start again at **Step 2**.

At this point, you can print the document or you can leave Backstage view and return to the document in Print Layout view. In the following steps, you should print the document only if your instructor asks you to. If you will be printing the document, make sure your printer is turned on and contains paper.

To leave Backstage view or to print the document:

(1). **Click** the **Back** button at the top of the navigation pane to leave Backstage view and return to the document in Print Layout view, or **Click** the **Print** button. Backstage view closes, and the letter prints if you clicked the Print button.

(2). **Click** the **File** tab, and then **Click Close** in the navigation pane to close the document without closing Word.

Next, David asks you to create an envelope he can use to send a water quality report to an environmental engineering publication.

Creating an Envelope

Before you can create the envelope, you need to open a new, blank document. To create a new document, you can start with a blank document—as you did with the letter to Flomo Tunde—or you can start with one that already contains formatting and generic text commonly used in a variety of professional documents, such as a fax cover (not common in Liberia, Nigeria and in many African countries) sheet or a memo. These preformatted files are called **Templates**. You could use a template to create a formatted envelope, but to create a basic envelope for a business letter, it's better to start with a new, blank document.

To create a new document for the envelope:

(1). **Click** the **File tab**, and then **Click New** in the navigation pane. The New screen is similar to the one you saw when you first started Word, with a blank document in the upper-left corner, along with a variety of templates. Kindly see **Figure 2-29** on the next page.

(2). **Click Blank Document**. A new document named Document2 opens in the document window, with the Home tab selected on the ribbon.

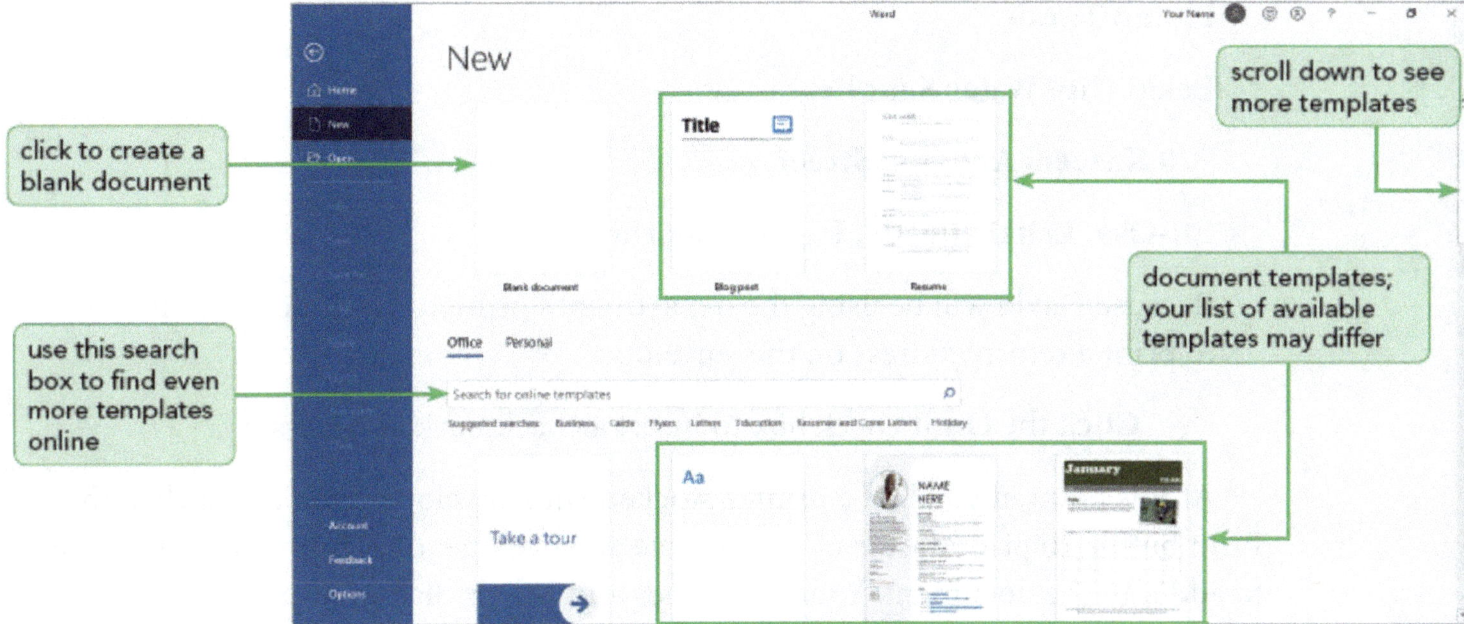

Figure 2-28 New options in Backstage view

(3). If necessary, change the Zoom level to **120%**, and display nonprinting characters and the rulers.

(4). **Save** the new document as **Chapter 2 Word Envelope** in the location specified by your instructor.

To create the envelope:

(1). On the ribbon, **Click** the **Mailings Tab**. The ribbon changes to display the various Mailings options.

Are You Having Trouble? If you like, you can select/highlight the inside address before clicking the **Mailing Tab**. By doing this, the address will auto populate.

(2). In the Create group, **Click** the **Envelopes** button. The Envelopes and Labels dialog box opens, with the Envelopes tab displayed. The insertion point appears in the Delivery address box, ready for you to type the recipient's address. Depending on how your computer is set up, and whether you are working on your own computer or a school computer, you might see an address in the Return address box.

(3). In the Delivery address box, type the following address, **Pressing ENTER** to start each new line:

Flomo Tunde

Lekki City Water Supplies

970 Kazeem Abogum Street

Eti-Osa, Lekki 106103, Lagos, Nigeria

Because David will be using the department's printed envelopes, you don't need to print a return address on this envelope.

(4). **Click** the **Omit** check box to insert a checkmark, if necessary.

At this point, if you had a printer stocked with envelopes, you could click the Print button to print the envelope. To save an envelope for printing later, you need to add it to the document. Your Envelopes and Labels dialog box should match the one in **Figure 2-29**.

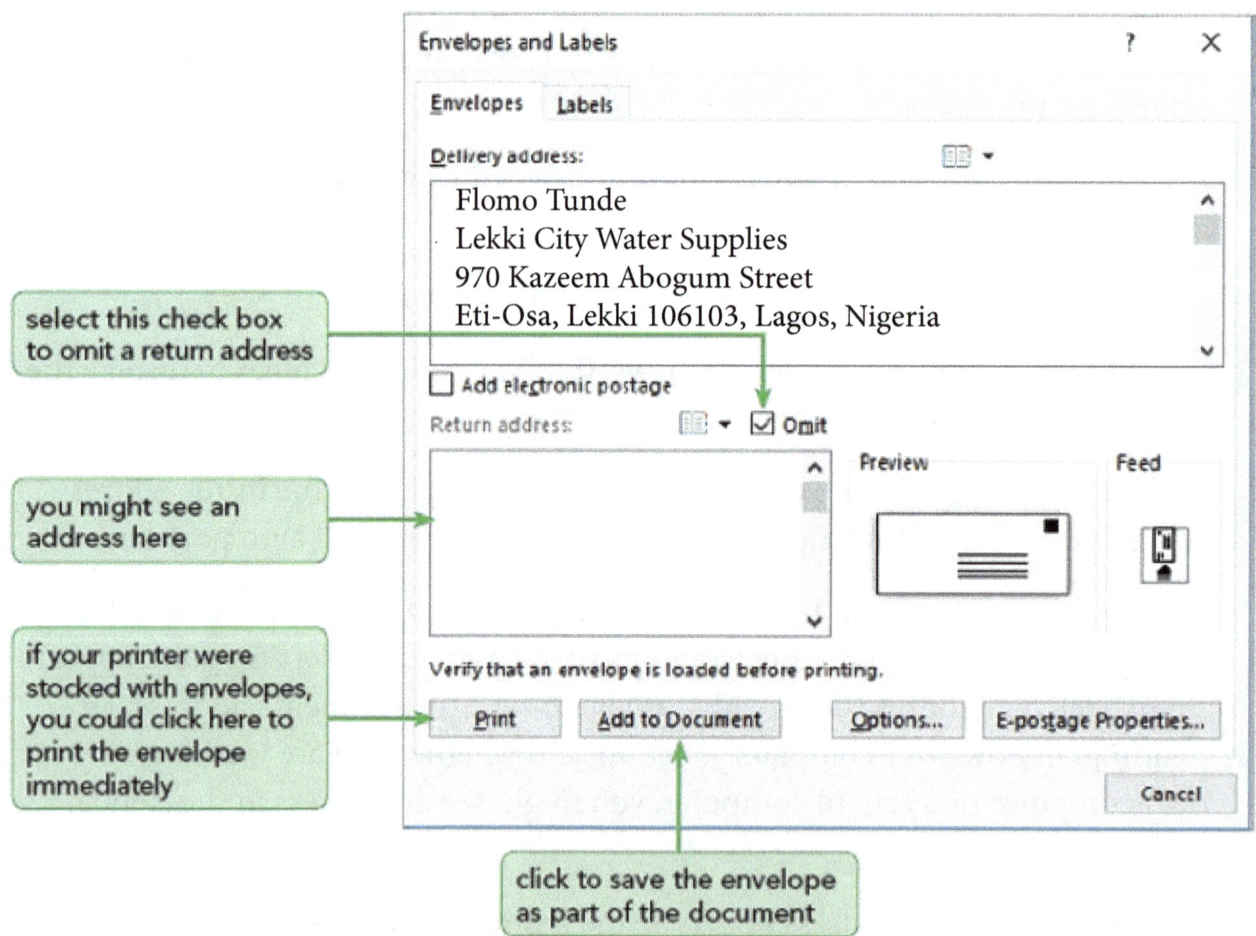

Figure 2-29 Envelopes and Labels dialog box

(5). **Click Add to Document**. The dialog box closes, and you return to the document window. The envelope is inserted at the top of your document, with 1.0 line spacing. The double line with the words "Section Break (Next Page)" is related to how the envelope is formatted and will not be visible when you print the envelope. The envelope will print in the standard business envelope format. In this case, you added the envelope to a blank document, but you could also add an envelope to a completed letter, in which case Word adds the envelope as a new page before the letter.

(6). **Save** the document. David will print the envelope later, so you can close the document now.

(7). **Click** the **File Tab**, and then **Click Close** in the navigation pane. The document closes, but Word remains open

You're finished creating the cover letter and the envelope. In the next session, you will modify a flyer by formatting the text and adding a photo.

SMARTSKILLS **Creating Documents with Templates**

Microsoft offers predesigned templates for all kinds of documents, including calendars, reports, and thank-you cards. You can use the scroll bar on the right of the New screen (shown earlier in **Figure 2-28**) to scroll down to see more templates, or you can use the Search for online templates box in the New screen to search among thousands of other options available at Office.com. When you open a template, you actually open a new document containing the formatting and text stored in the template, leaving the original template untouched. A typical template includes placeholder text that you replace with your own information.

Templates allow you to create stylish, professional-looking documents quickly and easily. To use them effectively, however, you need to be knowledgeable about Word and its many options for manipulating text, graphics, and page layouts. Otherwise, the complicated formatting of some Word templates can be more frustrating than helpful. As you become a more experienced Word user, you'll learn how to create your own templates.

Quick Review

1. What Word feature automatically inserts dates and other regularly used items for you?
2. Explain how to display nonprinting characters.
3. In a block-style letter, does the inside address appear above or below the date?
4. Explain how to use a hyperlink in a Word document to open a new email message.
5. Define the term "paragraph spacing."
6. Explain how to display a shortcut menu with options for correcting a word with a wavy red underline.

OPENING AN EXISTING DOCUMENT

In this session, you'll complete a flyer encouraging community members to join a citizen advisory panel. David has already typed the text of the flyer, inserted a photo into it, and saved it as a Word document. He would like you to check the document for spelling and grammar errors, format the flyer to make it eye-catching and easy to read, and then replace the current photo with a new one. You'll start by opening the document.

To open the flyer document:

(1). On the ribbon, **Click** the **File Tab** to open Backstage view, and then **Click Open** in the navigation pane. On the left side of the Open screen is a list of places you can go to locate other documents, and on the right is a list of recently opened documents.

Are You Having Trouble? If you closed Word at the end of the previous session, start Word now, click Open in the navigation pane in Backstage view, and then begin with **Step 2**.

(2). **Click the Browse** button. The Open dialog box opens.

Are You Having Trouble? If your instructor asked you to store your files to your OneDrive account, click OneDrive, and then log in to your account.

(3). Navigate to the **Word** > **Chapter 2** folder included with your Data Files, **Click Chapter 2 Flyer.docx** in the file list, and then **Click Open**. The document opens with the insertion point blinking in the first line of the document.

Before making changes to David's document, you will save it with a new name. Saving the document with a different file name creates a copy of the file and leaves the original file unchanged in case you want to work through the module again.

To save the document with a new name:

(1). On the ribbon, click the **File Tab**.

(2). In the navigation pane in Backstage view, **click Save As**. Save the document as **Chapter 2 Flyer Final.docx** in the location specified by your instructor. Backstage view closes, and the document window appears again with the new file name in the title bar. The original **Chapter 2 Flyer.docx** document closes, remaining unchanged.

SMARTSKILLS Decision Making: Creating Effective Documents

Before you create a new document or revise an existing document, take a moment to think about your audience. Ask yourself these questions:

- Who is your audience?
- What do they know?
- What do they need to know?

How can the document you are creating change your audience's behavior or opinions?

Every decision you make about your document should be based on your answers to these questions. To take a simple example, if you are creating a flyer to announce an upcoming seminar on college financial aid, your audience would be students and their parents. They probably all know what the term "financial aid" means, so you don't need to explain that in your flyer. Instead, you can focus on telling them what they need to know—the date, time, and location of the seminar. The behavior you want to affect, in this case, is whether your audience will show up for the seminar. By making the flyer professional looking and easy to read, you increase the chance that they will.

You might find it more challenging to answer these questions about your audience when creating more complicated documents, such as corporate reports. But the focus remains the same—connecting with the audience. As you are deciding what information to include in your document, remember that the goal of a professional document is to convey the information as effectively as possible to your target audience.

Before revising a document for someone else, it's a good idea to familiarize yourself with its overall structure.

To review the document:

(1). Verify that the document is displayed in Print Layout view and that nonprinting characters and the rulers are displayed. For now, you can ignore the wavy underlines that appear in the document.

(2). Change the Zoom level to **120%**, if necessary, and then scroll down, if necessary, so that you can read the last line of the document.

You will start by correcting the spelling and grammar errors.

USING THE EDITOR PANE

As you type, Word marks possible spelling and grammatical errors, as well as wordiness, with underlines so you can quickly go back and correct those errors. A more thorough way of checking the spelling in a document is to use the Editor pane to check a document word by word for a variety of errors. You can customize the spelling and grammar settings to add or ignore certain types of errors.

David asks you to use the Editor pane to check the flyer for mistakes. Before you do, you'll review the various Spelling and Grammar settings.

To review the Spelling and Grammar settings:

(1). On the ribbon, **Click** the **File tab**, and then **Click Options** in the navigation pane. The Word Options dialog box opens. You can use this dialog box to change a variety of settings related to how Word looks and works.

(2). In the left pane, click **Proofing**.

Note the four selected options in the "When correcting spelling and grammar in Word" section. These options tell you that Word will check for misspellings, grammatical errors, and frequently confused words as you type, marking them with wavy underlines as necessary.

(3). In the "When correcting spelling and grammar in Word" section, **Click Settings**. The Grammar Settings dialog box opens. Here you can control the types of grammar errors Word checks for. All of the boxes in the Grammar section are selected by default, which is what you want. See **Figure 2-30**.

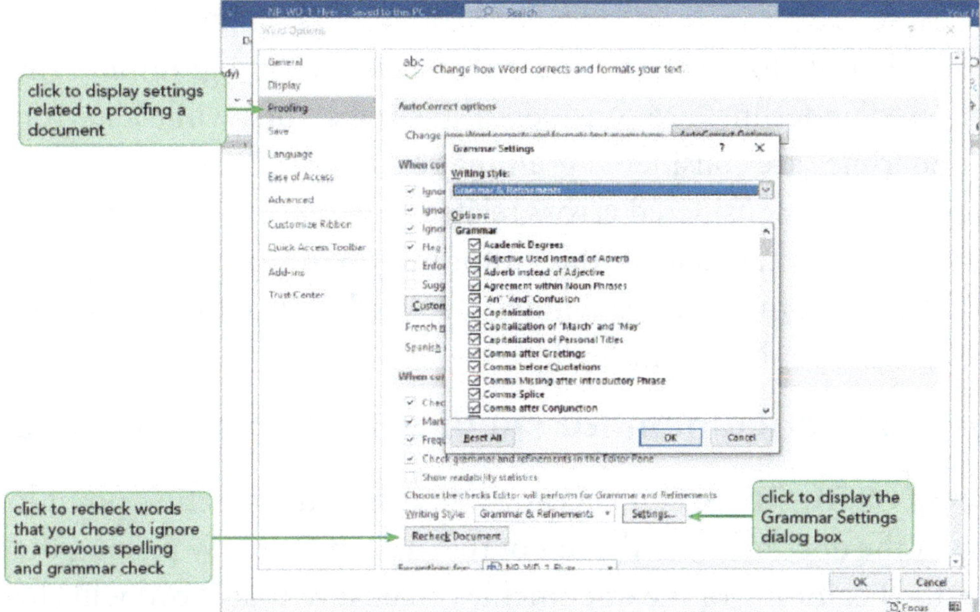

Figure 2-30 Grammar Settings dialog box

(4). Scroll down in the Grammar Settings dialog box to display the Conciseness settings. By default, only Nominalizations and Wordiness are selected.

(5). **Click Cancel** to close the Grammar Settings dialog box and return to the Word Options dialog box.

Note that the results of the Spelling and Grammar checker are sometimes hard to predict. For example, in some documents Word will mark a misused word or duplicate punctuation as an error and then fail to mark the same items as errors in another document. Also, if you choose to ignore a misspelling in a document and then, without closing Word, type the same misspelled word in another document, Word will probably not mark it as an error. Sometimes, if you change a document's line or paragraph spacing, Word will mark text as errors that it previously did not.

These issues can be especially problematic when working on a document typed by someone else. So to ensure that you get the best possible results, it's a good idea to click Recheck Document in the Word Options dialog box before you use the Spelling and Grammar checker.

(6). **Click** the **Recheck Document** button, and then **Click Yes** in the warning dialog box.

(7). In the Word Options dialog box, **Click OK** to close the dialog box. You return to the document.

Now you are ready to check the document's spelling and grammar. All errors marked with red underlines are considered spelling errors, while all errors marked with blue underlines are considered grammatical errors. Errors marked with purple dotted underlines are considered errors related to a lack of conciseness. To begin checking the document, you'll use the Editor button in the Proofing group on the Review tab. Note that in some installations of Word, this button might be called the "Spelling & Grammar" button instead.

To check the document for spelling and grammatical errors:

(1). **Press CTRL+HOME**, if necessary, to move the insertion point to the beginning of the document, to the left of the "W" in "We." By placing the insertion point at the beginning of the document, you ensure that Word will check the entire document from start to finish, without having to go back and check an earlier part.

(2). On the ribbon, **Click** the **Review Tab**. The ribbon changes to display reviewing options.

(3). In the Proofing group, **Click** the **Editor** button. The Editor pane opens on the right side of the Word window, indicating that the document contains three spelling errors, one grammar error, and no clarity and conciseness errors.

Are You Having Trouble? If you see the Spelling & Grammar button instead of the Editor button, click the Spelling & Grammar button.

(4). Near the top of the Editor pane, **Click Total Consideration 4**. Now the Editor pane displays information about the first error. As in the document, the word "prottect" is underlined in red as a possible spelling error. To the right of the sentence in the Editor pane is a speaker icon, which you can click to hear the sentence read aloud. Below, in the Consider box, the correctly spelled word "protect" appears along with its definition. You might also see some other suggestions/considerations. The incorrectly spelled word "prottect" is also highlighted in gray in the document. See **Figure 2-31**.

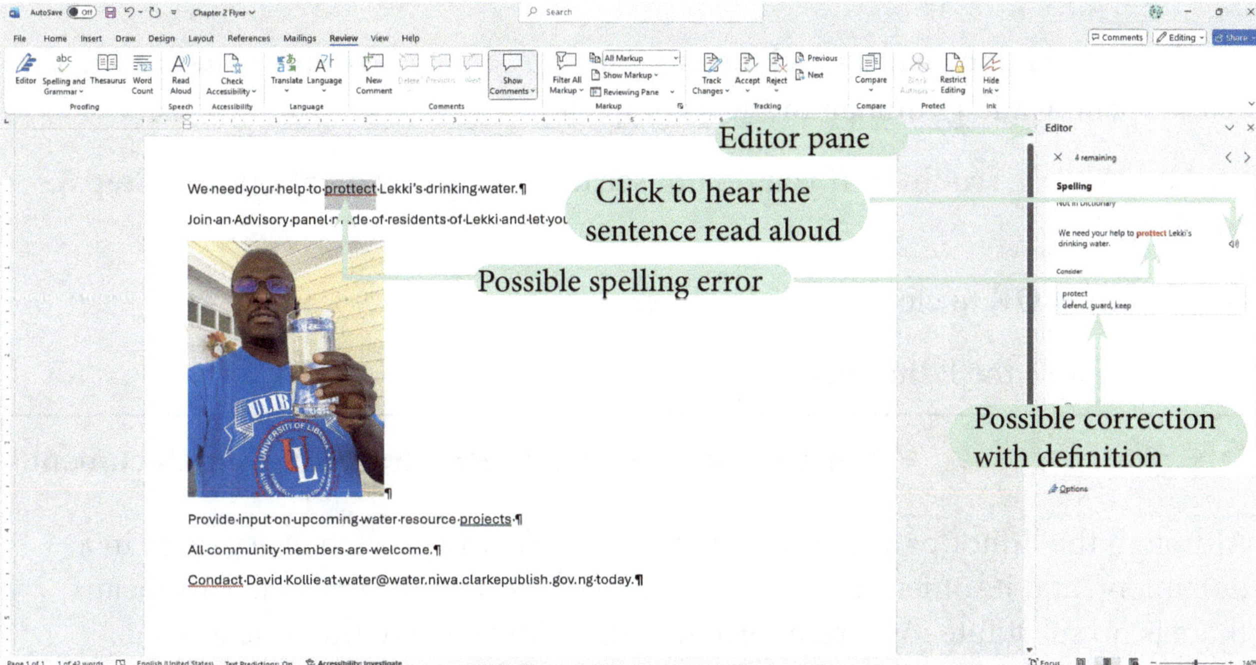

Figure 2-31 Editor pane

Are You Having Trouble? If you don't see "Total Consider 4" at the top of the Editor pane, that's fine. Everything else in **Step 4** should still match what you see on your screen.

(5). In the Editor pane, **Click** the **protect** suggestion. The misspelled word "prottect" is replaced with "protect."

Next Word highlights the last word in the second sentence, indicating another possible error. The explanation near the top of the pane indicates that Word has detected a redundant punctuation mark—that is, an extra period.

(6). In the Suggestions list, **Click heard**. (with one period).

The first word of the last sentence is now highlighted in the document.

(7). In the Editor pane, **Click** the **Contact suggestion**. (Make sure to click

"Contact," with an uppercase "C"). David's last name is now highlighted in the document. Although the Editor pane doesn't recognize "Kollie" as a word, it is spelled correctly, so you can ignore it. Note that if his name appeared repeatedly in the document, you could click Ignore All to ignore all instances of it.

Are You Having Trouble? If you see "Resume" in the Editor pane instead of "Resume checking all results," click "Resume" instead.

(8). In the Editor pane, **Click Ignore Once**. A dialog box opens, indicating that the spelling and grammar check is complete.

Are You Having Trouble? If you do not see the dialog box mentioned in **Step 8**, skip **Step 9**.

(9). **Click OK** to close the dialog box.

(10). **Close** the Editor pane.

SMARTSKILLS Written Communication: Proofreading Your Document

Although the Editor pane is a useful tool, it won't always catch every error in a document, and it sometimes flags "errors" that are actually correct. This means there is no substitute for careful proofreading. Always take the time to read through your document to check for errors the Editor pane might have missed. Keep in mind that the Editor pane cannot pinpoint inaccurate phrases or poorly chosen words. You'll have to find those yourself. To produce a professional document, you must read it carefully several times. It's a good idea to ask one or two other people to read your documents as well; they might catch something you missed.

At this point you still need to proofread the document. You'll do that next.

To proofread the document:

(1). Review the document text for any remaining errors. In the second paragraph, change the lowercase "p" in "panel" to an uppercase "P."

(2). In the last line of text, replace "David Kollie" with your first and last names, and then save the document. Including your name in the document will make it easier for you to find your copy later if you print it on a shared printer.

Now you're ready to begin formatting the document. You will start by turning the page so it is wider than it is tall. In other words, you will change the document's **Orientation**.

CHANGING PAGE ORIENTATION

Portrait Orientation, with the page taller than it is wide, is the default page orientation for Word documents because it is the orientation most commonly used for letters, reports, and other formal documents. However, David wants you to format the flyer in **Landscape Orientation**—that is, with the page turned so it is wider than it is tall—to better accommodate the photo. You can accomplish this task by using the Orientation button located on the Layout tab on the ribbon. After you change the page orientation, you will select narrower margins so you can maximize the amount of color on the page.

To change the page orientation:

(1). Change the document **Zoom level** to **One Page** so that you can see the entire document.

(2). On the ribbon, **Click** the **Layout Tab**. The ribbon changes to display options for formatting the overall layout of text and images in the document.

(3). In the Page Setup group, **Click** the **Orientation** button, and then **Click** **Landscape** on the menu. The document changes to landscape orientation.

(4). In the Page Setup group, **Click** the **Margins** button, and then **Click** the **Narrow** option on the menu. The margins shrink from 1 inch to .5 inch on all four sides. See **Figure 2-32** on the next page.

The flyer in Figure 2-32 on the next page appears zoomed out in the Word window so that the entire page is displayed. The Orientation button in the Page Setup group on the Layout Tab was used to change the layout from portrait to landscape orientation. The page appears wider than it is tall. The margins were made narrow, and now are .5 inch instead of the 1 inch default.

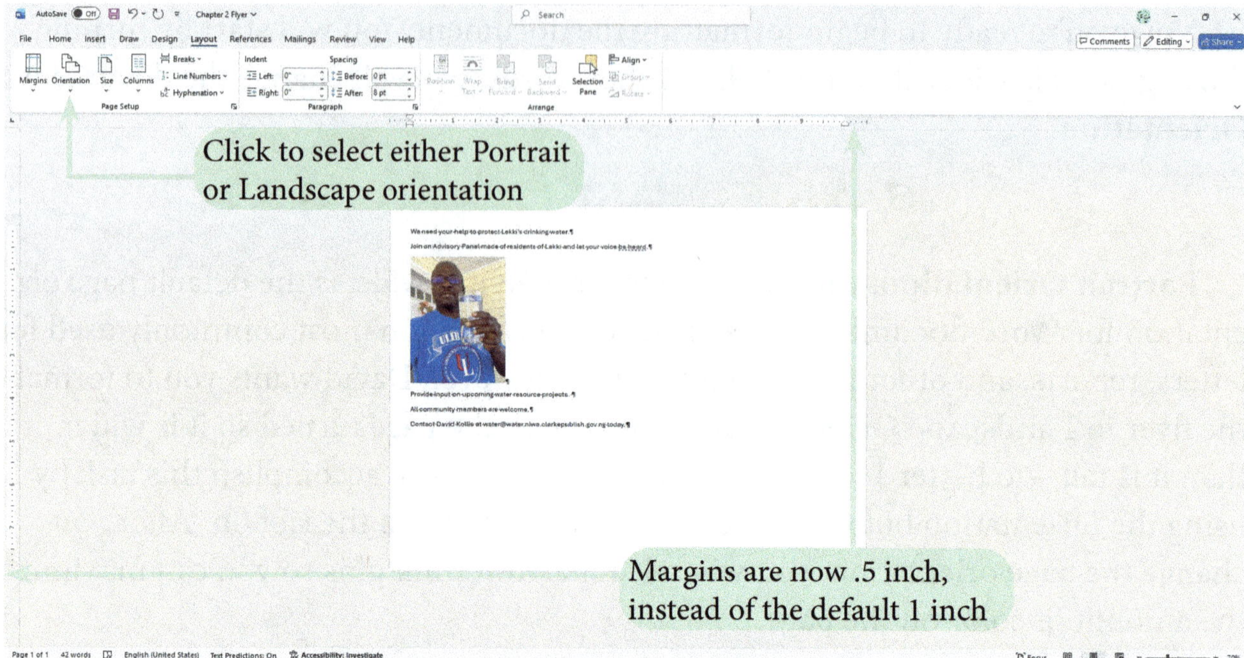

Figure 2-32 Document in landscape orientation

CHANGING THE FONT AND FONT SIZE

David typed the document in the default font size, 12 point, and the default font, Aptos, but he would like to switch to the Arial font instead. Also, he wants to increase the size of all five paragraphs of text. To apply these changes, you start by selecting the text you want to format. Then you select the options you want in the Font group on the Home tab.

To change the font and font size:

(1). **Change** the document **Zoom** level to **120%**.

(2). On the ribbon, **Click** the **Home Tab**.

(3). To verify that the insertion point is located at the beginning of the document, **Press CTRL+HOME**.

(4). **Press** and **Hold SHIFT**, and then **Click** to the right of the second paragraph marker, at the end of the second paragraph of text. The first two paragraphs of text are selected, as shown in **Figure 2-33** on the next page.

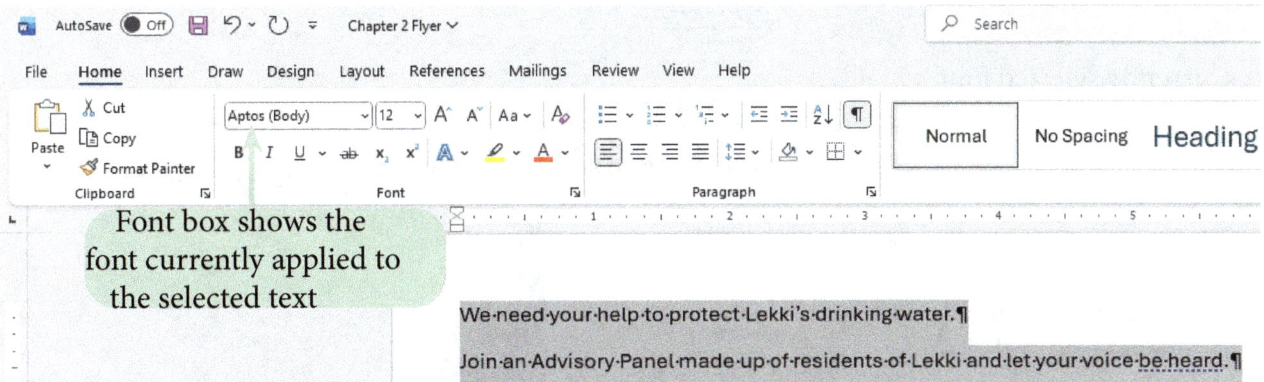

Figure 2-33 Selected text, with default font displayed in Font box

The Font box in the Font group displays the name of the font applied to the selected text, which in this case is Aptos. The word "Body" next to the font name indicates that the Aptos font is intended for formatting body text. **Body Text** is ordinary text, as opposed to titles or headings.

(5). In the Font group on the Home tab, **Click** the **Font Arrow**. A list of available fonts appears, with Aptos at the top of the list. Aptos is highlighted in gray, indicating that this font is currently applied to the selected text. The word "Headings" next to the font name "Aptos" indicates that Aptos is intended for formatting headings.

Below Aptos, you might see a list of fonts that have been used recently on your computer, followed by a complete alphabetical list of all available fonts. (You won't see the list of recently used fonts if you just installed Word.) You need to scroll the list to see all the available fonts. Each name in the list is formatted with the relevant font. For example, the name "Arial" appears in the Arial font. See **Figure 2-34** on the next page.

The document in Figure 2-34 is zoomed in closer, and the first two paragraphs in the document are selected. The Font arrow in the Font group on the Home tab was clicked to display a menu. Aptos (body) is the font currently applied to the selected, and appears selected on the menu. A scroll box on the menu can be dragged to scroll the font list. Users might see a list of recently used fonts on the menu. Each font name in the list is an example of that font.

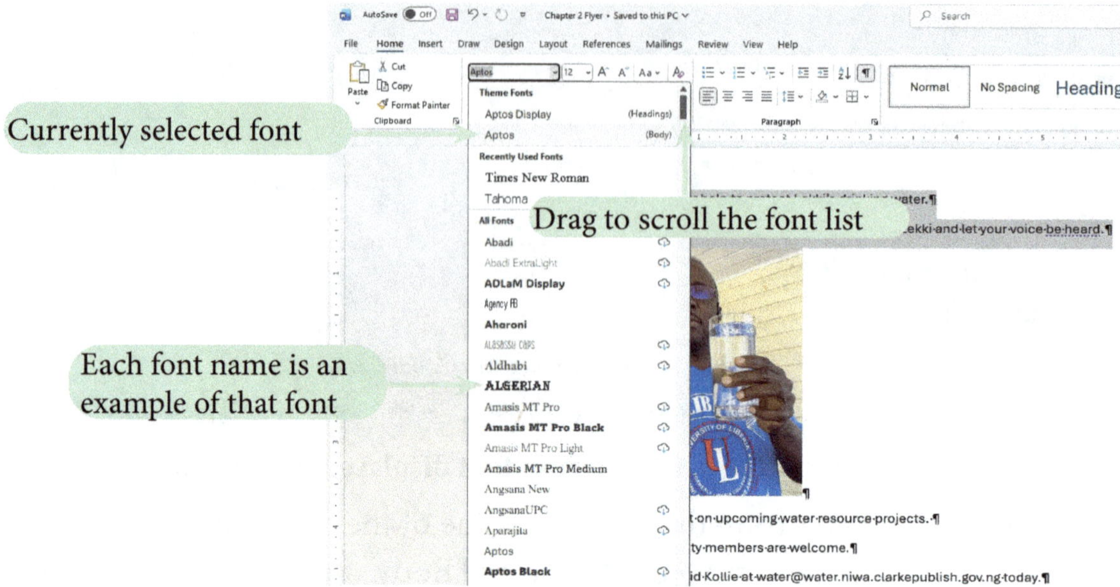

Figure 2-34 Font list

(6). Without clicking, move the pointer over a dramatic-looking font in the font list, such as Algerian or Arial Black, and then move the pointer over another font.

The selected text in the document changes to show a Live Preview of the font the pointer is resting on. **Live Preview** shows the results that would occur in your document if you clicked the option you are pointing to.

(7). When you are finished reviewing the Font list, **click Arial**. The Font menu closes, and the selected text is formatted in Arial.

Next, you will make the text more eye-catching by increasing the font size. The Font Size box currently displays the number "11," indicating that the selected text is formatted in 11-point font.

(8). Verify that the first two paragraphs are still selected, and then click the **Font Size arrow** in the Font group to display a menu of font sizes. As with the Font menu, you can move the pointer over options in the Font Size menu to see a Live Preview of that option.

(9). On the Font Size menu, **Click 22**. The selected text increases significantly in size, and the Font Size menu closes.

(10). Select the three paragraphs of text below the photo, format them in the

Arial font, and then increase the paragraph's font size to 22 points.

(11). **Click** a blank area of the document to deselect the text, and then save the document.

Keep in mind that to restore selected text to its default appearance, you can click the Clear All Formatting button in the Font group on the Home tab.

David examines the flyer and decides he would like to apply more character formatting, which affects the appearance of individual characters, in the middle three paragraphs. After that, you can turn your attention to paragraph formatting, which affects the appearance of the entire paragraph.

Applying Text Effects, Font Colors, and Font Styles

For formal, professional documents, you typically only need to use **bold** or **italic** to make a word or paragraph stand out. Occasionally you might need to underline a word. To apply these forms of character formatting, select the text you want to format, and then click the Bold, Italic, or Underline button in the Font group on the Home tab. To really make text stand out, you can use text effects. You access these options by clicking the Text Effects and Typography button in the Font group on the Home tab. Keep in mind that text effects can be very dramatic.

David suggests applying text effects to the second paragraph.

To apply text effects to the second paragraph:

(1). Scroll up, if necessary, to display the beginning of the document, and then click in the selection bar to the left of the second paragraph. The entire second paragraph is selected.

(2). In the Font group on the Home tab, **Click** the **Text Effects and Typography** button A.

A gallery of text effects appears. Options that allow you to fine-tune a particular text effect, perhaps by changing the color or adding an even more pronounced shadow, are listed below the gallery. A **Gallery** is a menu or grid that shows a visual representation of the options available when you click a button.

(3). In the middle of the bottom row of the gallery, place the pointer over the blue letter "A." This displays a ScreenTip with the text effect's full name: Fill: Blue, Accent color 5; Outline: White, Background color 1; Hard Shadow: Blue, Accent color 5. A Live Preview of the effect appears in the document. See **Figure 2-35**.

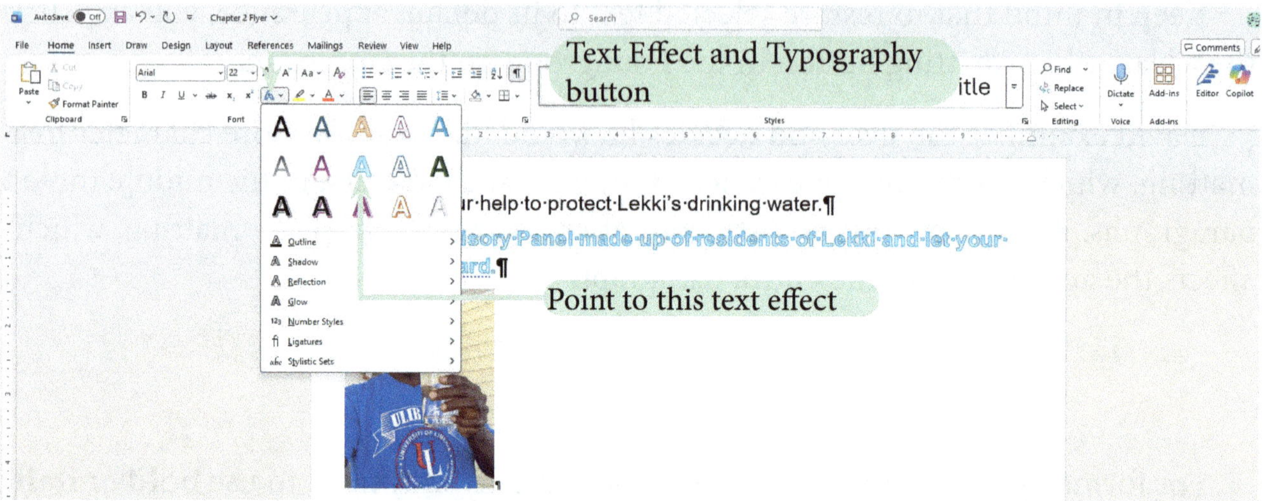

Figure 2-35 Live Preview of a text effect

(4). In the bottom row of the gallery, click the blue letter "A." The text effect is applied to the selected paragraph, and the Text Effects gallery closes. The second paragraph is formatted in blue. On the ribbon, the Bold button in the Font group is now highlighted because bold formatting is part of this text effect.

Next, to make the text stand out a bit more, you'll increase the font size. This time, instead of using the Font Size button, you'll use a different method.

(5). In the Font group, **Click** the **Increase Font Size** button. The font size increases from **22 Points** to **24 Pints**, which is the next higher font size on the Font menu.

(6). **Click** the **Increase Font Size** button again. The font size increases to **26 Points**, which is the next higher font size on the Font menu. If you need to decrease the font size of selected text, you can use the Decrease Font Size button. Each time you click the Decrease Font Size button, the font decreases to the next lower font size on the Font menu.

David asks you to emphasize the third and fourth paragraphs by adding bold and a blue font color.

To apply a font color and bold:

(1). **Select** the **third** and **fourth paragraphs** of text, which contain the text "Provide input on upcoming water resource projects. All community members are welcome." Font Color gallery showing a Live Preview

(2). In the Font group on the Home tab, **Click** the **Font Color arrow** [A ▾]. A gallery of font colors appears. Black is the default font color and appears at the top of the Font Color gallery, with the word "Automatic" next to it.

The options in the Theme Colors section of the menu are complementary colors that work well when used together in a document. The options in the Standard Colors section are more limited. For more advanced color options, you could use the More Colors or Gradient options. David prefers a simple blue.

Are You Having Trouble? If the third and fourth paragraphs turned red, you clicked the Font Color button [A ▾] instead of the arrow next to it. On the Quick Access Toolbar, click the Undo button [↶], and then repeat **Step 2**.

(3). In the Theme Colors section, place the pointer over the square that's second from the right in the top row. A ScreenTip with the color's name, "Blue, Accent 5," appears. A Live Preview of the color appears in the document, where the text you selected in **Step 1** now appears formatted in blue. See **Figure 2-36**.

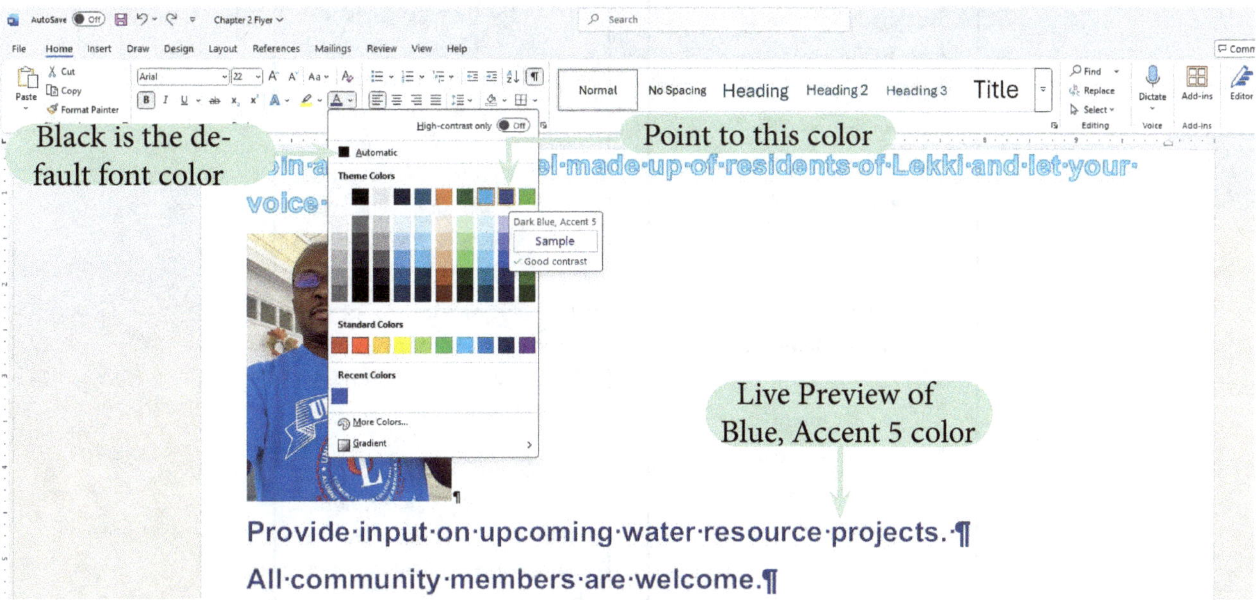

Figure 2-36 Font Color gallery showing a Live Preview

(4). **Click** the **Blue, Accent 5** square. The Font color gallery closes, and the selected text is formatted in blue. On the Font Color button, the bar below the letter "A" is now blue, indicating that if you select text and click the Font Color button, the text will automatically change to blue.

(5). In the Font group, **Click** the **Bold** button **B**. The selected text is now formatted in bold, with thicker, darker lettering.

Next, you will complete some paragraph formatting, starting with paragraph alignment.

ALIGNING TEXT

Alignment refers to how text and graphics line up between the page margins. By default, text is **Left-Aligned** in Word. That is, the text is flush with the left margin, with the text along the right margin ragged, or uneven. By contrast, **Right-Aligned** text is aligned along the right margin and is ragged along the left margin. **Centered** text is positioned evenly between the left and right margins and is ragged along both the left and right margins. Finally, with **Justified Alignment**, full lines of text are spaced between both the left and the right margins, and no text is ragged. Text in newspaper columns is often justified. See **Figure 2-37**.

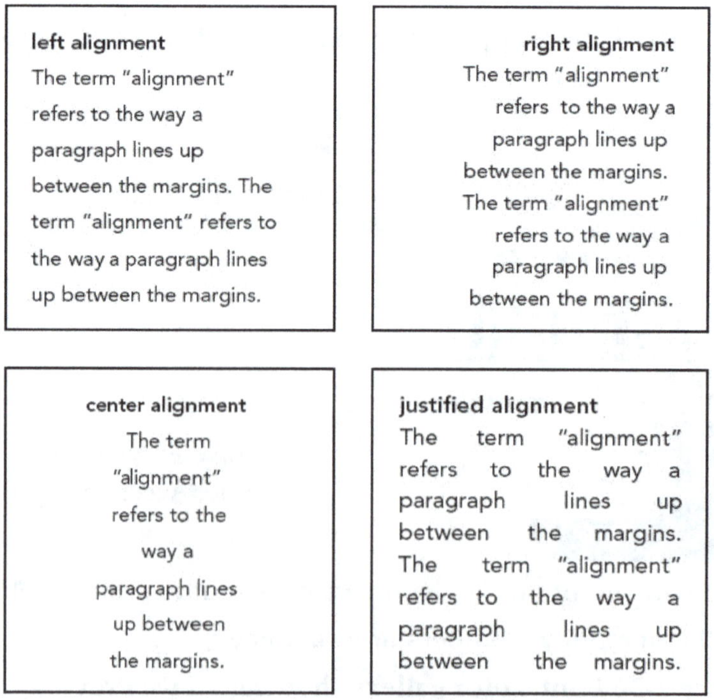

left alignment
The term "alignment" refers to the way a paragraph lines up between the margins. The term "alignment" refers to the way a paragraph lines up between the margins.

right alignment
The term "alignment" refers to the way a paragraph lines up between the margins. The term "alignment" refers to the way a paragraph lines up between the margins.

center alignment
The term "alignment" refers to the way a paragraph lines up between the margins.

justified alignment
The term "alignment" refers to the way a paragraph lines up between the margins. The term "alignment" refers to the way a paragraph lines up between the margins.

Figure 2-37 Varieties of text alignment

The Paragraph group on the Home tab includes a button for each of the four major types of alignment described in **Figure 2-37**: the Align Left button, the Center button, the Align Right button, and the Justify button. To align a single paragraph, click anywhere in that paragraph, and then click the appropriate alignment button. To align multiple paragraphs, select the paragraphs first, and then click an alignment button.

You need to center all the text in the flyer now. You can center the photo at the same time.

To center-align the text:

(1). Make sure the Home tab is still selected, and **Press CTRL+A** to select the entire document.

(2). In the Paragraph group, **Click** the **Center** button , and then **Click** a **Blank Area** of the document to deselect the selected paragraphs. The text and photo are now centered on the page.

(3). **Save** the document.

ADDING A PARAGRAPH BORDER AND SHADING

A **Paragraph Border** is an outline that appears around one or more paragraphs in a document. You can choose to apply only a partial border—for example, a bottom border that appears as an underline under the last line of text in the paragraph—or an entire box around a paragraph. You can select different colors and line weights for the border as well, making it more or less prominent as needed. You apply paragraph borders using the Borders button in the Paragraph group on the Home tab. **Shading** is background color that you can apply to one or more paragraphs and that can be used in conjunction with a border for a more defined effect. You apply shading using the Shading button in the Paragraph group on the Home tab.

Now you will apply a border and shading to the first paragraph. Then you will use the Format Painter to copy this formatting to the last paragraph in the document.

To add shading and a paragraph border:

(1). Scroll up if necessary and select the first paragraph. Be sure to select the paragraph mark at the end of the paragraph.

(2). On the Home tab, in the Paragraph group, click the **Borders Arrow**. A gallery of border options appears, as shown in **Figure 2-38**. To apply a complete outline around the selected text, you use the Outside Borders option.

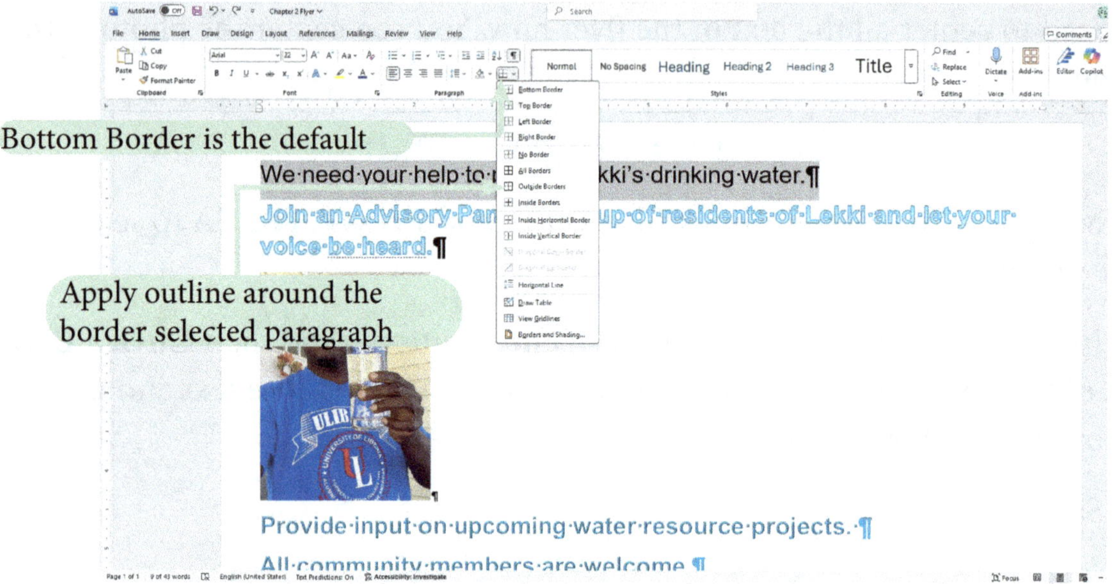

Figure 2-38 Border Gallery

Are You Having Trouble? If the gallery does not open and instead the paragraph becomes underlined with a single underline, you clicked the Borders button instead of the arrow next to it. On the Quick Access Toolbar, click the Undo button, and then repeat **Step 2**.

(3). In the Border gallery, **Click Outside Borders**. The menu closes and a black border appears around the selected paragraph, spanning the width of the page. In the Paragraph group, the Borders button changes to show the Outside Borders option.

Are You Having Trouble? If the border around the first paragraph doesn't extend all the way to the left and right margins and instead encloses only the text, you didn't select the paragraph mark as directed in **Step 1**. Click the Undo button repeatedly to remove the border, and begin again with **Step 1**.

(4). In the Paragraph group, click the **Shading arrow**. A gallery of shading options opens, divided into Theme Colors and Standard Colors. You will use a shade of dark blue in the fifth column from the left.

(5). In the bottom row in the Theme Colors section, move the pointer over the square in the fifth column from the left to display a ScreenTip that reads "Blue, Accent 1, Darker 50%." A Live Preview of the color appears in the document. See **Figure 2-39**.

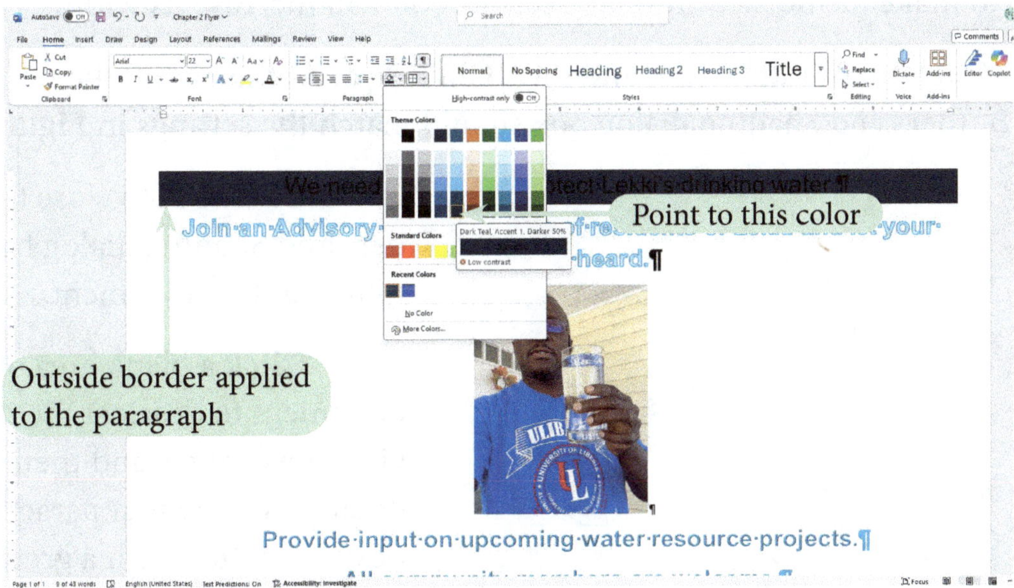

Figure 2-39 Shading gallery with a Live Preview

(6). **Click** the **Blue, Accent 1, Darker 50%** square to apply the shading to the selected text.

On a dark background like the one you just applied, a white font creates a striking effect. David asks you to change the font color for this paragraph to white.

(7). Make sure the Home tab is still selected.

(8). In the Font group, click the **Font Color Arrow** [A ▾] to open the Font Color gallery, and then **Click** the **White** square in the top row of the Theme Colors. The Font Color gallery closes. The paragraph is now formatted with white font.

The black paragraph border is hard to see with the dark blue shading, so you will change the border to a different color. To make more advanced changes to borders or paragraph shading, you need to use the Borders and Shading dialog box.

(9). **Click** the Borders arrow [⊞ ▾] and then, at the bottom of the menu, **Click**

Borders and Shading. The Borders and Shading dialog box opens with the Borders tab displayed.

(10). **Click** the **Color Arrow** to open the Color gallery, and then **Click** the **Green, Accent 6** square, which is the right-most square in the top row of the Theme Colors section.

Next, to make the border more noticeable, you will increase its width.

(11). **Click** the Width arrow, and then **Click 3 pt**. At this point, the settings in your Borders and Shading dialog box should match the settings in **Figure 2-40**.

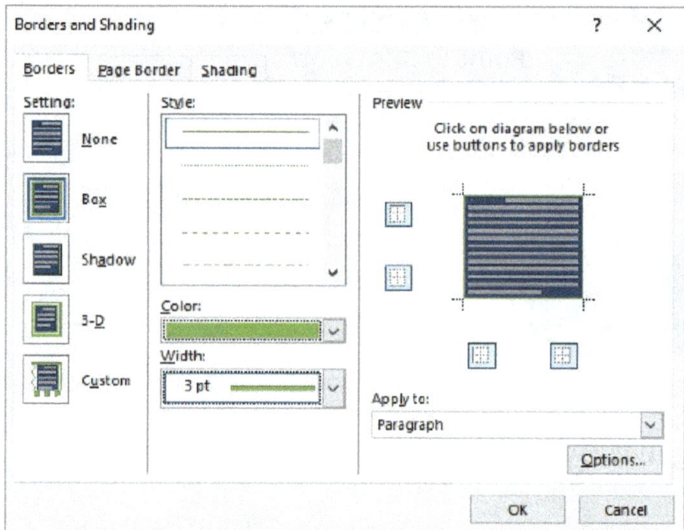

Figure 2-40 Borders and Shading dialog box

(12). **Click OK** to close the Borders and Shading dialog box and return to the document.

(13). **Click** a blank area of the document to deselect the text, review the change, and then save the document. The first paragraph is now formatted with a green border, a dark blue background, and white text.

Figure 2-40 shows the Borders and Shading dialog box with the Borders tab displayed. The Box option is selected in the Setting section. The Color box shows a green color. The Width box shows that the border will be 3 pt. and green.

To add balance to the flyer, David suggests formatting the last paragraph in the document with the same shading, border, and font color as the first paragraph. You'll do that next.

Copying Formatting with the Format Painter

You could select the last paragraph and then apply the border, shading, and font color one step at a time. But it's easier to copy all the formatting from the first paragraph to the last paragraph using the Format Painter button in the Clipboard group on the Home tab.

Smart Tips

Using the Format Painter

Below are some smart tips to help you use the Format Painter in Word.

- Select the text whose formatting you want to copy.
- On the Home tab, in the Clipboard group, click the Format Painter button, or to copy formatting to multiple sections of nonadjacent text, double-click the Format Painter button.
- The pointer changes to the Format Painter pointer, the I-beam pointer with a paint-brush.
- Click the words you want to format, or drag to select and format entire paragraphs.
- When you are finished formatting the text, click the Format Painter button again to turn off the Format Painter.

For this part of the lab, you'll use the Format Painter now.

To use the Format Painter:

(1). Change the document Zoom level to One Page so you can easily see both the first and last paragraphs.

(2). **Select** the first paragraph, which is formatted with the dark blue shading, the green border, and the white font color.

(3). On the ribbon, **Click** the **Home** tab.

(4). In the Clipboard group, **Click** the **Format Painter** button to activate, or turn on, the Format Painter.

(5). **Move** the **Format Painter** pointer over the document. The pointer changes to the Format Painter pointer when you move the pointer near an item that can be formatted. See **Figure 2-41** on the next page.

(6). **Click** and **Drag** the **Format Painter** pointer to select the last paragraph in the document. The paragraph is now formatted with dark blue shading, a green border, and white font. The pointer returns to its original I-beam shape.

Are You Having Trouble? If the text in the newly formatted paragraph wrapped to a second line, replace your full name with your first name, or, if necessary, use only your initials so the paragraph is only one-line long.

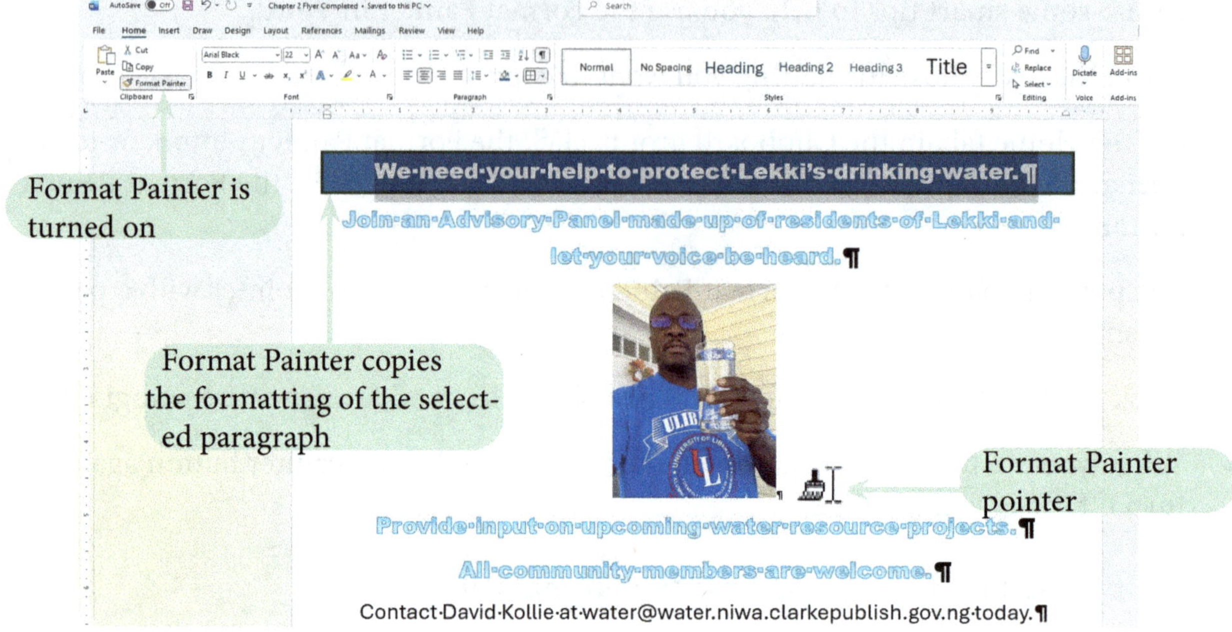

Figure 2-41 Format Painter

(7). **Click** anywhere in the document to deselect the text, review the change, and then save the document.

Your next task is to increase the paragraph spacing below the first paragraph and above the last paragraph. This will give the shaded text even more weight on the page. To complete this task, you will use the settings on the Layout tab, which offer more options than the Line and Paragraph Spacing button on the Home tab.

To increase the paragraph spacing below the first paragraph and above the last paragraph:

(1). **Click** anywhere in the first paragraph, and then **Click** the **Layout Tab**. On this tab, the Paragraph group contains settings that control paragraph spacing. Currently, the paragraph spacing for the first paragraph is set to the default 0 points before the paragraph and 8 points after.

(2). In the Paragraph group, **Click** the **After** box to select the current setting, **Type 42**, and then **Press ENTER**. The added space causes the second paragraph to move down 42 points.

(3). **Click** anywhere in the last paragraph.

(4). On the **Layout Tab**, in the Paragraph group, **Click** the **Before** box to select the current setting, **Type 42**, and then **Press ENTER**. The added space causes the last paragraph to move down 42 points.

SMARTSKILLS Formatting Professional Documents

In more formal documents, use color and special effects sparingly. The goal of letters, reports, and many other types of documents is to convey important information, not to dazzle the reader with fancy fonts and colors. Such elements only serve to distract the reader from your main point. In formal documents, it's a good idea to limit the number of colors to two and to stick with left alignment for text. In a document like the flyer you're currently working on, you have a little more leeway because the goal of the document is to attract attention. However, you still want it to look professional.

Next, David wants you to replace the photo with one that will look better in the document's new landscape orientation. You'll replace the photo with a you own photo with you in your school-branded shirt holding a glass of water just like one of the authors did with a picture of him wearing a branded University Liberia T-Shirt with a glass of water in his hands, or you can use a photo that promotes your local drink water source, and then you'll resize it so that the flyer fills the entire page.

INSERTING A PICTURE AND ADDING ALT TEXT

A **Picture** is a photo or another type of image that you insert into a document. To work with a picture, you first need to select it. Once a picture is selected, a contextual tab—the Picture Format tab—appears on the ribbon, with options for editing the picture and adding effects such as a border, a shadow, a reflection, or a new shape. A **Contextual Tab** appears on the ribbon only when an object is selected. It contains commands related to the selected object so that you can manipulate, edit, and format the selected object. You can also use the mouse to resize or move a selected picture. To insert a new picture, you use the Pictures button in the Illustrations group on the Insert tab.

To delete the current photo and insert a new one:

(1). **Click** the photo to select it.

The circles, called **Sizing Handles**, around the edge of the photo to indicate that the photo is selected. The Layout Options button, to the right of the photo, gives you access to options that control how the document text flows around the photo. You don't need to worry about these options now. Finally, note that the Picture Format tab appeared on the ribbon when you selected the photo. See **Figure 2-42**.

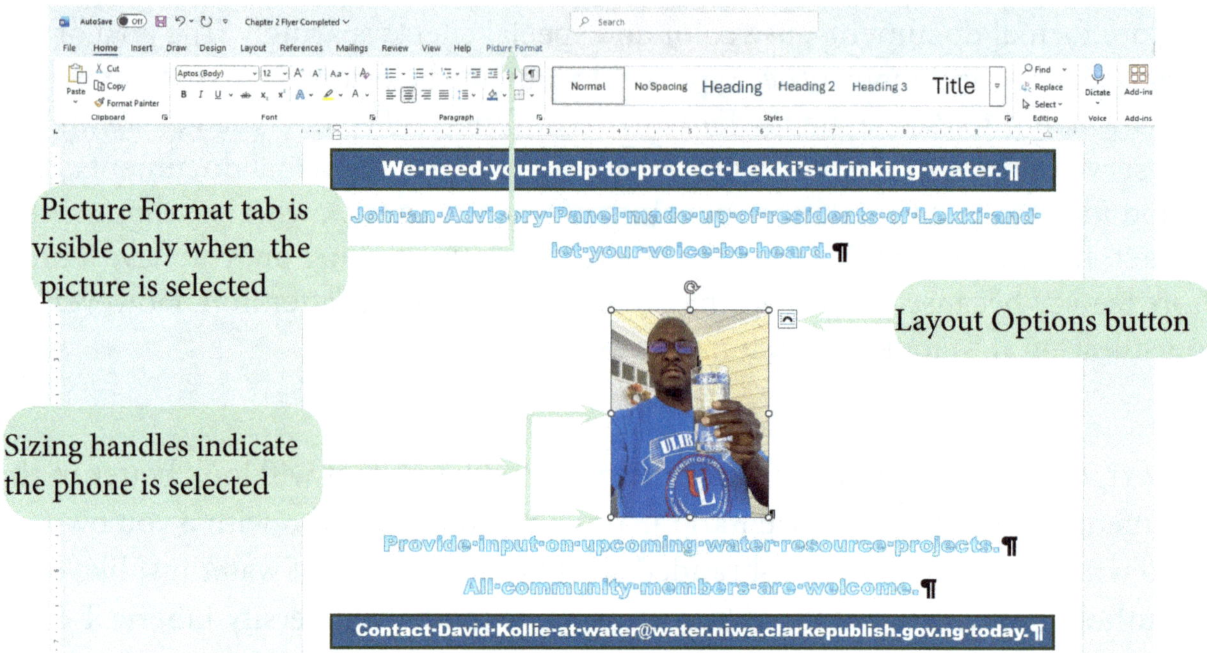

Figure 2-42 Selected Phone

(2). **Press DEL**. The photo is deleted from the document. The insertion point blinks next to the paragraph symbol.

Now you are ready to insert the new photo in the paragraph containing the insertion point. When you do, you will briefly see a gray box at the bottom of the photo containing a description of the image. This description, which is called **Alternative Text** (or **alt text**, for short), makes it possible for a screen-reading device to read a description of the image aloud. This is useful for people with vision impairment, who would otherwise find it difficult or impossible to read a document.

Word automatically creates alt text for most photos, although it is often too generic to be really helpful (for example, "Two people"). To refine alt text created by Word so that it accurately describes an image, click the Alt Text button in the Accessibility group on the Picture Format tab. This opens the Alt Text pane, where you can edit the existing alt text. As you become a more experienced Word user, you'll have the chance to create new alt text for charts, tables, and other items. Before you can use automatic alt text, you need to be using Word 365 and make sure the automatic alt text option is turned on in the Word Options dialog box. To add alt text in an older version of Word (e.g., 2016 or earlier), right-click the image, select Format Picture, and find the Alt Text or Layout & Properties option in the dialog box or side panel.

To turn on automatic alt text, insert a new photo, and edit its alt text:

******To complete this section or module, you will need to take and save an image of yourself in the Word Folder, into the Chapter 2 Folder. The picture can be that of you with you college branded shirt or T-shirt with you holding a bottle of water in your hand, with the water pouring from the bottle into a glass, or you can take and save a picture of your local government water source.***

(1). **Click File**, and then **Click Options** to open the Word Options dialog box with the General tab displayed.

(2). In the navigation pane, **Click Ease of Access**. In the "Automatic Alt Text" section, **Click** the **Automatically generate alt text for me** check box to insert a checkmark, if necessary, and then **Click OK** to close the Word Options dialog box.

(3). On the ribbon, **Click** the **Insert Tab**. The ribbon changes to display the Insert options.

(4). In the Illustrations group, **Click** the **Pictures** button. The Insert Picture From gallery opens.

(5). **Click This Device**. The Insert Picture dialog box opens.

(6). Navigate to the **Word** > **Chapter 2** folder included with your Data Files, and then **Click** the picture that you saved that folder. The name of the selected file appears in the File name box.

(7). **Click** the **Insert** button to close the Insert Picture dialog box and insert the photo. The image that you saved with water pouring from a bottle into a glass appears in the document, below the second paragraph. The photo is selected, as indicated by the sizing handles on its border, and the Picture Format tab is displayed. After a pause, a gray box with the text "Alt Text: A picture containing . . ." appears as shown in **Figure 2-43**, remains on the screen for about five seconds, and then disappears.

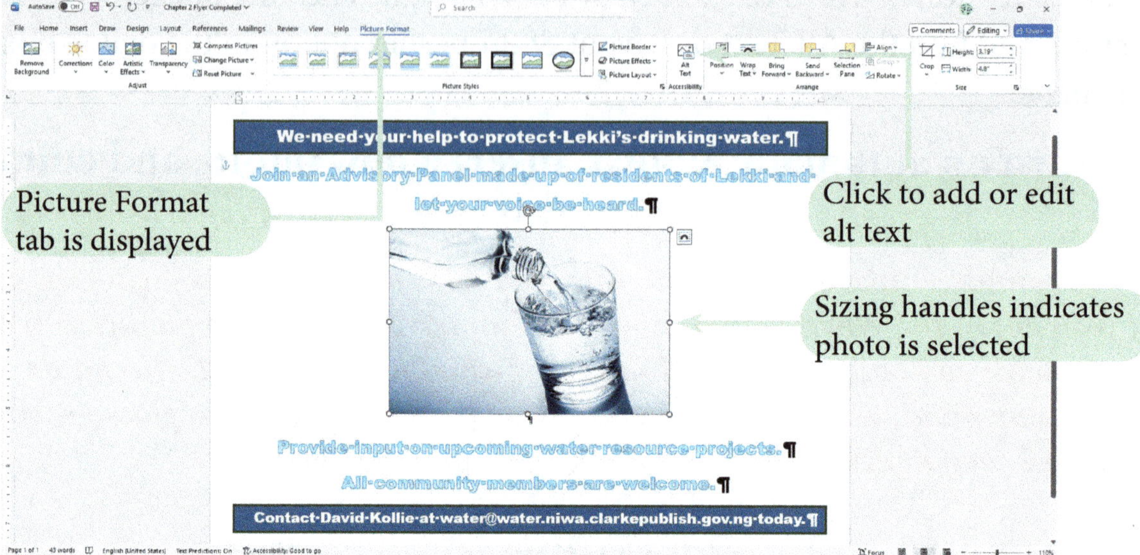

Figure 2-43 Selected Phone

Are You Having Trouble? If you see a blue message box explaining how alt text works, **Click Got It** to close the message box.

(8). In the Accessibility group on the ribbon, **Click** the **Alt Text** button to display the Alt Text pane, which displays the current alt text, as well as a note indicating the description was automatically generated. See **Figure 2-44**.

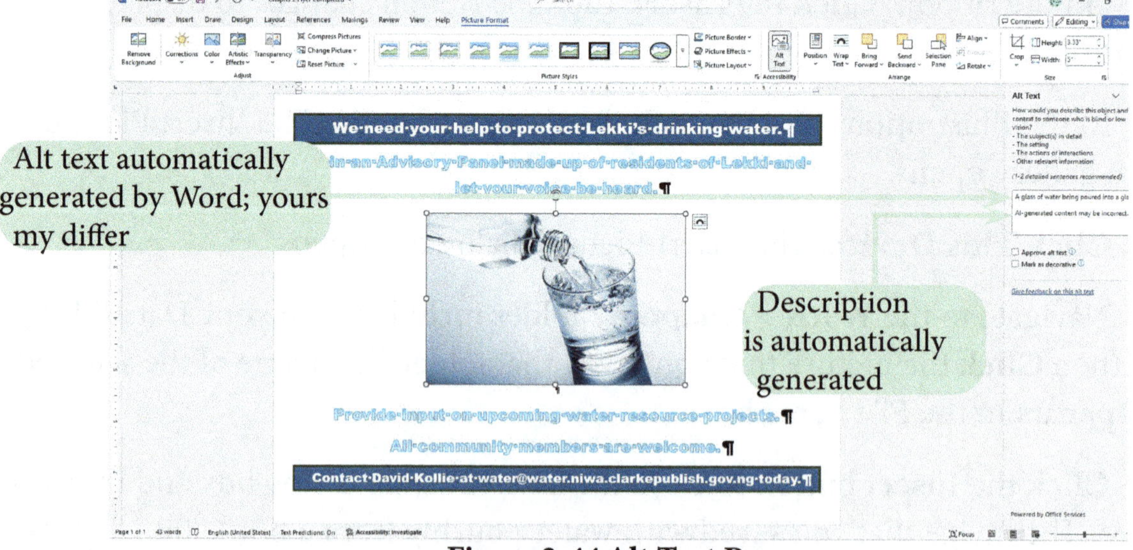

Figure 2-44 Alt Text Pane

Are You Having Trouble? If the Alt Text pane does not contain alt text, click the Generate a description for me button.

(9). In the Alt Text pane, in the white box, select all the text, including the phrase "Description automatically generated."

(10). **Type Water pouring from a bottle into a glass** and then click the Close button ☒ to close the Alt Text pane.

Now you need to resize the photo so it fills more space on the page. You could do so by clicking one of the picture's corner sizing handles, holding down the mouse button, and then dragging the sizing handle to resize the picture. But using the Shape Height and Shape Width boxes on the Picture Format tab gives you more precise results. For a screen reading image, select the Mark as decorative check box.

To resize the photo:

(1). Make sure the Picture Format tab is still selected on the ribbon.

(2). In the Size group on the far-right edge of the ribbon, locate the Shape Height box, which indicates that the height of the selected picture is currently . The Shape Width box indicates that the width of the picture is . As you'll see in the next step, when you change one of these measurements, the other changes accordingly, keeping the overall shape of the picture the same. See **Figure 2-45**.

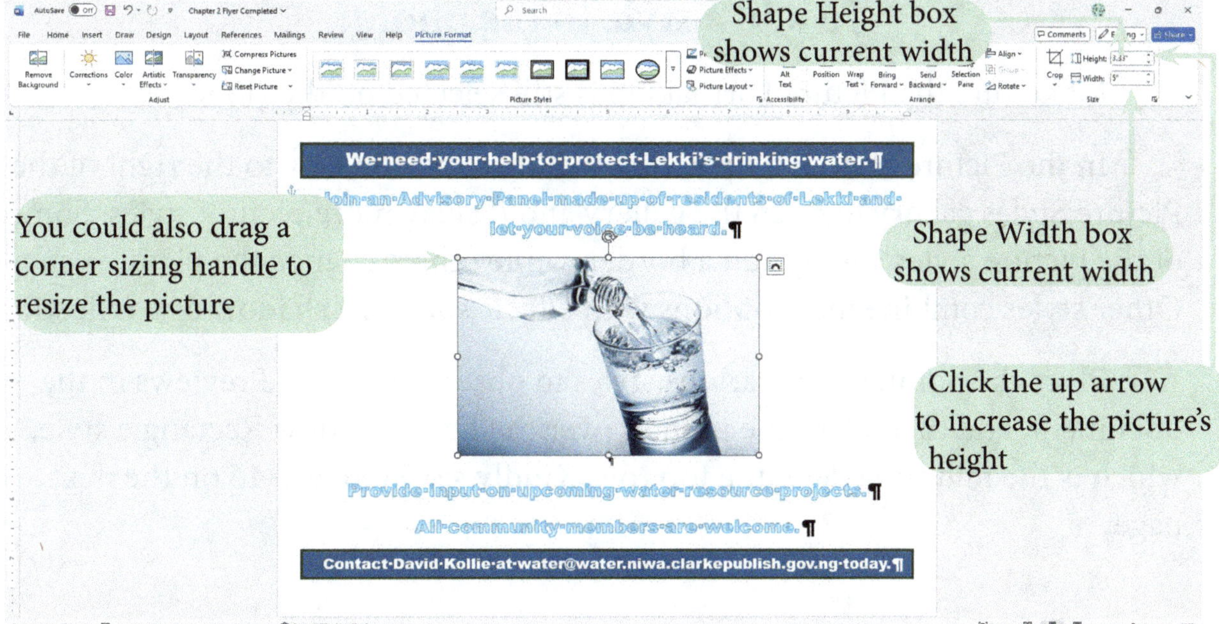

Figure 2-45 Shape Height and Shape Width Boxes

(3). **Click** the **Up Arrow** in the Shape Height box in the Size group. The photo increases in size slightly. The measurement in the Shape Height box increases to , and the measurement in the Shape Width box increases to .

(4). **Click** the **Up Arrow** in the Shape Height box repeatedly until the picture is 3.33" tall and 5" wide.

Smart Tips

Adjusting the Height and Width of an image

Depending on the size of the image you inserted, you should adjust the height and width of the image for this lab. Our recommended size for the height is 3.33" and width is 5". You are free to adjust the size of the image to the one recommended by your instructor or to your like. We encourage students to be creative, stylish, and innovative while observing school and department's guidelines.

Finally, to make the photo more noticeable, you can add a **Picture Style**, which is a collection of formatting options, such as a frame, a rounded shape, and a shadow. You can apply a picture style to a selected picture by clicking the style you want in the Picture Styles gallery on the Picture Format tab. Note that to return a picture to its original appearance, you can click the Reset Picture button in the Adjust group on the Picture Format tab. In the following steps, you'll start by displaying the Picture Styles gallery.

To add a style to the photo:

(1). Make sure the Picture Format tab is still selected on the ribbon.

(2). In the Picture Styles group, **Click** the **More** button to the right of the Picture Styles gallery to open the gallery and display more picture styles. Some of the picture styles simply add a border, while others change the picture's shape. Other styles combine these options with effects such as a shadow or a reflection.

(3). Place the pointer over various styles to observe the Live Previews in the document, and then place the pointer over the Drop Shadow Rectangle style, which is the middle style in the top row. Kindly see **Figure 2-46** on the next page.

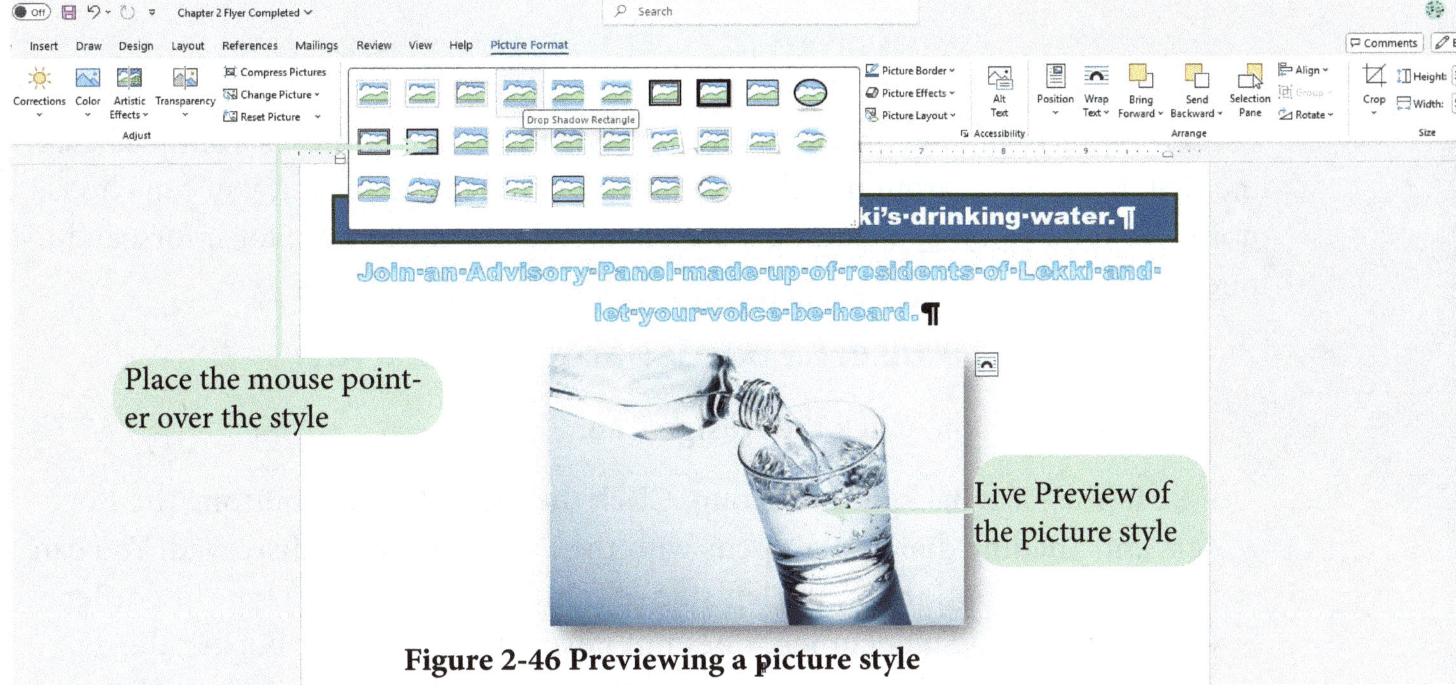

Figure 2-46 Previewing a picture style

(4). In the gallery, click the **Drop Shadow Rectangle** style to apply it to the photo and close the gallery. The photo will be formatted with a shadow on the bottom and right sides.

(5). **Click** anywhere outside the photo to deselect it, and then save the document. Congratulations, you are almost done with this hands on lab exercise.

SMARTSKILLS Working with Inline Pictures

By default, when you insert a picture in a document, it is treated as an inline object, which means its position changes in the document as you add or delete text. Also, because it is an inline object, you can align the picture just as you would align text, using the alignment buttons in the Paragraph group on the Home tab. Essentially, you can treat an inline picture as just another paragraph.

When you become a more advanced Word user, you'll learn how to wrap text around a picture so that the text flows around the picture—with the picture maintaining its position on the page no matter how much text you add to or delete from the document. The alignment buttons don't work on pictures that have text wrapped around them. Instead, you can drag the picture to the desired position on the page.

ADDING A PAGE BORDER

As with a paragraph border, the default style for a page border is a simple black line that forms a box around each page in the document. However, you can choose more elaborate options, including a dotted line, double lines, and, for informal documents, a border of graphical elements, such as stars or trees.

To insert a border around the flyer:

(1). On the ribbon, **Click** the **Design Tab**.

(2). In the Page Background group, **Click** the **Page Borders** button. The Borders and Shading dialog box opens with the Page Border tab displayed. You can use the Setting options on the left side of this tab to specify the type of border you want. Because a document does not normally have a page border, the default setting is None. The Box setting is the most professional and least distracting choice, so you'll select that next.

It's important to select the Box setting before you select other options for the border. Otherwise, when you click OK, your document won't have a page border, and you'll have to start over.

(3). In the Setting section, **Click** the **Box** setting. Selecting this option would add a simple line page border, but David prefers a different line style.

(4). In the Style box, scroll down and **Click** the D**ouble-Line Style**. Now you can select a different line color, just as you did when creating a paragraph border.

(5). **Click** the **Color Arrow** to open the Color gallery, and then **Click** the **Green**, **Accent 6 square**, which is the right-most square in the top row of the Theme Colors section. The Color gallery closes and the Green, Accent 6 color is displayed in the Color box. At this point, you could change the line width as well, but David prefers the default setting. Kindly see **Figure 2-47** on the next page.

(6). In the lower-right corner of the Borders and Shading dialog box, **Click** the **Options** button. The Border and Shading Options dialog box opens.

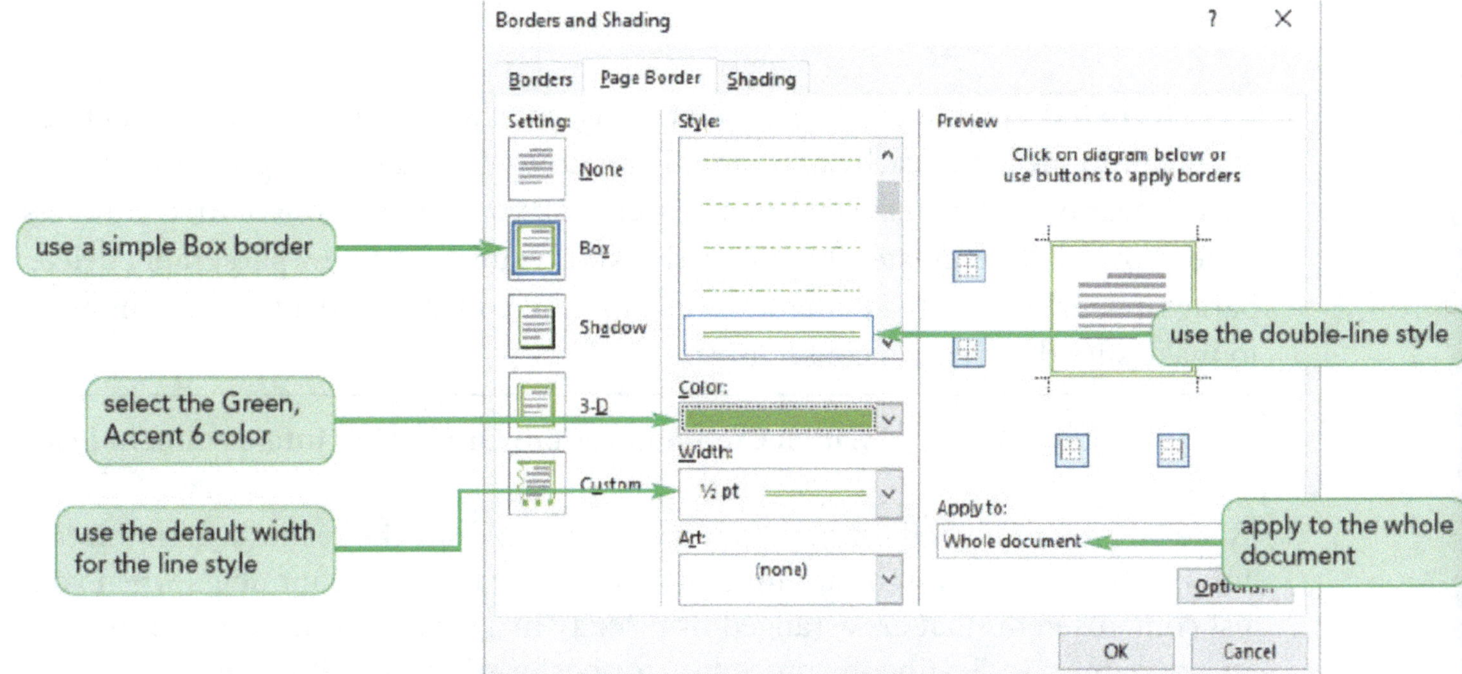

Figure 2-47 Adding a border to the flyer

By default, the border is positioned 24 points from the edges of the page. If you plan to print your document on an older printer, it is sometimes necessary to change the Measure from setting to Text, so that the border is positioned relative to the outside edge of the text rather than the edge of the page. Alternatively, you can increase the settings in the Top, Bottom, Left, and Right boxes to move the border closer to the text. For most modern printers, however, the default settings are fine.

(7). In the Border and Shading Options dialog box, **Click Cancel**, and then **Click OK** in the Borders and Shading dialog box. The flyer now has a double-line green border.

(8). **Save** the document.

(9). **Close** the document without closing Word.

David needs your help with one last task—adding bulleted and numbered lists to a document containing the minutes of the Citizen Advisory Panel's May meeting. After you finish formatting the document, David can make the minutes available to the residents of Lekki through the department's website.

CREATING BULLETED AND NUMBERED LISTS

A **Bulleted List** is a group of related paragraphs with a black circle or other character to the left of each paragraph. For a group of related paragraphs that have a particular order (such as steps in a procedure), you can use consecutive numbers instead of bullets to create a **Numbered List**. If you insert a new paragraph, delete a paragraph, or reorder the paragraphs in a numbered list, Word adjusts the numbers to make sure they remain consecutive.

SMARTSKILLS Written Communication: Organizing Information in Lists

Bulleted and numbered lists are both great ways to draw the reader's attention to information. But it's important to know how to use them. Use numbers when your list contains items that are arranged by priority in a specific order. For example, in a document reviewing the procedure for performing CPR, it makes sense to use numbers for the sequential steps. Use bullets when the items in the list are of equal importance or when they can be accomplished in any order. For example, in a resume, you could use bullets for a list of professional certifications.

To add bullets to a series of paragraphs, you use the Bullets button in the Paragraph group on the Home tab. To create a numbered list, you use the Numbering button in the Paragraph group instead. Both the Bullets button and the Numbering button have arrows you can click to open a gallery of bullet or numbering styles.

David asks you to add two bulleted lists and a numbered list to the minutes of the last meeting of the Citizen Advisory Panel.

To apply bullets to paragraphs:

(1). Open the document **Chapter 2 Meeting.docx** > **Chapter 2** folder, and then **Save** the document as **Minutes** in the location specified by your instructor.

(2). Verify that the document is displayed in **Print Layout View** and that the rulers and nonprinting characters are displayed. Make sure the Zoom level is set to 120%.

(3). On page 1, **Select** the complete list of members in attendance, starting with Oludele Awodele and concluding with Shagun Abara.

(4). On the ribbon, **Click** the **Home tab**, if necessary.

(5). In the Paragraph group, **Click** the **Bullets** button. Black circles appear as bullets before each item in the list. Also, the bulleted list is indented, and the paragraph spacing between the items is reduced.

After reviewing the default, round bullet in the document, David decides he would prefer square bullets.

(6). In the Paragraph group, **Click** the **Bullets** arrow. A gallery of bullet styles opens. See **Figure 2-48**.

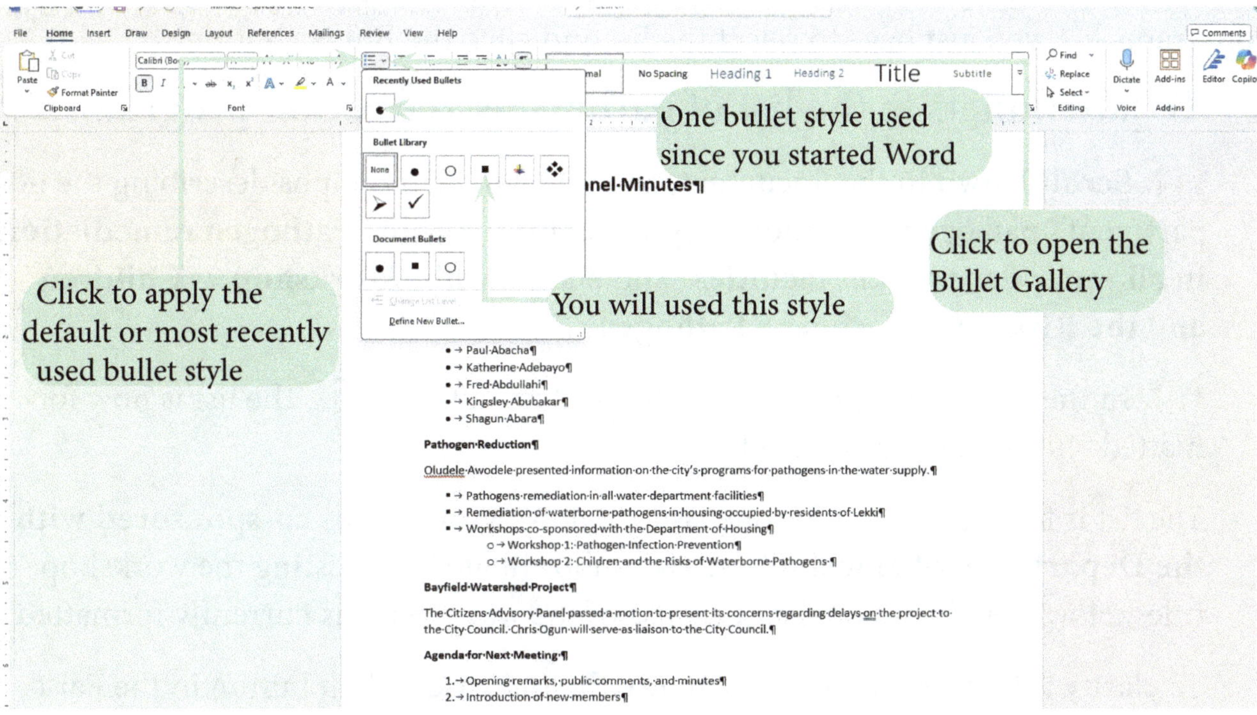

Figure 2-48 Bullets gallery

The Recently Used Bullets section appears at the top of the gallery of bullet styles; it displays the bullet styles that have been used since you started Word, which, in this case, is just the round black bullet style that was applied by default when you clicked the Bullets button. The **Bullet Library**, which offers a variety of bullet styles, is shown below the Recently Used Bullets. To create your own bullets from a picture file or from a set of predesigned symbols including diamonds, hearts, or Greek letters, **Click Define New Bullet**, and then **Click Symbol** or **Picture** in the Define New Bullet dialog box.

(7). **Move** the **Pointer** over the bullet styles in the Bullet Library to see a Live Preview of the bullet styles in the document. David prefers the black square style.

(8). In the Bullet Library, **Click** the **Black Square**. The round bullets are replaced with square bullets.

Next, you need to format the list of pathogen-reduction programs with square bullets. When you first start Word, the Bullets button applies the default, round bullets you saw earlier. But after you select a new bullet style, the Bullets button applies the last bullet style you used. So, to add square bullets to the pathogen-reduction programs list, you just have to select the list and click the Bullets button.

To add bullets to the list of pathogens-reduction programs:

(1). **Scroll Down** in the document, and select the paragraphs describing the department's pathogen-reduction programs, starting with "**Pathogen remediation in all water department facilities**" and ending with "**Workshop 2: Children and the Risks of Waterborne Pathogens** ."

(2). In the Paragraph group, **Click** the **Bullets** button . The list is now formatted with square black bullets.

The list is finished except for one issue. Below "**Workshops co-sponsored with the Department of Housing**" are two subordinate items listing the workshop titles. However, that's not clear because of the way the list is currently formatted.

To clarify this information, you can use the Increase Indent button in the Paragraph group to indent the last two bullets. When you do this, Word inserts a different style bullet to make the indented paragraphs visually subordinate to the bulleted paragraphs above.

To indent the last two bullets:

(1). In the list of pathogen-reduction programs, **Select** the last two paragraphs.

(2). In the Paragraph group, **Click** the Increase Indent button . The two paragraphs move to the right, and the black square bullets are replaced with open circle bullets. Note that to remove the indent from selected text, you could click the Decrease Indent button in the Paragraph group.

Next, you will format the agenda items as a numbered list.

To apply numbers to the list of agenda items:

(1). **Scroll down**, if necessary, until you can see the last paragraph in the document.

(2). **Select** all the paragraphs below the "**Agenda for Next Meeting**" heading, starting with "Opening remarks, public comments, and minutes" and ending with "**Report on upcoming projects**. . . ."

(3). In the Paragraph group, **click** the **Numbering** button. Consecutive numbers appear in front of each item in the list. See **Figure 9-43.**

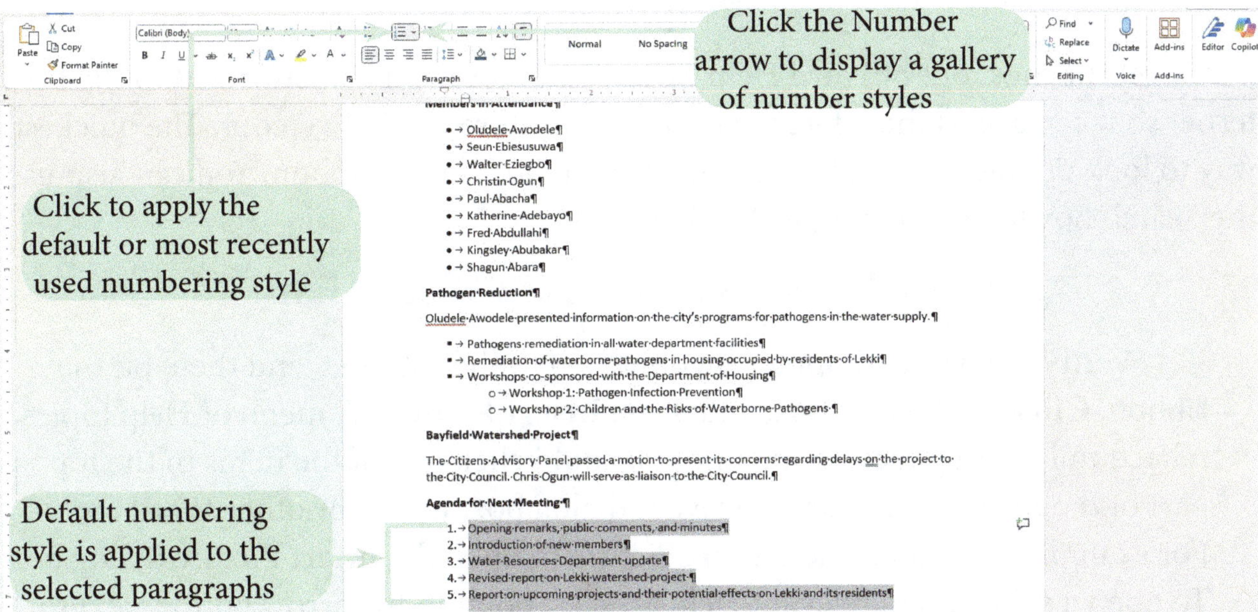

Figure 2-49 Numbered list

(4). **Click** anywhere in the document to deselect the numbered list.

(5). **Save** the document.

Submit your completed lab to your instructor. Congratulations on completing.

As with the Bullets arrow, you can click the Numbering arrow, and then select from a library of numbering styles. You can also indent paragraphs in a numbered list to create an outline, in which case the indented paragraphs will be preceded by lowercase letters instead of numbers. To apply a different list style to the outline (for example, with Roman numerals and uppercase letters), select the list, click the Multilevel List button in the Paragraph group, and then click a multilevel list style. Keep in mind that you can always add items to a bulleted or numbered list by moving the insertion point to the end of the last item in the list and pressing ENTER. The Bullets button is a **Toggle Button**, which means you can click it to add or remove bullets from selected text. The same is true of the Numbering button.

The document is complete and ready for David to post to the department's website. Because David is considering creating a promotional brochure that would include numerous photographs, he asks you to look up more information about inserting pictures. You can do that using Word Help.

Getting Help

To get the most out of Word Help, your computer must be connected to the Internet so it can access the reference information stored at Office.com. The quickest way to look up information is to use the Search box on the ribbon. You can also use the Search box to quickly access Word features.

To look up information in Word Help:

(1). Verify that your computer is connected to the Internet, and then, on the ribbon, **Click** the **Search** box, and **Type Insert Picture**. A menu of Help topics related to inserting pictures opens. You could click one of the items in the top part of the menu to access the relevant dialog box, menu, or other Word tool. For example, you could click Insert Picture to open the Insert Picture dialog box. To open a submenu of relevant Help articles, point to the arrow button next to the "insert picture" command in the Get Help section. If you prefer to expand your search to the entire web, you could click the More search results for "insert picture" command at the bottom of the menu to open the Search pane with links to articles from Wikipedia and other sources. See **Figure 2-50** on the next page.

To search the web for information on a word or phrase in a document, select the text, click the reference tab, and then click the search button in the Research group.

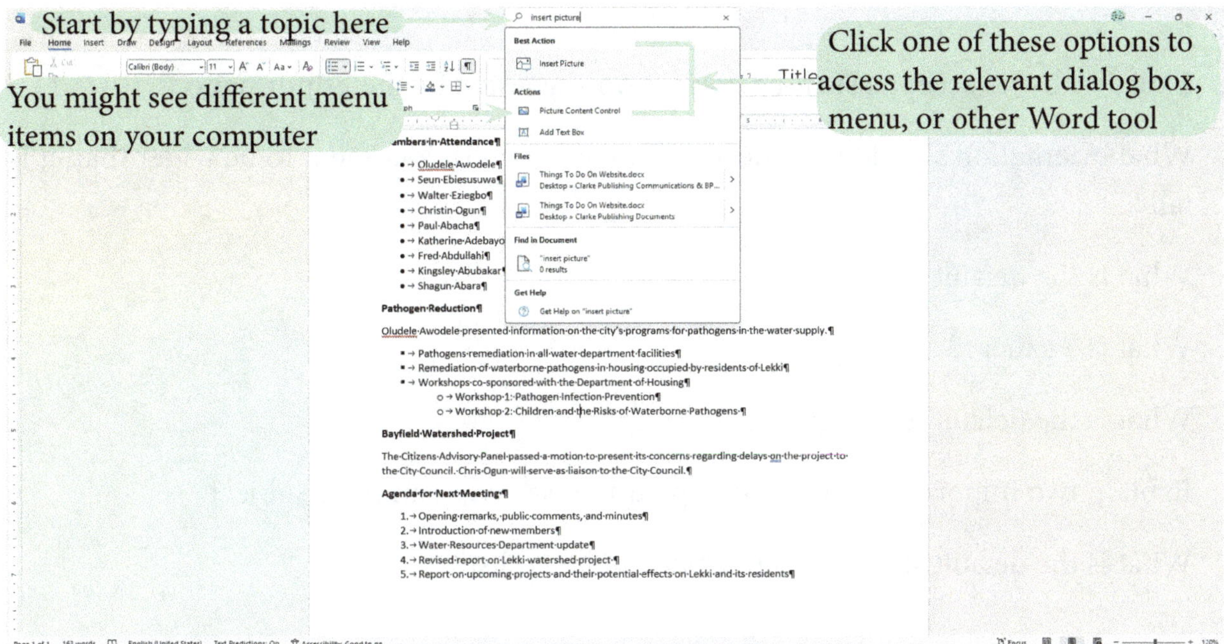

Figure 2-50 Word Help menu

(2). **Point** to the **arrow** button next to "insert picture" in the Get Help on section. A menu of articles about inserting pictures appears.

(3). **Click** the first item in the menu to open the Help pane with information about inserting pictures.

(4). **Scroll down** in the Help pane to read all the information. Note that you can use the Search help box at the top of the Help pane to look up information on other topics.

(5). **Click** the **Close** button in the upper-right corner to close the Help pane.

(6). **Click** the **File** tab, and then **Click Close** in the navigation pane to close the document without closing Word.

Word Help is a great way to learn about and access Word's many features. Articles and videos on basic skills provide step-by-step guides for completing tasks, while more elaborate, online tutorials walk you through more complicated tasks. Be sure to take some time on your own to explore Word Help so you can find the information and features you want when you need it.

Quick Review

1. Explain how to accept a spelling correction suggested by the Editor pane.
2. What orientation should you choose if you want your document to be wider than it is tall?
3. What is the default font size?
4. What is a gallery?
5. What is the default text alignment?
6. Explain two important facts about a picture inserted as an inline object.
7. What is the default shape for bullets in a bulleted list?

CHAPTER REVIEW

Lab Project 1

CHAPTER REVIEW LABS

Instructions: Data Files needed for the Review Assignments: ChumaLetter.docx, Chapter 2 Water Usage.docx, Glass.png

David asks you to write a cover letter to Chuma Gayflor at the New Day Neighborhood Center to accompany a pamphlet on water quality that will be used in an upcoming workshop. After that, he wants you to create an envelope for the letter, and then format a flyer announcing free educational tours of Lekki's water resource facilities. Finally, he needs you to add bulleted and numbered list to the minutes for the Citizen Advisory Panel's July meeting. Change the Zoom level as necessary while you are working. Complete the following steps:

(1). **Open** a **New, Blank** document and then **Save** the document as **Chuma-Letter** in the location specified by your instructor.

(2). Type the date **February 15, 2027** using AutoComplete for "February."

(3). **Press ENTER** twice, and then type the following inside address, using the default paragraph spacing and pressing **ENTER** once after each line:

CHAPTER REVIEW

Lab Project 1

CHAPTER REVIEW LABS

Chuma Gayflor

New Day Neighborhood Center

2 Akin Leigh Cres, Lekki Phase 1

Lagos 106104, Lagos, Nigeria

(4). **Type Dear Chuma:** as the salutation, **Press ENTER**, and then type the following two paragraphs as the body of the letter:

Enclosed you will find the water quality pamphlet we discussed. I hope the young people taking part in your sustainability workshop find this information useful. Additional data on our city's water supply is available at www.water.niwa.clarkepublish.gov.ng.

Keep in mind that we also offer free educational tours of our water resources facilities. We can accommodate groups as large as thirty.

(5). **Press ENTER, Type Sincerely yours,** as the complimentary closing, **Press ENTER** three times, **Type David Kollie** as the signature line, insert a manual line break, and **type Communications Director** as his title.

(6). **Press ENTER, type** your initials, insert a **Manual Line Break**, and then use the **Undo** button to make your initials all lowercase, if necessary.

(7). **Type Enclosure** and **Save** the document.

(8). **Scroll** to the beginning of the document and proofread your work. Remove any wavy underlines by using a shortcut menu or by typing a correction yourself. **Remove** the hyperlink formatting from the **web address**.

(9). **Remove** the paragraph spacing from the first three lines of the inside address.

CHAPTER REVIEW

Lab Project 1

CHAPTER REVIEW LABS

(10). **Change** the top **Margin** to **2.75 inches**. Leave the other margins at their default settings.

(11). **Save** your changes to the letter, preview it, print it if your instructor asks you to, and then close it.

(12). **Create** a **New, Blank** document, and then create an envelope. Use Chuma Gayflor's address (from Step 3) as the delivery address. Use your school's name and address for the return address. Add the envelope to the document. If you are asked if you want to save the return address as the new return address, click No.

(13). **Save** the document as **Chuma-Envelope.docx** in the location specified by your instructor, and then close the document.

(14). **Open** the document **Chapter 2 Lekki Water.docx**, located in the **Chapter 2 > Labs folder** included with your Data Files, and then check your screen to make sure your settings match those in the chapter.

(15). Save the document as **Lekki Drinking Water.docx** in the location specified by your instructor.

(16). Use the **Recheck Document** button in the Word Options dialog box to reset the Spelling and Grammar checker, and then use the Editor pane to correct any errors. Ignore any items marked as errors that are in fact correct, and accept any suggestions regarding clarity and conciseness. If the Editor pane does not give you the opportunity to correct all the errors marked in the document, close the Editor pane and correct the errors using shortcut menus.

(17). **Proofread** the document and correct any other errors. Be sure to change "Today" to "today " in the last paragraph.

(18). Change the page orientation to **Landscape** and the margins to **Narrow**.

(19). Format the document text in **22-point Times New Roman** font.

CHAPTER REVIEW

Lab Project 1

CHAPTER REVIEW LABS

(20). **Center** the text and the photo.

(21). Format the first paragraph with an outside border using the default style, and change the border color to **Gold, Accent 4**, and the border width to **11/2 pt**. Add blue shading to the paragraph, using the Blue, Accent 5 color in the Theme Colors section of the Shading gallery. Format the paragraph text in **white**.

(22). Format the last paragraph in the document using the same formatting you applied to the first paragraph.

(23). **Increase** the paragraph **spacing** after the first paragraph to **42 points**. Increase the paragraph spacing before the last paragraph in the document to 42 points.

(24). **Format** the **second paragraph** with the **Gradient Fill: Gold, Accent color 4; Outline: Gold, Accent color 4** text effect. **Increase** the paragraph's font size to **26 points**.

(25). **Format** the **text** in the **third** and **fourth paragraphs** (the first two paragraphs below the photo) using the **Blue, Accent 5** font color, and then add **bold** and **italic**.

(26). **Delete** the **photo** and replace it with the a photo of your choice with a water pouring into a glass, located in the **Chapter 2** > **Labs** folder.

(27). **Delete** the existing **alt text** and the text indicating that the description was automatically generated, and then **type Water pouring into a glass.** (Do not include the period after "glass.")

(28). **Resize** the new photo so that it is tall, and then add the Soft Edge Rectangle style in the Pictures Styles gallery.

(29). **Add** a **page border** using the **Box setting**, a double-line style, the default width, and the **Gold, Accent 4** color.

CHAPTER REVIEW

Lab Project 1

Chapter Review Labs

(30). **Save** your changes to the flyer, preview it, and then close it.

(31). **Open** the document **Chapter 2 Meeting.docx**, located in the **Chapter 2** > **Labs** folder, and then check your screen to make sure your settings match those in the chapter.

(32). **Save** the document as **July-Minutes** in the location specified by your instructor.

(33). **Format** the list of members in attendance as a bulleted list with **square bullets**, and then format the list of lawn-care initiatives with square bullets (starting with "Alternate-day watering . . ." and ending with "Workshop 2: Drought-Tolerant Gardening"). Indent the paragraphs for Workshop 1 and Workshop 2 so they are formatted with open circle bullets.

(34). **Format** the five paragraphs below the "Agenda for Next Meeting" heading as a numbered list.

(35). Use Word Help to look up the topic work with pictures. Read the first article, return to the Help home page, and then close Help.

Self Challenge Lab

Lab Project 2

Self Challenge Lab

There are no Data Files needed for this Self Driven Problem.

Monrovia City Real Estate, Inc You are a real estate agent at Monrovia City Real Estate, Inc., located in Paynesville City, Liberia. You recently sold a building on Ashmun Street in downtown Monrovia and need to forward an extra key to the building's new owner. Create a cover letter to accompany the key by completing the following steps.

CHAPTER REVIEW

SELF CHALLENGE LAB

Lab Project 2

Because your office is currently out of letterhead, you'll start the letter by typing a return address. As you type the letter, remember to include the appropriate number of blank paragraphs between the various parts of the letter. Complete the following steps:

(1). **Open** a **new, blank** document, and then save the document as **Toe-Letter** in the location specified by your instructor. If necessary, change the **Zoom level** to **120%**.

(2). **Type** the following return address, using the default paragraph spacing and replacing [Your Name] with your first and last names:

[Your Name]

Monrovia City Real Estate, Inc.

15 AB Tolbert Road

1000 Paynesville, 10 Liberia

(3). **Type November 9, 2025** as the date, leaving a blank paragraph between the last line of the return address and the date.

(4). **Type** the following inside address, using the default paragraph spacing and leaving the appropriate number of blank paragraphs after the date:

Musa Saye Toe

Brewerville Properties Management, LLC.

36 Hotel Africa Road

1000 Brewerville, 10 Liberia

CHAPTER REVIEW

SELF CHALLENGE LAB

Lab Project 2

(5). Type **Dear Mr. Toe:** as the salutation.

(6). To begin the body of the letter, type the following two paragraphs: **Enclosed please find the extra office key for the apartment building you recently purchased at 85 Ashmun Street. The previous owner found it when he was cleaning out his desk and asked me to send it to you.**

It was a pleasure working with you. In order to improve our service, I would be grateful if you would review the following questions, and then email me your answers at toe@mcrealestate.clarkepublish.com.

(7). **Remove** the hyperlink formatting from the email address.

(8). **Add** the following questions as separate paragraphs, using the default paragraph spacing:

Did you find our staff helpful and well-informed?

Were you satisfied with the service provided during your real estate transaction?

Would you recommend Monrovia City Real Estate to others?

Can you suggest any ways to improve our service?

(9). **Insert** a **new paragraph** after the last question, and then type the complimentary closing Sincerely, (including the comma).

(10). Leave the appropriate amount of space for your signature, type your full name, insert a manual line break, and then type Licensed Real Estate Agent.

(11). **Type Enclosure** in the appropriate place.

(12). Use the Editor pane to correct any errors. Ignore any items marked as errors that are in fact correct (such as the word "Musa, Saye, or Ashmun"), and

CHAPTER REVIEW

Self Challenge Lab

Lab Project 2

accept any suggestions regarding clarity and conciseness. Instruct the Editor pane to ignore the recipient's name. If the Editor pane does not give you the opportunity to correct all the errors marked in the document, close the Editor pane and correct the errors using shortcut menus.

(13). **Italicize** the four paragraphs containing the questions.

(14). **Format** the list of questions as a bulleted list with square bullets.

(15). **Remove** the paragraph spacing from the first three lines of the return address. Do the same for the first three paragraphs of the inside address.

(16). **Center** the four paragraphs containing the return address, format them in 16-point font, and then add the Fill: Blue, Accent color 1; Shadow text effect.

(17). **Deselect** any selected text, and then create an envelope in the current document. Use Musa Saye Toe's address (from Step 4) as the delivery address. Edit the delivery address as necessary to remove any incorrect text. Use the return address shown in Step 2. Add the envelope to the Toe-Letter.docx document. If you are asked if you want to save the return address as the default return address, **Click No**.

(18). **Save** the document, preview it, and close it.

CHAPTER 3 HIGHLIGHT

Excel: Tracking Miscellaneous Expenses for a Conference or a Program

After completing this chapter, you will be able to:

1. Open and close a workbook
2. Navigate through a workbook and worksheet
3. Select cells and ranges
4. Plan and create a workbook
5. Insert, rename, and move worksheets
6. Enter text, dates, and numbers
7. Undo and redo actions
8. Resize columns and rows
9. Open and close a workbook
10. Navigate through a workbook and worksheet
11. Select cells and ranges
12. Plan and create a workbook
13. Insert, rename, and move worksheets
14. Enter text, dates, and numbers
15. Undo and redo actions
16. Resize columns and rows

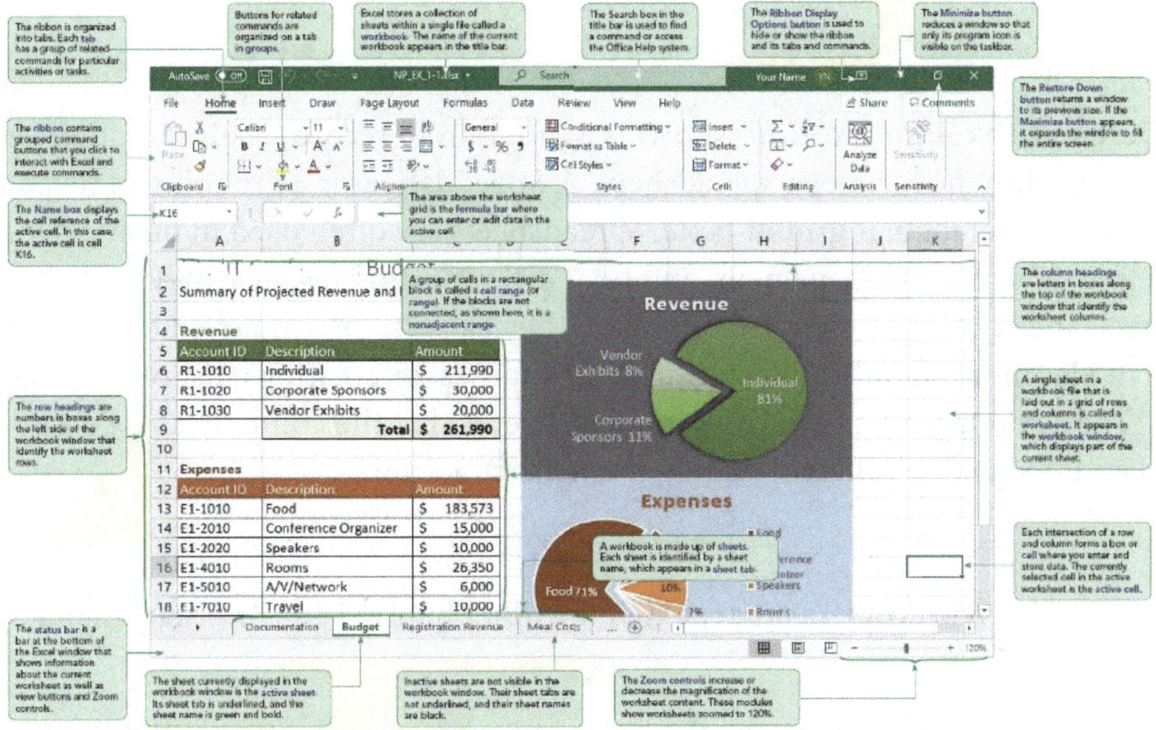

This overview screenshot shows an example of an Excel Workbook. At the top of the workbook is the title bar, which include the name of the current file (Excel stores a collection of sheets within a single file called workbook). Below the title bar is the Ribbon, which is organized into tabs. Each tab has a group of related commands for particular activities or tasks. The Tell Me box on the ribbon is used to find a command or access the Office Help system. The ribbon contains grouped command buttons that you click to interact with Excel and execute commands. Buttons for related commands are organized on a tab in groups. At the right end of the ribbon are commands related to how the workbook is displayed: the Ribbon Display Options button (used to hide or show the ribbon and its tabs and commands); the Minimize button (reduces a window so that only its program icon is visible on the taskbar); the Restore Down Button (returns a window to its previous size); and the Maximize button (if this appears instead of the Restore Down button, it expands the window to fill the entire screen). Below the ribbon, three text boxes appear. The first box, the Name box, displays the cell reference of the active cell and has a drop-option to select the cell. "K16" appears in the box, so the active cell is K16. The third box above the worksheet grid is the formula bar where you can enter or edit data in the active cell. A workbook is made up of sheets (identified by a sheet name in the sheet tab at the bottom of the workbook). A single sheet in a workbook file that is laid out in a grid of rows and columns is called a worksheet.

CHAPTER 3 | MICROSOFT EXCEL

INTRODUCING EXCEL AND SPREADSHEETS

Microsoft Excel (or **Excel** for short) is a program to record, analyse, and present data arranged in the form of a spreadsheet. A **Spreadsheet** is a grouping of text and numbers in a rectangular grid or table. Spreadsheets are often used in business for budgeting, inventory management, and financial reporting because they unite text, numbers, and charts within one document. They can also be employed for personal use in planning a family budget, tracking expenses, or creating a list of personal items. The advantage of an electronic spreadsheet is that the content can be easily edited and updated to reflect changing financial conditions.

To start Excel:

(1). On the **Windows Taskbar**, **Click** the **Start** button ⊞. The Start menu opens.

(2). On the **Start Menu**, scroll through the list of apps, and then **Click Excel**. Excel starts in Backstage view. See **Figure 3-1**.

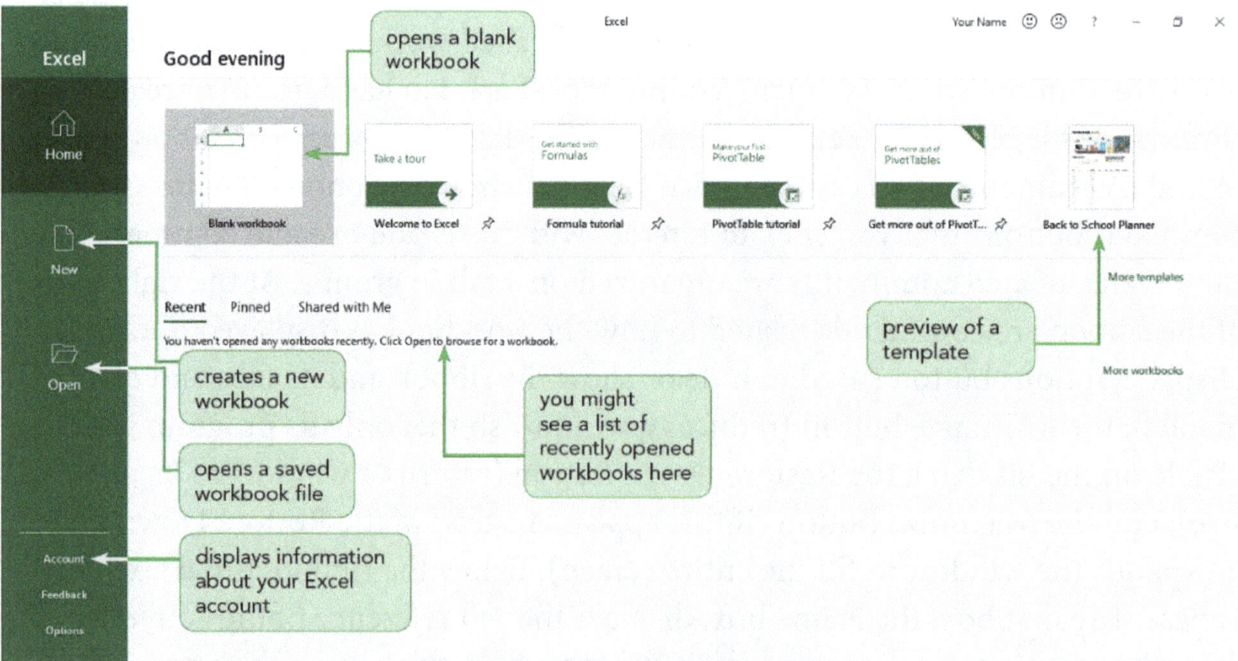

Figure 3-1 Backstage view

Backstage View, the File tab of the ribbon, contains various screens with commands that allow you to manage files and options for Excel. Excel documents are called workbooks. From Backstage view, you can open a blank workbook, open an existing workbook, or create a new workbook based on a template. A **Template** is a preformatted workbook that contains the document design and some content already entered into the document. Templates can speed up the process of creating a workbook because much of the effort in designing the workbook and entering its data and formulas is already done for you.

Mardia Obiamiwe created an Excel workbook containing information on the budget for an upcoming Monrovia conference. You'll open that workbook now.

To open the Conference workbook:

(1). In the navigation bar in Backstage view, **Click Open**. The Open screen is displayed and provides access to different locations where you might store files.

(2). **Click Browse**. The Open dialog box appears.

(3). **Navigate** to the **Excel** > **Chapter 3** folder included with your Data Files.

Are You Having Trouble? If you don't have the starting Data Files, you need to get them before you can proceed. Your instructor will either give you the Data Files or ask you to obtain them from a specified location (such as a network drive). If you have any questions about the Data Files, see your instructor or technical support person for assistance.

(4). **Click Chapter 3 CP Conference.xlsx** in the file list to select it.

(5). **Click** the **Open** button. The workbook opens in Excel.

Are You Having Trouble? If you don't see the full ribbon as shown in the Session 1.1 Visual Overview, the ribbon may be partially or fully hidden. To pin the ribbon so that the tabs and groups are fully displayed and remain visible, click the Ribbon Display Options button, and then click Show Tabs and Commands.

(6). If the Excel window doesn't fill the screen, **Click** the **Maximize** button in the upper-right corner of the title bar. Kindly see **Figure 3-2** on the next page.

Before reviewing the contents of this workbook, you first should understand how to work with the Excel interface.

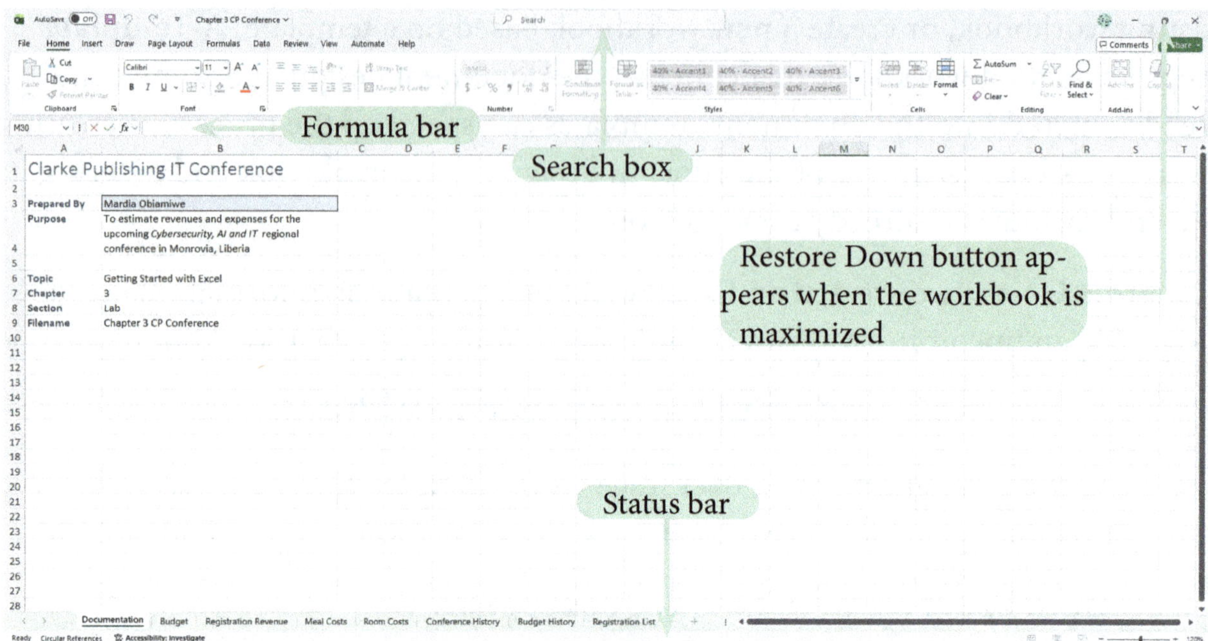

Figure 3-2 Conference workbook

Getting Help

Excel is an extensive and powerful program supporting a wide variety of commands and tools. If you are unsure about the function of an Excel command or you want information about how to accomplish a task, you can use the Help system. To access Excel Help, you press F1. You can also enter a phrase or keyword into the Search box next to the file name in the title bar. From this box, you can get quick access to detailed information on all of the Excel features.

Using Keyboard Shortcuts to Work Faster

There are several ways of accessing an Excel command. Perhaps the most efficient method is entering the command through your device's keyboard through the use of keyboard shortcuts. A **Keyboard Shortcut** is a key or combination of keys that you press to access a feature or perform a command. Excel provides keyboard shortcuts for many commonly used commands. For example, CTRL+S is the keyboard shortcut for the Save command, which means you hold down CTRL while you press S to save the workbook. (Note that the plus sign is not pressed; it is used to indicate that an additional key is pressed.) When available, a keyboard shortcut is listed next to the command's name in a ScreenTip. A **ScreenTip** is a label that appears next to an object, providing information about that object or giving a link to associated help topics. **Figure 3-3** on the next page lists some of the keyboard shortcuts commonly used in Excel.

Press	To	Press	To
ALT	Display the Key Tips for the commands and tools on the ribbon	CTRL+V	Paste content that was cut or copied
CTRL+A	Select all objects in a range	CTRL+W	Close the current workbook
CTRL+C	Copy the selected object(s)	CTRL+X	Cut the selected object(s)
CTRL+G	Go to a location in the workbook	CTRL+Y	Repeat the last command
CTRL+N	Open a new blank workbook	CTRL+Z	Undo the last command
CTRL+O	Open a saved workbook file	F1	Open the Excel Help window
CTRL+P	Print the current workbook	F5	Go to a location in the workbook
CTRL+S	Save the current workbook	F12	Save the current workbook with a new name or to a new location

Figure 3-3 Excel keyboard shortcuts

You can also use the keyboard to quickly select commands on the ribbon. First, you display the **KeyTips**, which are labels that appear over each tab and command on the ribbon when ALT is pressed. Then you press the key or keys indicated to access the corresponding tab, command, or button while your hands remain on the keyboard.

Using Excel in Touch Mode

If your computer has a touchscreen, another way to interact with Excel is in **Touch Mode** in which you use your finger or a stylus to tap objects on the touchscreen to invoke a command or tool. In Touch Mode, the ribbon increases in height, the buttons are bigger, and more space appears around each button so you can more easily use your finger or a stylus to tap the button you need.

The figures in these modules show the screen in **Mouse Mode**, in which you use a computer mouse to interact with Excel and invoke commands and tools. If you plan on doing some of your work on a touch device, you'll need to switch between Touch Mode and Mouse Mode. You should turn Touch Mode on only if you are working on a touch device.

To switch between Touch Mode and Mouse Mode:

(1). On the Quick Access Toolbar, click the **Customize Quick Access Toolbar** button . A menu opens, listing buttons you can add to the Quick Access Toolbar as well as other options for customizing the toolbar.

Are You Having Trouble? If the Touch/Mouse Mode command on the menu has a checkmark next to it, press ESC to close the menu, and then skip **Step 2**.

(2). From the Quick Access Toolbar menu, **Click Touch/Mouse Mode**. The Quick Access Toolbar now contains the Touch/Mouse Mode button , which you can use to switch between Mouse Mode and Touch Mode.

(3). On the Quick Access Toolbar, **Click** the T**ouch/Mouse Mode** button . A menu opens listing Mouse and Touch, and the icon next to Mouse is shaded to indicate that it is selected.

Are You Having Trouble? If the icon next to Touch is shaded, press ESC to close the menu and continue with **Step 5**.

(4). **Click Touch**. The display switches to Touch Mode with more space between the commands and buttons on the ribbon. See **Figure 3-4** on the next page.

Figure 3-4 on the next page displays the ribbon of the workbook with the Home tab is displayed in Touch Mode. The buttons appear larger and have more space around them. . On the left side of the title bar is the Customize Quick Access Toolbar button. To the left of that button is the Touch/Mouse Mode button.

Next, you will switch back to Mouse Mode. If you are working with a touchscreen and want to use Touch Mode, skip **Steps 5** and **6**.

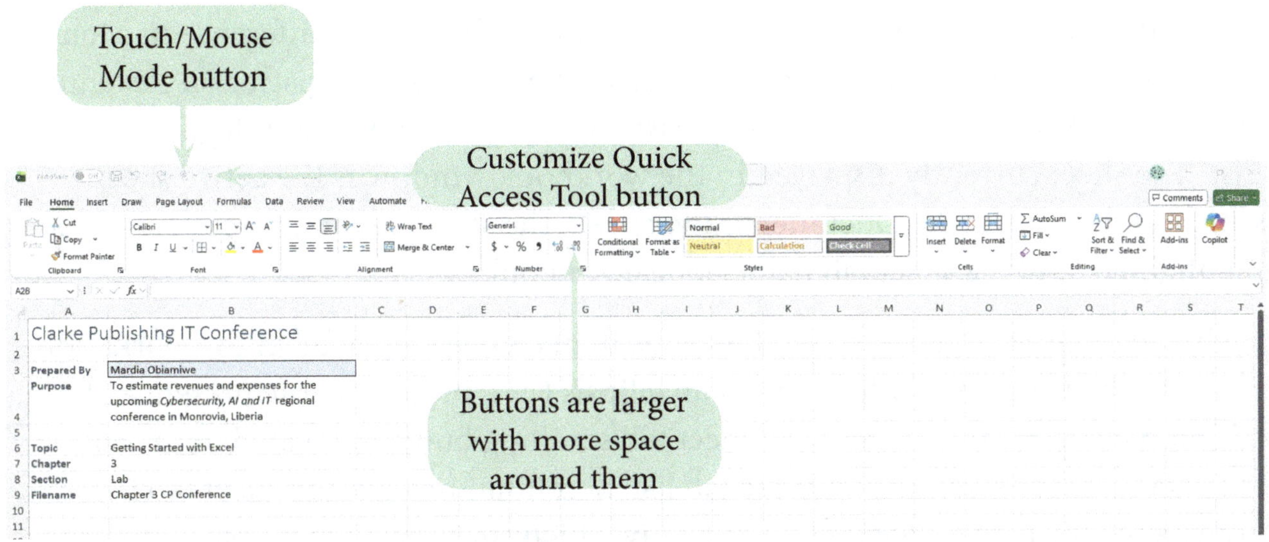

Figure 3-4 Excel displayed in Touch Mode

(5). On the Quick Access Toolbar, **Click** the **Touch/Mouse Mode** button , and then click Mouse. The ribbon returns to Mouse Mode, as shown earlier in Figure 3-2.

(6). On the Quick Access Toolbar, **Click** the **Customize Quick Access Toolbar** button , and then **Click Touch/Mouse Mode** to deselect it. The Touch/Mouse Mode button is removed from the Quick Access Toolbar.

Now that you've seen how to interact with the Excel program, you are ready to explore the workbook that Mardia has prepared.

EXPLORING A WORKBOOK

The contents of a workbook are shown in the workbook window, which is below the ribbon. Workbooks are organized into separate pages called sheets. Excel supports two types of sheets: worksheets and chart sheets. A worksheet contains a grid of rows and columns into which you can enter text, numbers, dates, and formulas. Worksheets can also contain graphical elements such as charts, maps, and clip art. A chart sheet is a sheet that contains only a chart that is linked to data within the workbook. A **Chart Sheet** can also contain other graphical elements like clip art, but it doesn't contain a grid for entering data values.

CHANGING THE ACTIVE SHEET

Worksheets and chart sheets are identified by the sheet tabs at the bottom of the workbook window. The workbook for the Clarke Publishing Conference in Monrovia contains eight sheets labeled Documentation, Budget, Registration Revenue, Meal Costs, Room Costs, Conference History, Budget History, and Registration List. The sheet currently displayed in the workbook window is the active sheet, which in this case is the Documentation sheet. The sheet tab of the active sheet is highlighted, and the sheet tab name appears in bold.

If a workbook contains more sheet tabs than can be displayed in the workbook window, the list of tabs will end with an ellipsis (…), indicating the presence of additional sheets. You can use the sheet tab scrolling buttons, located to the left of the sheet tabs, to scroll through the tab list. Scrolling through the sheet tab list does not change the active sheet; it changes only which sheet tabs are visible within the workbook window.

You will view the contents of the Conference workbook by clicking the tabs for each sheet.

To change the active sheet:

(1). **Click** the **Budget Sheet Tab**. The Budget worksheet becomes the active sheet, and its name is in bold green. See **Figure 3-5**.

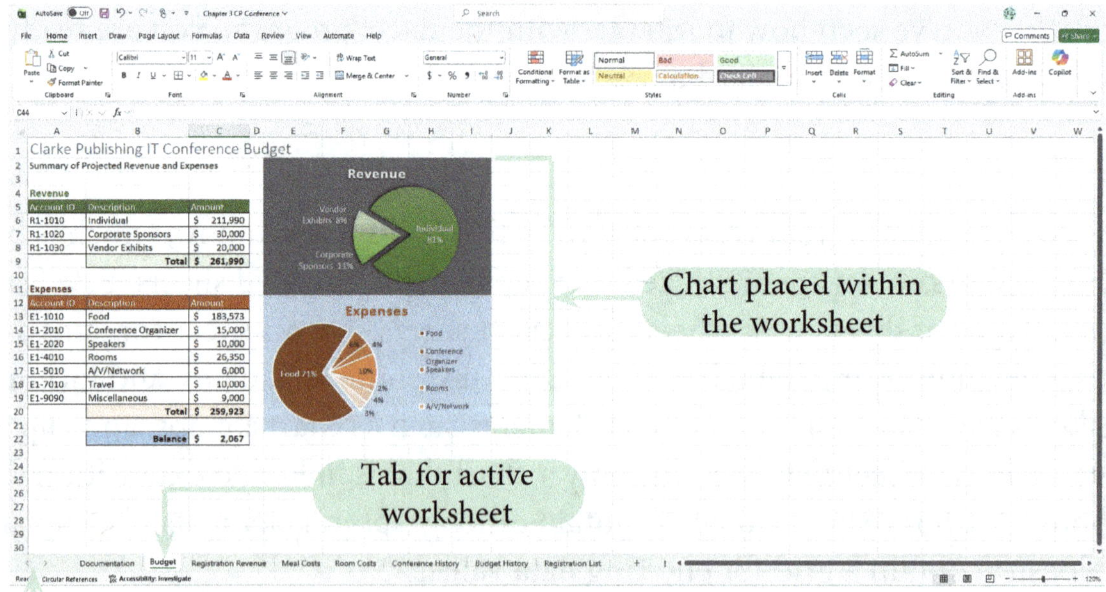

Figure 3-5 Budget worksheet

The Budget sheet contains estimates of the conference's revenue and expenses. The sheet also contains charts of the revenue and expense categories. From the charts, it's easily apparent that the major source of revenue for the conference comes from individual registrations and the major expense comes from feeding all the attendees over the three conference days.

(2). **Click** the **Registration Revenue Sheet Tab** to make it the active sheet. The Registration Revenue tab provides a more detailed breakdown of the revenue estimates for the conference.

(3). **Click** the **Meal Costs, Room Costs**, and **Conference History Sheet Tabs** to view each worksheet. **Figure 3-6** shows the contents of the Conference History chart sheet. Because this is a chart sheet, it contains only the Excel chart and not the rows and columns of text and numbers you saw in the worksheets.

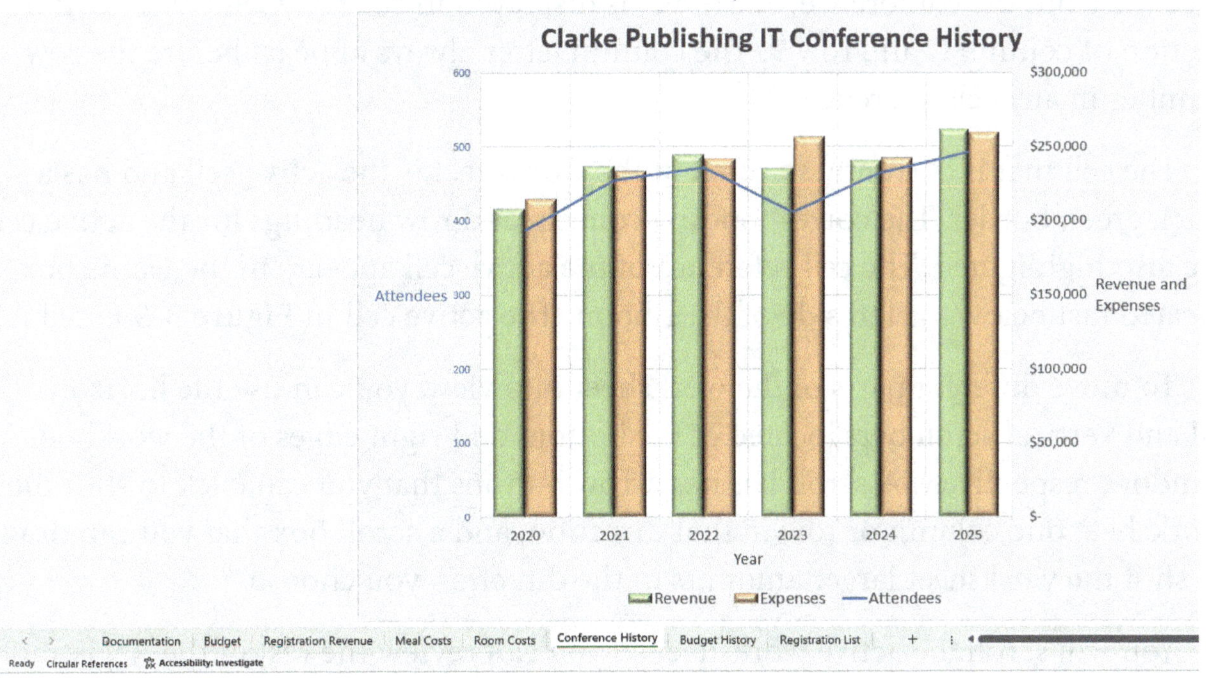

Figure 3-6 Conference History chart sheet

(4). **Click** the **Budget History** and **Registration List Sheet** tabs to view the remaining two worksheets in the workbook.

Are You Having Trouble? If you can't see the sheet tabs for the last few sheets in the workbook, click the sheet tab scrolling buttons to scroll through the tab list.

Now you know how to switch between the eight sheets in the workbook. Next, you will move around the individual worksheets so you can review their contents.

NAVIGATING WITHIN A WORKSHEET

A worksheet is organized into a grid of cells, which are formed by the intersection of rows and columns. Rows are identified by numbers, and columns are identified by letters. Row numbers range from 1 to 1,048,576. Column labels start with the letters A to Z. After Z, the next column headings are labeled AA, AB, AC, and so forth. The last possible column label is XFD, which means there are 16,384 columns available in a worksheet. The total number of possible cells in a single Excel worksheet is more than 17 billion, providing an extremely large worksheet for reports.

Each cell is identified by a **Cell Reference**, which indicates the column and row in which the cell is located. For example, as shown in **Figure 3-5**, the total expected revenue from the conference, $261,990, is displayed in cell C9. Cell C9 is the intersection of column C and row 9. The column letter always appears before the row number in any cell reference.

The cell that is currently selected in the worksheet is the active cell and has a thick green border. The corresponding column and row headings for the active cell are also highlighted. The cell reference of the active cell appears in the Name box, located just below the left side of the ribbon. The active cell in **Figure 3-5** is cell A3.

To move different parts of the worksheet into view, you can use the horizontal and vertical scroll bars located at the bottom and right edges of the workbook window, respectively. A scroll bar has arrow buttons that you can click to shift the worksheet one column or row in that direction, and a scroll box that you can drag to shift the worksheet larger amounts in the direction you choose.

You will scroll the active worksheet so you can review the rest of the Registration List worksheet.

To scroll through the Registration List worksheet:

(1). On the Registration List worksheet, **Click** the **down arrow** button ▾ on the vertical scroll bar to scroll down the worksheet until you see row 496 containing the last registration in the list.

(2). On the horizontal scroll bar, **click** the **right arrow** button ▸ three times.

The worksheet scrolls three columns to the right, moving columns A through C out of view.

(3). On the **horizontal scroll bar**, **drag** the **scroll box** to the **left** until you see column A.

(4). On the **vertical scroll bar**, **drag** the **scroll box up** until you see the top of the worksheet and cell A1.

Scrolling the worksheet does not change the location of the active cell. Although the active cell might shift out of view, you can always see the location of the active cell in the Name box. To make a different cell active, you can either click a new cell or use keyboard shortcuts to move between cells, as described in **Figure 1-7.**

Figure 3-7 Excel navigation keyboard shortcuts

Press	To move the active cell
↑↓ ←→	Up, down, left, or right one cell
HOME	To column A of the current row
CTRL+HOME	To cell A1
CTRL+END	To the last cell in the worksheet that contains data
ENTER	Down one row or to the start of the next row of data
SHIFT+ENTER	Up one row
TAB	One column to the right
SHIFT+TAB	One column to the left
PGUP, PGDN	Up or down one screen
CTRL+PGUP, CTRL+PGDN	To the previous or next sheet in the workbook

Keyboard shortcuts are especially useful in worksheets in which the data is spread across many rows or columns. For example, some financial worksheets can have tens of thousands of rows of data. You will use these shortcuts to move through the Registration List worksheet.

To change the active cell using keyboard shortcuts:

(1). On the **Registration List** worksheet, move the pointer over cell **C10** and then click the mouse button. The active cell moves from cell A2 to cell C10. A green border appears around cell C10 to indicate that it's now the active cell. The labels for row 10 and column C are highlighted, and the cell reference in the Name box is C10.

(2). **Press RIGHT ARROW**. The active cell moves one column to the right to cell D10.

(3). **Press PGDN**. The active cell moves down one full screen.

(4). **Press PGUP**. The active cell moves up one full screen, returning to cell D10.

(5). **Press CTRL+END**. The active cell is cell H496, the last cell containing data in the worksheet.

(6). **Press CTRL+HOME**. The active cell returns to the first cell in the worksheet, cell A1.

To change the active cell to a specific cell location, you can use the Go To dialog box or the Name box. These methods are especially helpful when you are working in worksheets with many rows or columns. You'll try both these methods now.

To change the active cell using the Go To dialog box and Name box:

(1). On the Home tab, in the Editing group, **Click** the **Find & Select** button, and then **Click Go To** on the menu that opens (or **Press CTRL+G** or **F5**). The Go To dialog box opens.

(2). **Type A100** in the Reference box. See **Figure 1-8**.

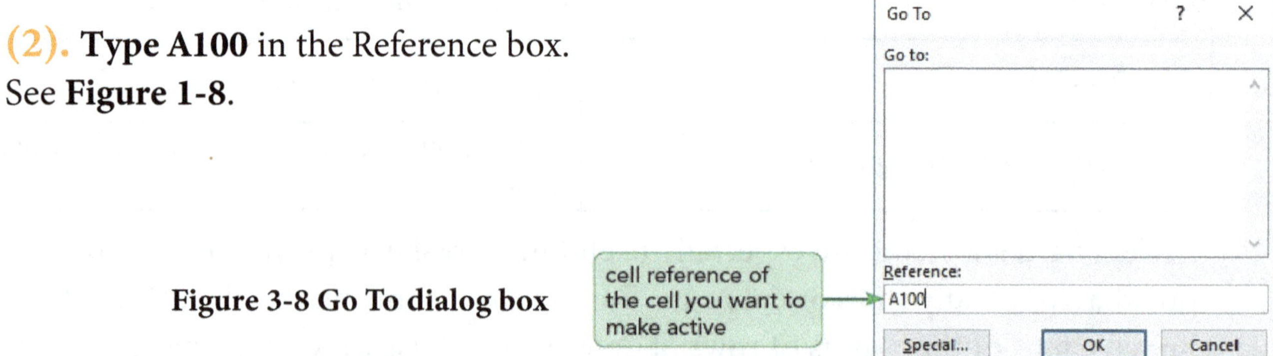

Figure 3-8 Go To dialog box

(3). **Click OK.** Cell A100 becomes the active cell showing the registration information for Saybah Flomo of 37 Clay Street, Monrovia, Liberia.

(4). Click the Name box, **Type A3**, and then **Press ENTER**. Cell A3 becomes the active cell in the worksheet.

SELECTING A CELL RANGE

Many tasks in Excel require you to work with a group of cells. A group of cells in a rectangular block is called a **Cell Range** (or simply a **Range**). Each range is identified with a range reference that includes the cell reference of the upper-left cell of the rectangular block and the cell reference of the lower-right cell separated by a colon. For example, the range reference A1:G5 refers to all the cells in the rectangular block from cell A1 through cell G5.

As with individual cells, you can select cell ranges using your mouse, the keyboard, or commands. You will select a range in the Budget worksheet.

To select a cell range in the Budget worksheet:

(1). **Click** the **Budget Sheet Tab**. The Budget worksheet becomes the active sheet.

(2). **Click** cell **A5** to select it and, without releasing the mouse button, drag down and right to cell C8.

(3). Release the mouse button. The range A5:C8 is selected. The selected cells are highlighted and surrounded by a green border. The first cell you selected in the range, cell A5, is the active cell in the worksheet. The Quick Analysis button appears next to the selected range, providing options for working with the range. See **Figure 3-9** on the next page.

A snippet of the Budget sheet is displayed. Cells A5 through C8 are selected, and appear highlighted. The Quick Analysis button next to the selected cells is called out. A5 appears in the Name box as the active cell in the selected range.

You can also select a cell range by typing its range reference in the Name box and pressing ENTER.

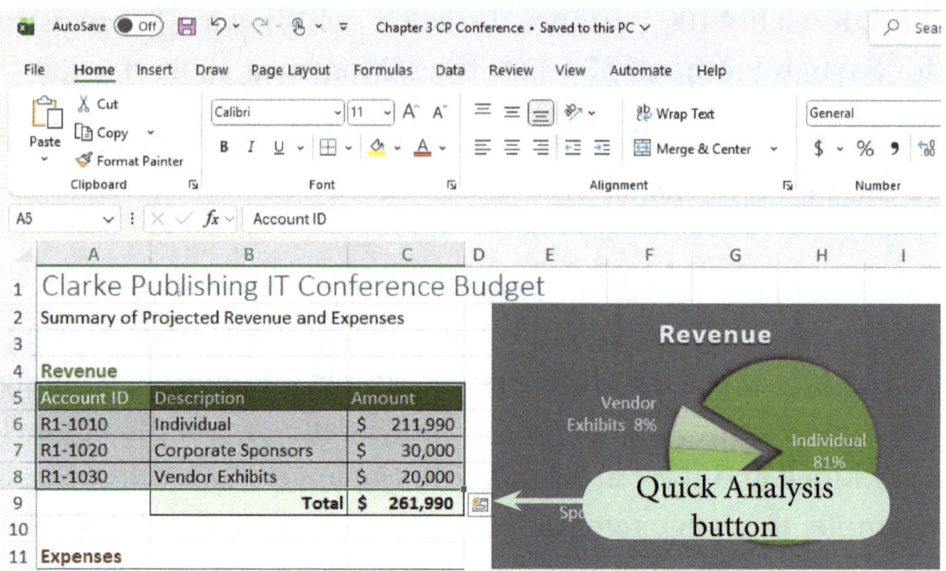

Figure 3-9 Range A5:C8 selected

(4). **Click** cell **A1** to deselect the range.

Another type of range is a nonadjacent range, which is a collection of separate rectangular ranges. The range reference for a nonadjacent range includes the range reference to each range separated by a comma. For example, the range reference A1:G5,A10:G15 includes two ranges—the first range is the rectangular block of cells from cell A1 to cell G5, and the second range is the rectangular block of cells from cell A10 to cell G15.

You will select a nonadjacent range in the Budget worksheet.

To select a nonadjacent range in the Budget worksheet:

(1). **Click** cell **A5**, hold down **SHIFT** as you **Click** cell **C8**, and then release **SHIFT** to select the range A5:C8.

(2). Scroll down the worksheet using the vertical scroll bar and then hold down **CTRL** as you drag to select the range **A12:C19** and then release **CTRL**. The two separate blocks of cells in the nonadjacent range A5:C8,A12:C19 are selected. See **Figure 3-10** on the next page.

You can also select a range by opening the Go To dialog box and entering the cell reference in the Reference box.

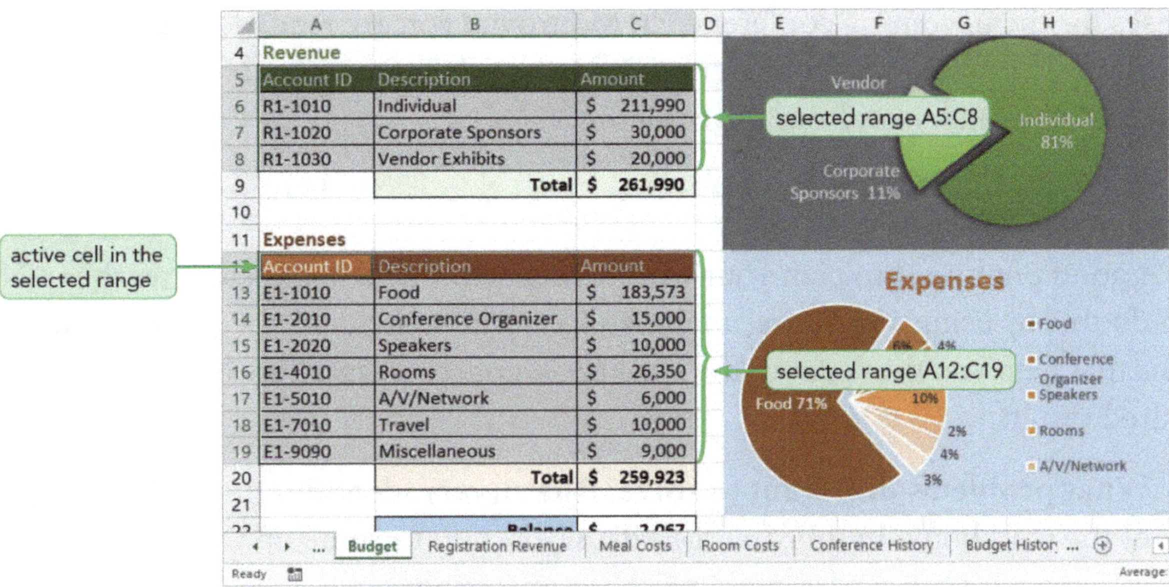

Figure 3-10 Range A5:C8 selected

You can also use the Name box and Go To dialog box to select entire columns or rows. Just enter the column letters or row numbers as the reference separated by a colon. For example, the range reference E:E selects all cells in column E and the range reference 5:5 selects all cells in row 5.

CLOSING A WORKBOOK

Once you are finished with a workbook, you can close it. When you close a workbook, a dialog box might appear, asking whether you want to save any changes you may have made to the workbook. If you have made changes that you want to keep, you should save the workbook. Because you have finished reviewing Mardia's workbook for Clark Publishing IT Conference Budget, you will close the workbook without saving any changes you may have inadvertently made to its contents.

To close Mardia's workbook:

(1). On the ribbon, **Click** the **File** tab to display Backstage view, and then **Click Close** in the navigation bar (or **Press CTRL+W**).

(2). If a dialog box opens, asking whether you want to save your changes to the workbook, **Click Don't Save**. The workbook closes without saving any changes. Excel remains opens, ready for you to create or open another workbook.

Now that you've reviewed Mardia's workbook estimating the revenues and expenses for the upcoming conference in Monrovia, you are ready to create a new workbook. Mardia wants you to create a workbook in which you will estimate the miscellaneous expenses from the conference.

PLANNING A WORKBOOK

A good practice is to plan your workbooks before you begin creating them. You can do this by using a planning analysis sheet, which includes the following questions that help you think about the workbook's purpose and how to achieve your desired results:

1. **What problems do I want to solve?** The answer identifies the goal or purpose of the workbook. In this case, Mardia wants you to come up with reasonable estimates for the conference's miscellaneous expenses.

1. **What data do I need?** The answer identifies the type of data that you need to collect for the workbook. Mardia needs a list of all the miscellaneous expenses and the estimated cost of each so that Clark Publishing IT Conference Budget will not be surprised by unexpected expenses. Miscellaneous expenses include printing brochures and schedules, decorations for the conference banquet, and gifts for the conference attendees and keynote speakers.

1. **What calculations do I need?** The answer identifies the formulas you need to apply to your data. Clark Publishing IT Conference Budget needs you to calculate the charge for each miscellaneous item, the total number of items ordered, the sales tax on all purchased items, and the total cost of all miscellaneous expenditures.

1. **What form should my solution take?** The answer impacts the appearance of the workbook content and how it should be presented to others. Mardia wants the estimates stored in a single worksheet that is easy to read and prints clearly.

You will create a workbook based on this plan. Mardia will then incorporate your projections for miscellaneous expenses into her full budget to ensure that the conference costs will not exceed the projected revenue.

SMARTSKILLS Written Communication: Creating Effective Workbooks

Workbooks convey information in written form. As with any type of writing, the final product creates an impression and provides an indicator of your interest, knowledge, and attention to detail. To create the best impression, all workbooks—especially those you intend to share with others such as coworkers and clients—should be well planned, well organized, and well written.

A well-designed workbook should clearly identify its overall goal and present information in an organized format. The data it includes—both the entered values and the calculated values—should be accurate. The process of developing an effective workbook includes the following steps:

1. Determine the workbook's purpose, content, and organization before you start.
2. Create a list of the sheets used in the workbook, noting each sheet's purpose.
3. Insert a documentation sheet that describes the workbook's purpose and organization. Include the name of the workbook's author, the date the workbook was created, and any additional information that will help others to track the workbook to its source.
4. Enter all the data in the workbook. Add labels to indicate what the values represent and, if possible, where they originated so others can view the source of your data.
5. Enter formulas for calculated items rather than entering the calculated values into the workbook. For more complicated calculations, provide documentation explaining them.
6. Test the workbook with a variety of values; edit the data and formulas to correct errors.
7. Save the workbook and create a backup copy when the project is completed. Print the workbook's contents if you need to provide a hard-copy version to others or for your files.
8. Maintain a history of your workbook as it goes through different versions, so that you and others can quickly see how the workbook has changed during revisions.

By including clearly written documentation, explanatory text, a logical organization, and accurate data and formulas, you will create effective workbooks that others can easily use.

STARTING A NEW WORKBOOK

You create new workbooks from the New screen in Backstage view. The New screen includes templates that you can use to preview different types of workbooks you can create with Excel. You will create a new workbook from the Blank workbook template, and then add all the content that Mardia wants for the miscellaneous expenses workbook.

To start a new, blank workbook for miscellaneous expenses:

(1). On the ribbon, **Click** the **File Tab** to display Backstage view.

(2). **Click New** in the navigation bar to display the New screen, which includes access to templates for a variety of workbooks.

(3). **Click Blank** workbook. A blank workbook opens.

In these modules, the workbook window is zoomed to 120% for better readability. If you want to zoom your workbook window to match the figures, complete Step 4. If you prefer to work in the default zoom of 100% or at another zoom level, read but do not complete Step 4; you might see more or less of the worksheet on your screen, but this will not affect your work in the modules.

(4). If you want your workbook window zoomed to 120% to match the figures, on the Zoom slider at the lower-right of the program window, click the **Zoom In** button [+] twice to increase the percentage to 120%. The 120% magnification increases the size of each cell but reduces the number of worksheet cells visible in the workbook window. See **Figure 3-11**.

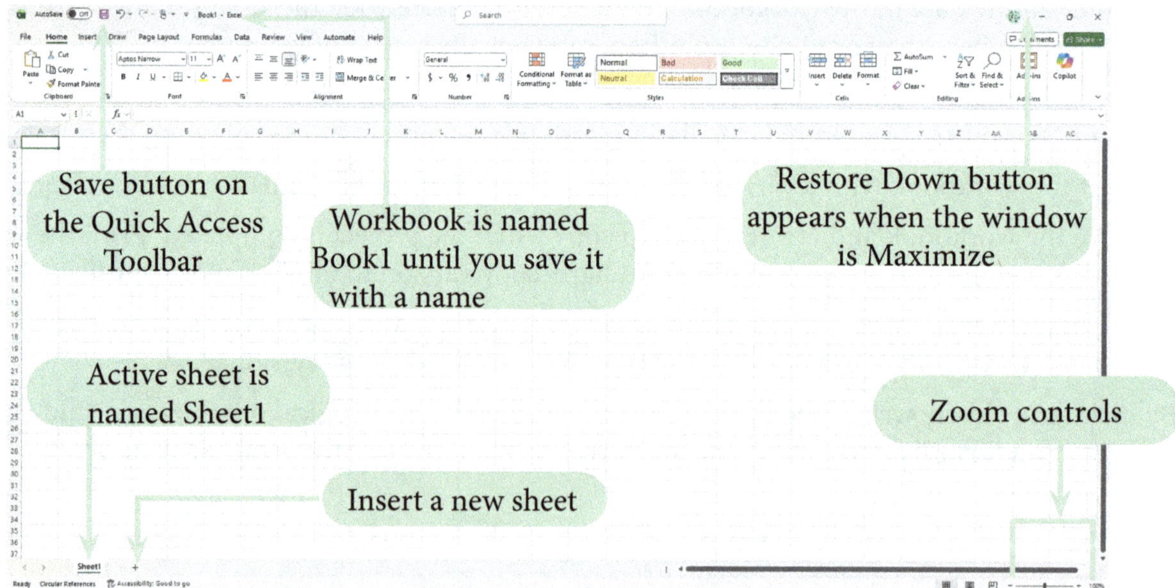

Figure 3-11 Blank workbook

The name of the active workbook, Book1, appears in the title bar. If you open multiple blank workbooks, they are named Book1, Book2, Book3, and so forth until you save them with a more descriptive name.

RENAMING AND INSERTING WORKSHEETS

Blank workbooks open with a single blank worksheet named Sheet1. It's a good practice to give your sheets more descriptive names that indicate the purpose and content of the sheet. Sheet names cannot exceed 31 characters, but they can contain blank spaces and include uppercase and lowercase letters.

Because Sheet1 is not a very descriptive name, Madia wants you to rename the worksheet as Miscellaneous Expenses.

To rename the Sheet1 worksheet:

(1). **Double-Click** the **Sheet1 Tab** to select the text of the sheet name.

(2). **Type Miscellaneous Expenses** as the new name, and then **Press ENTER** The width of the sheet tab expands to fit the longer sheet name.

Many workbooks include multiple sheets so that data can be organized in logical groups. A common business practice is to include a documentation worksheet that contains a description of the workbook, the name of the person who prepared the workbook, and the date it was created.

Mardia wants you to create two new worksheets. You will rename one worksheet as Documentation and the other worksheet as Site Information. The Site Information worksheet will be used to store information about the convention center that is hosting the conference.

To insert and name the Documentation and Site Information worksheets:

(1). To the right of the Miscellaneous Expenses sheet tab, **Click** the **New sheet** button ⊕ . A new sheet named Sheet2 is inserted to the right of the Miscellaneous Expenses sheet.

(2). **Double-Click the Sheet2** sheet tab, Type Documentation as the new name, and then press ENTER. The worksheet is renamed.

(3). To the right of the Documentation sheet, **Click** the **New Sheet** button ⊕ and then rename the inserted Sheet3 worksheet as **Site Information**.

(4). If you want these worksheets zoomed to 120% to match the figures, go to each worksheet, and then on the Zoom slider, **click** the **Zoom In** button [+] twice to increase the percentage to **120%**.

MOVING WORKSHEETS

Another good practice is to place the most important sheets at the beginning of the workbook (the leftmost sheet tabs) and the least important sheets at the end (the rightmost sheet tabs). To change the placement of sheets in a workbook, you drag them by their sheet tabs to the new location.

Mardia wants you to move the Documentation worksheet to the front of the workbook, so that it appears before the Miscellaneous Expenses sheet.

To move the Documentation worksheet:

(1). **Point** to the **Documentation Sheet Tab**. The sheet tab name changes to bold.

(2). **Press** and **Hold** the **Mouse** button. The pointer changes to , and a small arrow appears in the upper-left corner of the tab.

(3). **Drag** to the **Left** until the small arrow appears in the upper-left corner of the Miscellaneous Expenses sheet tab, and then **Release** the **Mouse** button. The Documentation worksheet is now the first sheet in the workbook.

You can copy a worksheet by holding down CTRL as you drag and drop the sheet tab. Copying the worksheet duplicates all of the worksheet data and its structure.

DELETING WORKSHEETS

In some workbooks, you will want to delete an existing sheet. The easiest way to delete a sheet is by using a **Shortcut Menu**, which is a list of commands related to an object that opens when you right-click the object. Mardia asks you to include site information on the Miscellaneous Expenses worksheet so all the information about the conference site and the miscellaneous expenses is on one sheet.

To delete a worksheet:

(1). **Right-Click** the **Site Information Sheet Tab**. A shortcut menu opens.

(2). **Click Delete**. The Site Information worksheet is removed from the workbook.

When you delete a sheet, you also delete any text and data it contains. So be careful that you do not remove important and irretrievable information.

SAVING A WORKBOOK

As you modify a workbook, you should save it regularly—every 10 minutes or so is a good practice. The first time you save a workbook, the Save As dialog box opens so you can name the file and choose where to save it. You can save the workbook in the location your instructor told you to, or you can save it on your computer or network, or even to your account on OneDrive.

You will save the miscellaneous expenses workbook that you just created.

To save the miscellaneous expenses workbook for the first time:

(1). On the Quick Access Toolbar, **Click** the **Save** button (or press **CTRL+S**). The Save this file dialog box opens.

(2). **Click More Options**. The Save As screen in Backstage view opens.

(3). **Click** the **Browse** button. The Save As dialog box opens.

(4). Navigate to the location specified by your instructor.

(5). In the File name box, **Select Book1** (the default name assigned to your workbook) if it is not already selected, and then **Type Chapter 2 Misc** as the new name.

(6). **Verify** that **Excel Workbook** appears in the Save as type box.

(7). **Click Save**. The workbook is saved, the dialog box closes, and the workbook window reappears with the new file name in the title bar.

As you modify the workbook, you will need to resave the file. Because you already saved the workbook with a file name, the next time you save, the Save command saves the changes you made to the workbook without opening the Save As dialog box.

Sometimes you will want to save a current workbook under a new file name. This is useful when you want to modify a workbook without losing its content and structure or when you want to save a copy of the workbook to a new location. To save a workbook with a new name, click the File tab to return to Backstage view, click Save As on the navigation bar, specify the new file name and location, and then click Save.

ENTERING TEXT, DATES, AND NUMBERS

Worksheet cells can contain text, numbers, dates, and times. **Text Data** is any combination of letters, numbers, and symbols. A **Text String** is a series of text data characters. **Numeric Data** is any number that can be used in a mathematical operation. **Date Data** and **Time Data** are values displayed in commonly recognized date and time formats. For example, Excel interprets the cell entry April 15, 2025, as a date and not as text. By default, text is left-aligned within worksheet cells, and numbers, dates, and times are right-aligned.

ENTERING TEXT

Text is often used in worksheets as labels for the numeric values and calculations displayed in the workbook. Mardia wants you to enter text content into the Documentation sheet.

To enter text in the Documentation sheet:

(1). **Go To** the Documentation sheet, and then **Press CTRL+HOME** to make sure cell **A1** is the active cell.

(2). **Type Clarke Publishing IT Conference Budget** in cell A1. As you type, the text appears in cell A1 and in the formula bar.

(3). **Press ENTER twice**. The text is entered into cell A1, and the active cell moves down two rows to cell **A3**.

(4). **Type Author** in cell **A3**, and then press TAB. The text is entered and the active cell moves one column to the right to cell B3.

(5). **Type your name** in cell **B3**, and then **press ENTER**. The text is entered and the active cell moves one cell down and to the left to cell **A4**.

(6). **Type Date** in cell **A4**, and then **press TAB**. The text is entered, and the active cell moves one column to the right to cell **B4**, where you would enter the date you created the worksheet. For now, you will leave the cell for the date blank.

(7). **Press ENTER** to make cell **A5** the active cell, **type Purpose** in the cell, and then **press TAB**. The active cell moves one column to the right to cell **B5**.

(8). **Type To estimate expenses at the Cybersecurity, AI and IT conference** in cell **B5**, and then **Press ENTER**. **Figure 1-12** shows the text entered in the Documentation sheet.

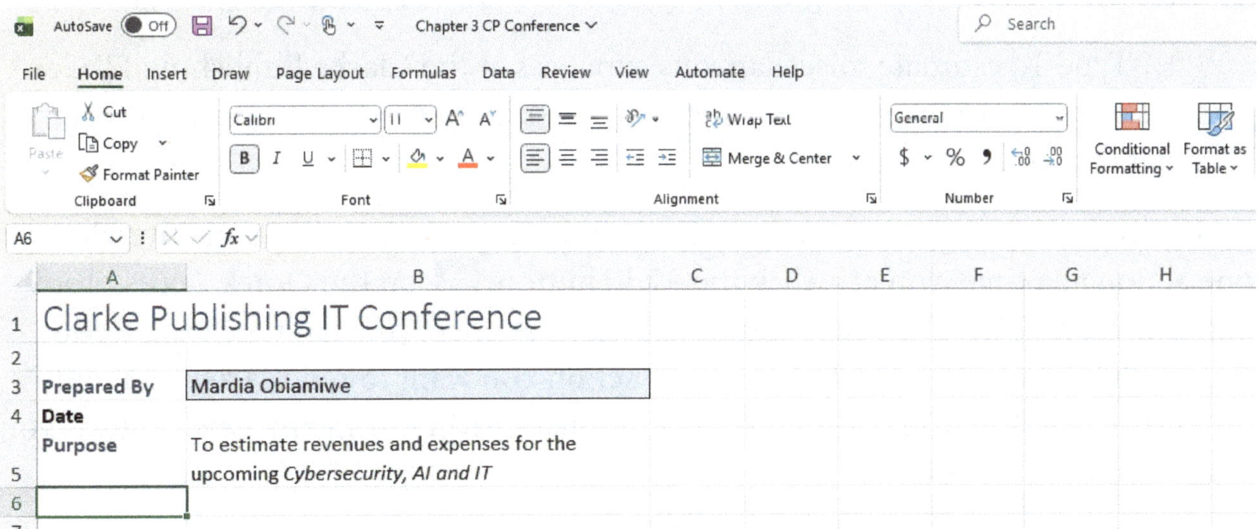

Figure 3-12 Text entered in the Documentation sheet

The text strings you entered in cells B3 and B5 are so long that they cover the adjacent cells. Any text that doesn't fit within a cell will cover the adjacent cells to the right if they are empty. If the adjacent cells contain data, only the text that fits into the cell is displayed and the rest of the text string is hidden. The complete text is still stored in the cell; it is just not displayed. (You will learn how to display all text in a cell in the next session.)

UNDOING AND REDOING AN ACTION

As you enter data in a workbook, you might need to undo a previous action. Excel maintains a list of the actions you performed in the workbook during the current session, so you can undo most of your actions. You can use the Undo button on the Quick Access Toolbar or press CTRL+Z to reverse your most recent actions one at a time. If you want to undo more than one action, you can click the Undo arrow and then select the earliest action you want to undo—all actions after that initial action will also be undone.

You will undo the most recent change you made to the Documentation sheet—the text you entered into cell B5. Then you will enter a different description of the workbook's purpose in cell B5.

To undo the text entry in cell B5:

(1). On the Quick Access Toolbar, click the **Undo** button (or press CTRL+Z). The last action is reversed, removing the text you entered in cell B5.

(2). Type To estimate miscellaneous expenses at the Clarke Publishing IT Conference in Monrovia in cell B5, and then **Press ENTER**. The new purpose statement is entered in cell B5.

If you want to restore actions you have undone, you can redo them. To redo one action at a time, you can click the Redo button on the Quick Access Toolbar or press **CTRL+Y**. To redo multiple actions at once, you can **Click** the **Redo Arrow** and then **Click** the earliest action you want to redo. After you undo or redo an action, Excel continues the action list starting from any new changes you make to the workbook.

EDITING CELL CONTENT

As you continue to create your workbook, you might find mistakes you need to correct or entries that you want to change. To replace all of a cell's content, you simply select the cell and then type the new entry to overwrite the previous entry. If you want to replace only part of a cell's content, you can switch to **Edit Mode** to make the changes directly in the cell. To switch to Edit mode, you double-click the cell. A blinking insertion point indicates where the new content you type will be inserted. In the cell or formula bar, the pointer changes to an I-beam, which you can use to select text in the cell. Anything you type replaces the selected content.

Because the meeting in Monrovia is a conference rather than a convention, Mardia wants you to edit the text in cell B5. You will do that in Edit mode.

To edit the text in cell B5:

(1). **Double-Click** cell **B5** to select the cell and switch to Edit mode. A blinking insertion point appears within the text of cell B5. The status bar displays Edit instead of Ready to indicate that the cell is in Edit mode.

(2). **Press LEFT ARROW** or **RIGHT ARROW** as needed to move the insertion point directly to the right of the word "**convention**" in the cell text.

(3). **Press BACKSPACE 10 Times** to delete the word "**convention,**" and then **type conference** in the entry. See **Figure 3-13**.

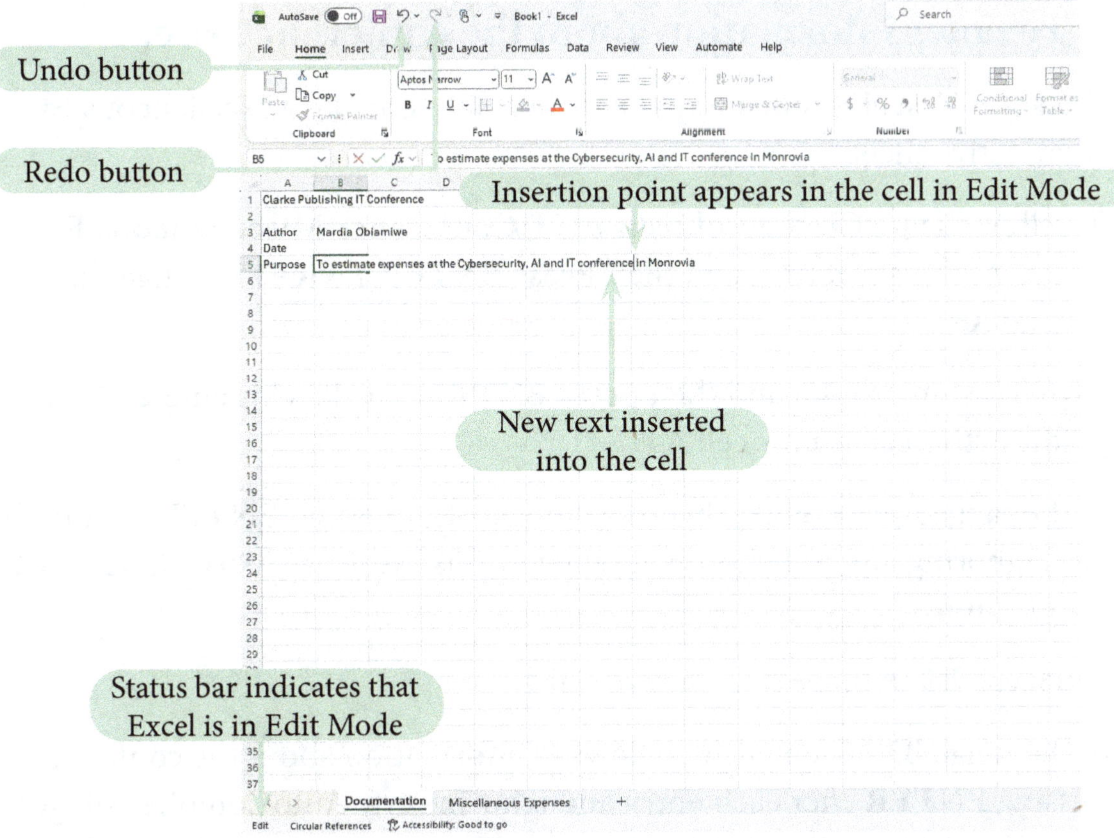

Figure 3-13 Edited text in the Documentation sheet

(3). **Press ENTER** to exit the cell and return to Ready mode.

Now that you have returned to Ready mode, you can continue to insert and edit content from other cells and sheets in the workbook.

UNDERSTANDING AUTOCOMPLETE

As you type text in the active cell, Excel tries to anticipate the remaining characters by displaying text that begins with the same letters as a previous entry in the same column. This feature, known as **AutoComplete**, helps make entering repetitive text easier and reduces data entry errors. To accept the suggested text, press TAB or ENTER. To override the suggested text, continue to type the text you want to enter in the cell. AutoComplete does not work with dates or numbers or when a blank cell is positioned between the previous entry and the text you are typing.

You will see AutoComplete entries as you enter descriptive text about the Monrovia conference in the Miscellaneous Expenses worksheet.

To enter information about the conference site:

(1). **Click** the **Miscellaneous Expenses** sheet tab to make Miscellaneous Expenses the active sheet.

(2). In cell **A1**, **Type Clarke Publishing IT Conference Miscellaneous Expenses** as the worksheet title, and then **Press ENTER Twice** to change the active cell to **A3**.

(3). **Type Host** in cell **A3**, and then **press ENTER**. The label is entered in the cell and the active cell is now cell **A4**.

(4). In the range A4:A8, enter the following labels, pressing **ENTER** after each entry and ignoring any AutoComplete suggestions: **Address, City, State, Postal Code,** and **Phone**.

(5). **Click** cell **B3** to make it the active cell.

(6). In the range B3:B8, enter the following information about the conference site, pressing **ENTER** after each entry and ignoring any AutoComplete suggestions: **231 Hotel Africa Road, 1000 Brewerville, 10 Liberia,** and **231-777-123-456**. See **Figure 3-14** on the next page.

Figure 3-14 on the next page, shows that the site information text has been entered into the Miscellaneous Expenses sheet. The text in cell A7 is partially hidden because not all of the text fits inside the cell and because there is text in cell A8.

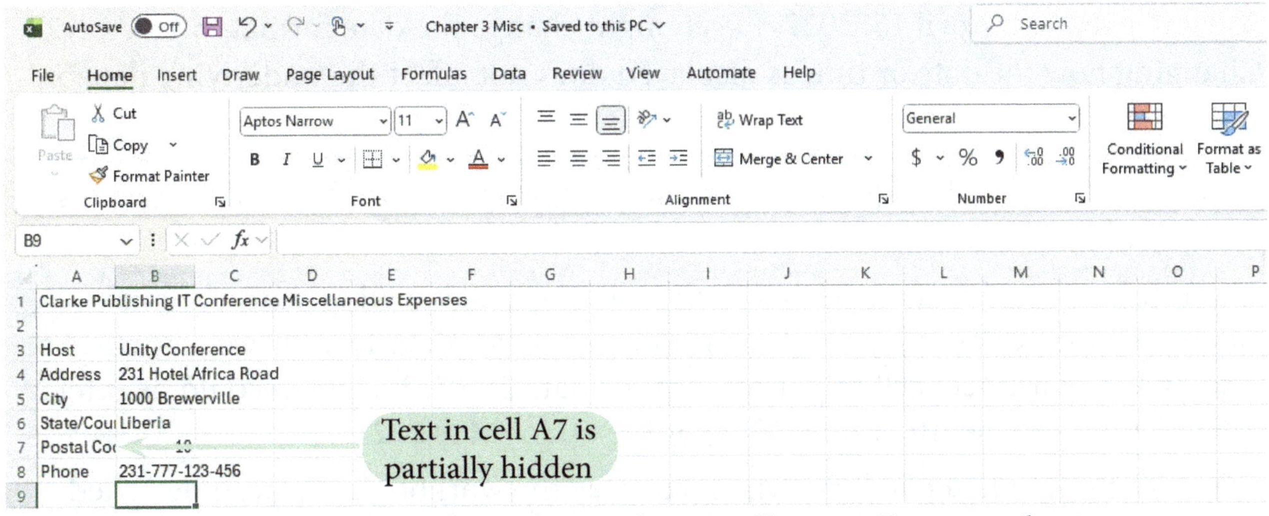

Figure 3-14 Site information in the Miscellaneous Expenses sheet

Entering Dates

Excel recognizes dates in any standard date formats. For example, in Excel, all the following entries represent the same date:

- 4/6/2027
- 4/6/27
- 4-6-2027
- April 6, 2027
- 6-Apr-27

Even though dates are entered as text, Excel stores the date as a number equal to the number of days between the specified date and January 0, 1900. Times are also entered as text and stored as fractions of a 24-hour day. For example, a date and time of April 15, 2027, 6:00 PM is stored as the number 45,762.75, which is 45,762 days after January 0, 1900, plus 3/4 of one day. Excel stores dates and times as numbers so they can be used for date and time calculations, such as determining the elapsed time between one date and another.

Based on how your computer displays dates, Excel might change the appearance of a date after you type it. For example, if you enter the date 4/15/27 into the active cell, Excel might display the date with the four-digit year value, 4/15/2027. If you enter the text April 15, 2027, Excel might change the date format to 15-Apr-27. Changing how the date or time is displayed does not affect the underlying date or time value.

SMARTSKILLS Written Communication: Creating Effective Workbooks

For international business transactions, you may need to adopt international standards for expressing dates, times, and currency values in your workbooks. For example, a worksheet cell might contain the date 06/05/25, which could be interpreted as either the 5th of June 2021, or the 6th of May 2021.

The interpretation depends on which country the workbook has been designed for. You can avoid this problem by entering the full date, as in June 5, 2025. However, this might not work with documents written in foreign languages, such as Japanese, that use different character symbols.

To solve this problem, many international businesses adopt ISO (International Organization for Standardization) dates in the format yyyy-mm-dd, where yyyy is the four-digit year value, mm is the two-digit month value, and dd is the two-digit day value. So, a date such as June 5, 2025, is entered as 2025/06/05. If you choose to use this international date format, make sure that everyone else using your workbook understands this format so they interpret dates correctly. You can include information about the date format in the Documentation sheet.

For your work, you will enter dates in the format mm/dd/yyyy, where mm is the two-digit month number, dd is the two-digit day number, and yyyy is the four-digit year number.

To enter a date into the Documentation sheet:

(1). **Click** the **Documentation Sheet Tab** to make the Documentation sheet the active worksheet.

(2). **Click** cell **B4** to make it the active cell, type the current date in the ***mm/dd/yyyy*** format, and then **Press ENTER**. The date is entered in the cell.

Are You Having Trouble? Depending on your system configuration, Excel might change the date to the date format dd-mm-yy. This difference will not affect your work.

(3). **Click** the **Miscellaneous Expenses Sheet Tab** to return to the Miscellaneous Expenses worksheet.

Expense Category	Subcategory	Description	Units	Cost per Unit
E2	9010	printing of brochures and conference materials	1600	$2.45
E2	9030	transportation shuttles	3	$335.75
E2	9020	decorations for banquet	1	$850.55
E2	9040	gift bags for conference attendees	525	$6.25
E2	9045	gifts for banquet speakers	6	$55.25

Figure 3-15 Miscellaneous expenses

You will enter the first three columns of this table into the worksheet.

To enter the first part of the table of miscellaneous expenses:

(1). In the Miscellaneous worksheet, **Click** cell **A10** to make it the active cell, **Type Expense Category** as the column label, and then **Press TAB** to move to cell **B10**.

(2). Type Subcategory in cell B10, **Press TAB** to move to cell **C10**, **Type Description** in cell **C10**, and then **Press ENTER**.

(3). In the range **A11:C15**, enter the Expense Category, Subcategory, and Description text for the five miscellaneous expenses listed in **Figure 3-15**, **Pressing TAB** to move from one cell to the next, and **Pressing ENTER** to move to a new row. Note that the text in some cells will be partially hidden; you will fix that problem shortly. See **Figure 3-16** on the next page.

Figure 3-16 on the next page shows the expense categories, subcategories, and descriptions text has been entered into the Miscellaneous Expenses sheet in cells A11 through C15. The text in cells A11 through A15 is left-aligned. The text in cells B11 through B15 is right-aligned. The text in cells A7, A10, and B10 is partially hidden because not all of the text fits inside those cells and because there is text in the cells to the right of each. The text in cells C11 through C15 overlaps the adjacent columns because not all of the text fits inside those cells and because there is no text to the right. Cell A16 is now the active cell.

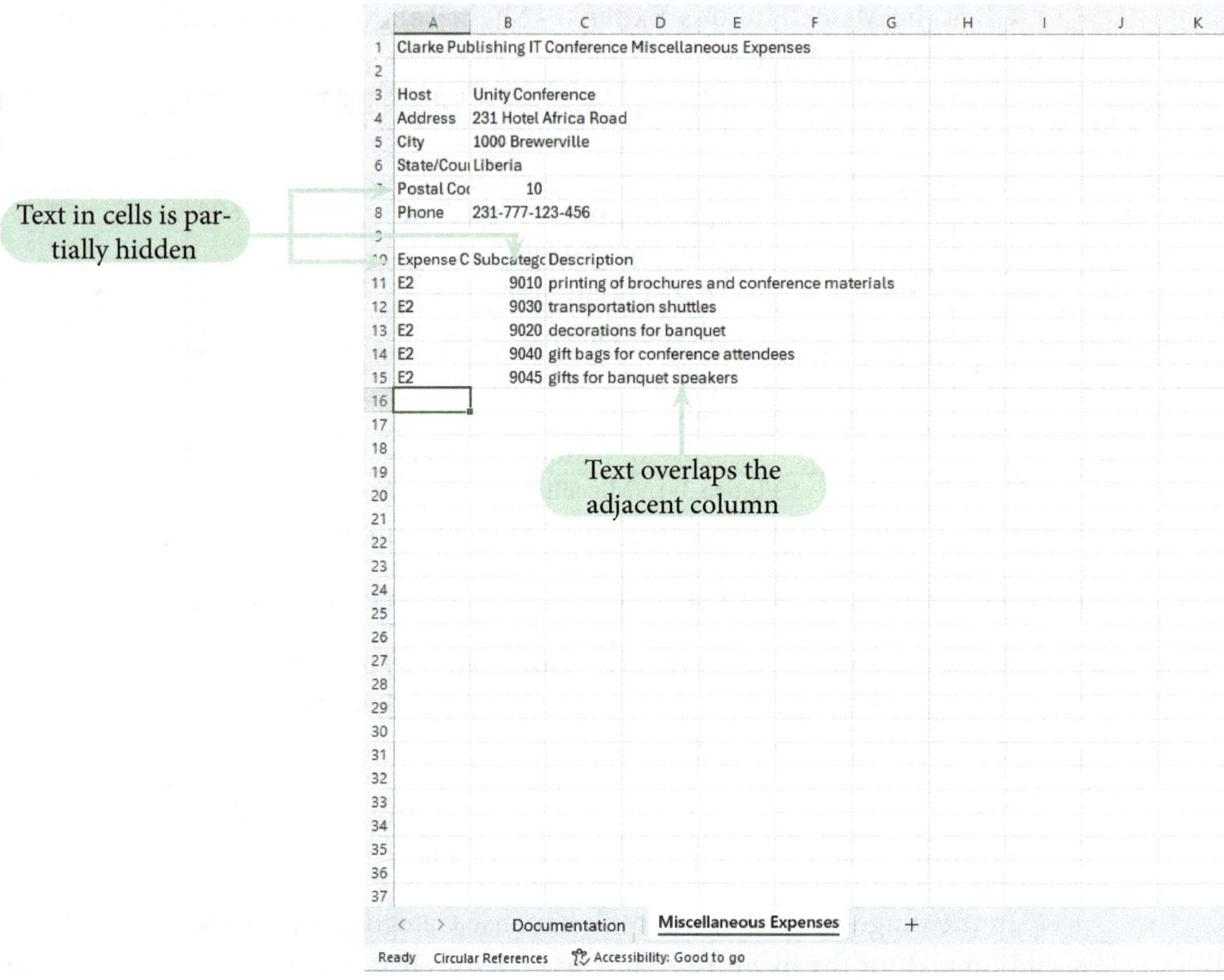

Figure 3-16 Miscellaneous expense

ENTERING NUMBERS

In Excel, numbers can be integers such as 234, decimals such as 1.95, or negative values such as –5.2. In the case of currency and percentages, you can include the currency symbol or the percent sign when you enter the value. Excel treats a currency value such as $87.25 as the number 87.25, and a percentage such as 95% as the decimal 0.95. Much like dates, currency and percentages are displayed with their symbols but stored as numbers.

You will complete the list of miscellaneous expenses by inserting their number of units and costs per unit.

To enter the miscellaneous expenses:

(1). **Click** cell **D10**, **Type Units** as the label, and then **Press TAB**. Cell E10 becomes the active cell.

(2). **Type Cost per unit** in cell **E10**, and then **Press ENTER**. Cell D11 becomes the active cell.

(3). **Type 1600** in cell **D11**, **Press TAB** to make cell E11 the active cell, **Type $2.45** in cell **E11**, and then **press ENTER**. Cell D12 becomes the active cell.

(4). In the range **D12:E15**, enter the number of units and cost per unit for the remaining four expense categories shown in **Figure 3-16**. See **Figure 3-17**.

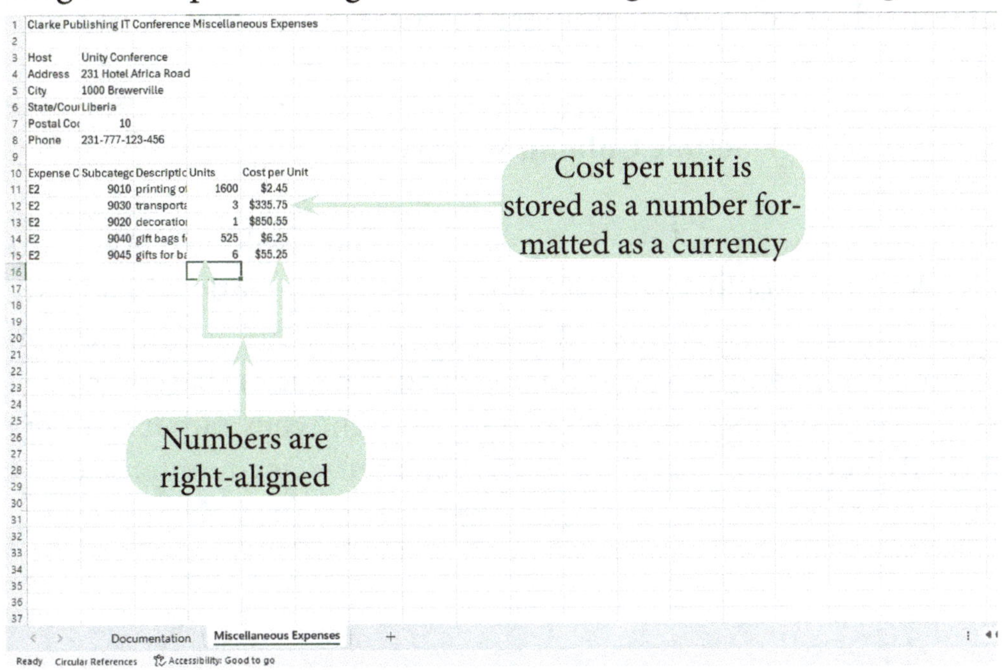

Figure 3-17 Miscellaneous expense units and costs per unit

(5). On the Quick Access Toolbar, **Click** the **Save** button (or **Press CTRL+S**) to save the workbook.

Much of the information from the miscellaneous expenses table is difficult to read because of the hidden text. You can display all the cell contents by changing the size of the columns and rows in the worksheet.

RESIZING COLUMNS AND ROWS

There are several ways to resize columns and rows, including changing column widths, wrapping text within cells, and changing row heights. You can use a combination of these in a worksheet to create the best fit for your data.

SETTING A COLUMN WIDTH

Column widths are expressed as the number of characters the column can contain. The default column width is 8.43 standard-sized characters. In general, this means that you can type eight characters in a cell. Any additional text is hidden or overlaps the adjacent cell. Column widths are also expressed in terms of pixels. A **Pixel** is an individual point on a computer monitor or printout. A column width of 8.43 characters is equivalent to 64 pixels.

SMARTSKILLS Setting Column Widths

On a computer monitor, pixel size is based on screen resolution. As a result, cell content that looks fine on one screen might appear differently when viewed on a screen with a different resolution. If you work on multiple computers or share your workbooks with others, you should set column widths based on the maximum number of characters you want displayed in the cells rather than pixel size. This ensures that everyone sees the cell contents the way you intended.

You will increase the width of column A so that all of the text labels within that column are completely displayed.

To increase the width of column A:

(1). Point to the **Right Border** of the column A heading until the pointer changes to ✛ .

(2). **Click** and **drag** to the **right** until the width of the column heading reaches 18 characters, but **do not release** the mouse **button**. The ScreenTip that appears as you resize the column shows the new column width in characters and in pixels. See **Figure 3-18**.

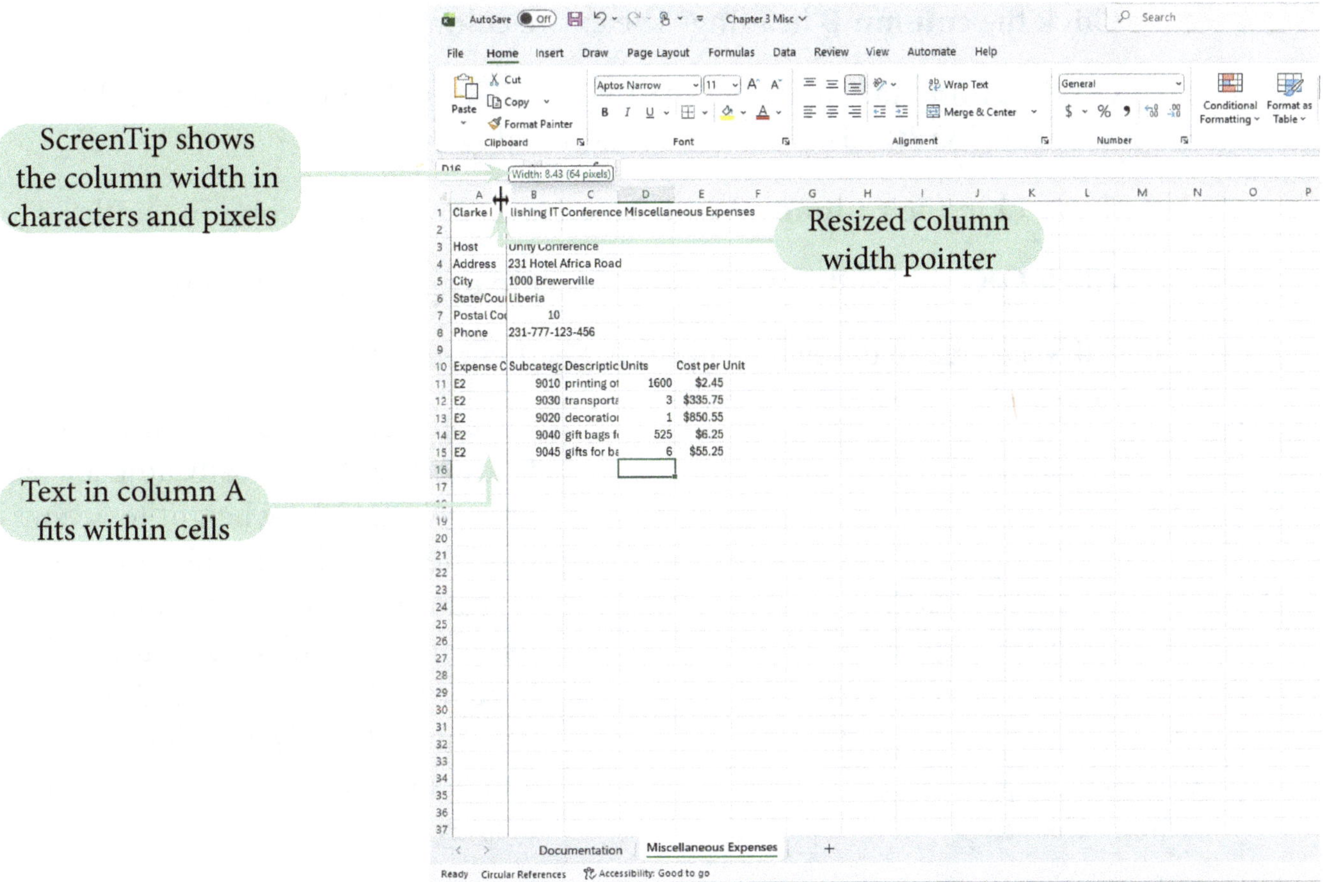

Figure 3-18 Width of column A increased to 18 characters

(3). **Release** the **mouse** button. The width of column **A** expands to **18 characters**, and all the text within that column is visible within the cells.

You can change the width of multiple columns at once. To select a range of columns, click the first column heading in the range, hold down SHIFT, and then click the heading of the last column or click and drag the pointer over the column headings. To select nonadjacent columns, hold down CTRL and click the heading of each column you want to select. When you change the width of one column, the widths of all the other columns that are selected also change.

Using the mouse to resize columns can be imprecise and a challenge to some users with special needs. The Format command on the Home tab gives you precise control over column width and row height settings. You will use the Format command to set the width of column B to exactly 12 characters so that the hidden text in cell B10 is completely visible.

To set the width of column B using the Format command:

(1). **Click** the **column B** heading. The entire column is selected.

(2). On the **Home tab**, in the Cells group, **Click** the **Format** button, and then **click Column Width**. The Column Width dialog box opens.

(3). **Type 12** in the **Column Width** box to specify the new column width.

(4). **Click OK**. The width of column B is set to exactly 12 characters.

(5). **Click** cell **A2** to deselect column B. See **Figure 3-19**.

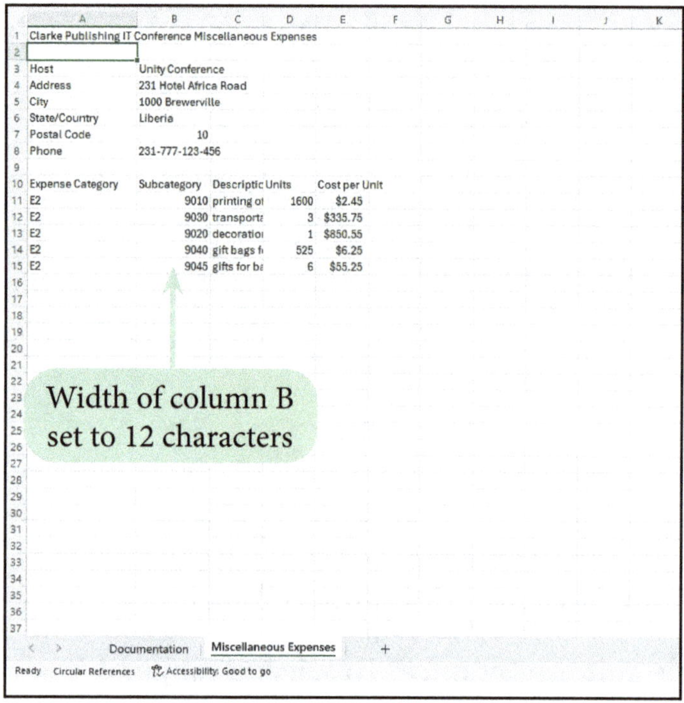

Figure 3-19 Width of column B

You can also use the **AutoFit** feature to automatically adjust a column width or row height to accommodate its widest or tallest entry. To AutoFit a column to the width of its contents, double-click the right border of the column heading. You'll use AutoFit to resize columns C and E so that all the content is fully displayed.

To use AutoFit to display all the contents of columns C and E:

(1). **Point** to the **right border** of column **C** until the pointer changes to the resize column width pointer ✛.

(2). **Double-Click** the **right border** of the column **C** heading. The width of column C increases to about **43** characters so that the longest item description is completely visible.

(3). **Double-Click** the **Right Border** of the column **E** heading. The width of column E increases to about **12** characters.

Sometimes when you use AutoFit, the column becomes wider than you want. Another way to display long text entries is to wrap the text within each cell.

WRAPPING TEXT WITHIN A CELL

When you wrap text within a cell, any content that doesn't fit on the first line is displayed on a new line in the cell. **Wrapping Text** increases the row height to display any additional new lines added in the cell. You can wrap only text within a cell; numbers, dates, or times do not wrap.

You'll reduce the width of column C to 30 characters, and then wrap the category descriptions so all of the text is visible.

To wrap text in column C:

(1). **Resize** the width of column **C** to **30** characters.

(2). **Select** the range **C11:C15**, which has the expense descriptions.

(3). On the **Home tab**, in the Alignment group, **Click** the **Wrap Text** button. The Wrap Text button is highlighted, indicating that it is applied to the selected range. Any text in the selected cells that exceeds the column width wraps to a new line in those cells.

(4). **Click** cell **A2** to make it the active cell. See **Figure 3-20**.

Another way to create a new line within a cell is to press ALT+ENTER where you want the new line to start. Subsequent characters will be on a new line.

Text wrapped to new line

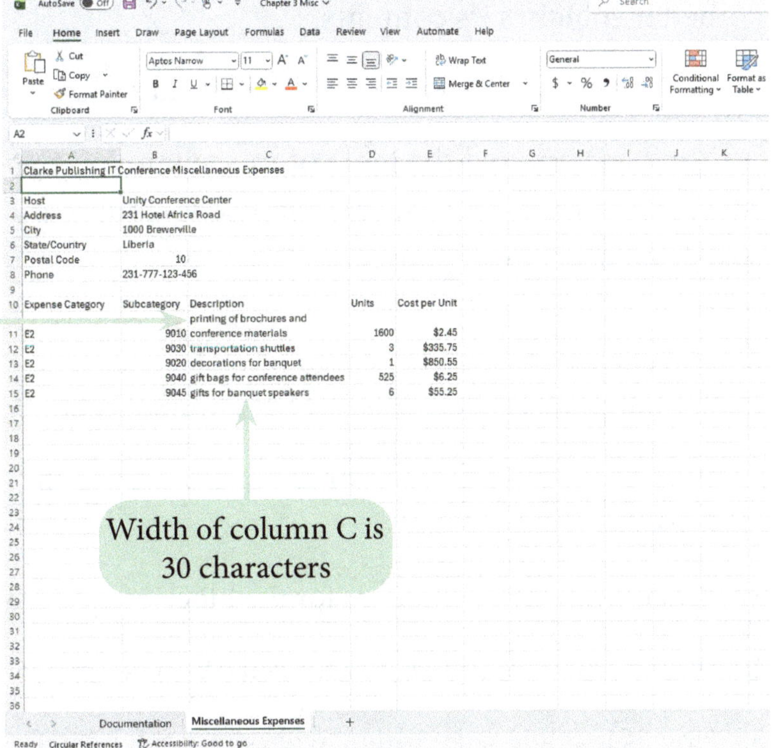

Figure 3-20 Text Wrap within a cell

Figure 3-20 Miscellaneous Expenses sheet shows that the width of columns C, D, and E have been increased. The width of column C is now 30 characters. The text in cell C11 is now wrapped on two lines because the text does not fit within the width of the cell. The Wrap Text button in the Alignment group is called out. Column E has been autofit so that the text in each cell in the column fits the width of the column. Cell A2 is the active cell.

CHANGING ROW HEIGHTS

Row heights are measured in points or pixels. A **Point** is approximately 1/72 of an inch. The default row height is 15 points or 20 pixels. There are several ways to set row heights. You can drag the bottom border of the row heading. You can click the Format button in the Cells group on the Home tab, and then click Row Height. Or, you can double-click the bottom border of the row heading to AutoFit a row to its tallest cell.

The height of row 14 is too tall for its contents. Mardia asks you to reduce it.

To change the height of row 14:

(1). **Point** to the **Bottom Border** of the row 14 heading until the pointer changes to the resize row height pointer .

(1). **Drag** the **Bottom Border** up until the height of the row is equal to **18** points (or **24 pixels**), and then release the mouse button. The height of row 14 better matches its contents.

(1). **Press CTRL+S** to save the workbook.

You have entered the table of miscellaneous expenses for the Monrovia conference. In the next session, you will use formulas and functions to calculate the total cost of all of those expenses.

Quick Review

1. How are chart sheets different from worksheets?
2. What is the cell reference for the cell located in the third column and fourth row of a worksheet?
3. What is the range reference for the block of cells D3 through E10?
4. What is the range reference for cells A1 through C5 and cells A8 through C12?
5. How is text aligned within a worksheet cell by default?

Quick Review

6. How would the number 00514 appear in a cell?
7. Cell B2 contains the entry May 3, 2025. Why doesn't Excel consider this a text entry?
8. How do you autofit a column to match its longest cell entry?

Calculating with Formulas

So far you have entered text, numbers, and dates in the worksheet. However, the main reason for using Excel is to perform calculations and analysis on data. For example, Mardia wants the workbook to calculate the number of items in the miscellaneous expense category and the total cost of those items. Such calculations are added to a worksheet using formulas and functions.

Entering a Formula

A **Formula** is an expression that returns a value. In most cases, this is a number—though it could also be text or a date. In Excel, every formula begins with an equal sign (=) followed by an expression containing the operations that return a value. If you don't begin the formula with the equal sign, Excel assumes that you are entering text or numbers.

A formula is written using **operators**, or mathematical symbols, that combine different values, resulting in a single value that is then displayed in the cell. The most common operators are arithmetic operators that perform mathematical calculations such as addition (+), subtraction (–), multiplication (*), division (/), and exponentiation (^). For example, the following formula adds 3 and 8, returning a value of 11:

=3+8

Most Excel formulas contain references to cells rather than specific values. This allows you to change the values used in the calculation without having to modify the formula itself. For example, the following formula returns the result of adding the values stored in cells C3 and D10:

=C3+D10.

If the value 3 is stored in cell C3 and the value 8 is stored in cell D10, this formula would also return a value of 11. If you later changed the value in cell C3 to 10, the formula would return a value of 18. **Figure 3-21** describes the different arithmetic operators and provides examples of formulas.

Operation	Arithmetic Operator	Example	Description
Addition	+	=B1+B2+B3	Adds the values in cells B1, B2, and B3
Subtraction	–	=C9-B2	Subtracts the value in cell B2 from the value in cell C9
Multiplication	*	=C9*B9	Multiplies the values in cells C9 and B9
Division	/	=C9/B9	Divides the value in cell C9 by the value in cell B9
Exponentiation	^	=B5^3	Raises the value of cell B5 to the third power

Figure 3-21 Arithmetic operators

If a formula contains more than one arithmetic operator, Excel performs the calculation based on the following order of operations, which is the sequence in which operators are applied in a calculation:

1. Calculate any operations within parentheses
2. Calculate any exponentiations (^)
3. Calculate any multiplications (*) and divisions (/)
4. Calculate any additions (+) and subtractions (–)

For example, the following formula returns the value 23 because multiplying 4 by 5 is done before adding 3:

=3+5*5

If a formula contains two or more operators with the same level of priority, the operators are applied in order from left to right. In the following formula, Excel first multiplies 4 by 10 and then divides that result by 8 to return the value 5:

=4*10/8

When parentheses are used, the value inside them is calculated first. In the following formula, Excel calculates (3+4) first, and then multiplies that result by 5 to return the value 35:

=(3+4)*5

Figure 3-22 shows how changes in a formula affect the order of operations and the result of the formula.

Formula	Order of Operations	Result
=50+10*5	10*5 calculated first and then 50 is added	100
=(50+10)*5	(50+10) calculated first and then 60 is multiplied by 5	300
=50/10–5	50/10 calculated first and then 5 is subtracted	0
=50/(10–5)	(10–5) calculated first and then 50 is divided by that value	10
=50/10*5	Two operators at same precedence level, so the calculation is done left to right with 50/10 calculated first and that value is then multiplied by 5	25
=50/(10*5)	(10*5) is calculated first and then 50 is divided by that value	1

Figure 3-22 Order of operations applied to formulas

Mardia wants the miscellaneous expenses workbook to calculate the total cost of each item. The total cost is equal to the number of units ordered multiplied by the cost per unit. You already entered this information in columns D and E. Now you will enter a formula in cell F11 to calculate the total cost of printing for the conference.

To enter a formula that calculates the total cost of printing:

(1). If you took a break after the previous lab, make sure the **Chapter 2 Misc. xlsx** workbook is open and the **Miscellaneous Expenses Worksheet** is active.

(2). **Click** cell **F10**, **Type Total** as the label, and then **Press ENTER**. The label is entered in cell F10 and cell F11 is the active cell.

(3). Type =D11*E11 (the number of units multiplied by the cost per unit). As you type the formula, a list of Excel function names appears in a ScreenTip, which provides a quick method for entering functions. The list will close when you complete the formula. You will learn more about Excel functions shortly. Also, Excel color codes each cell reference and its corresponding cell with the same color. See **Figure 3-23**.

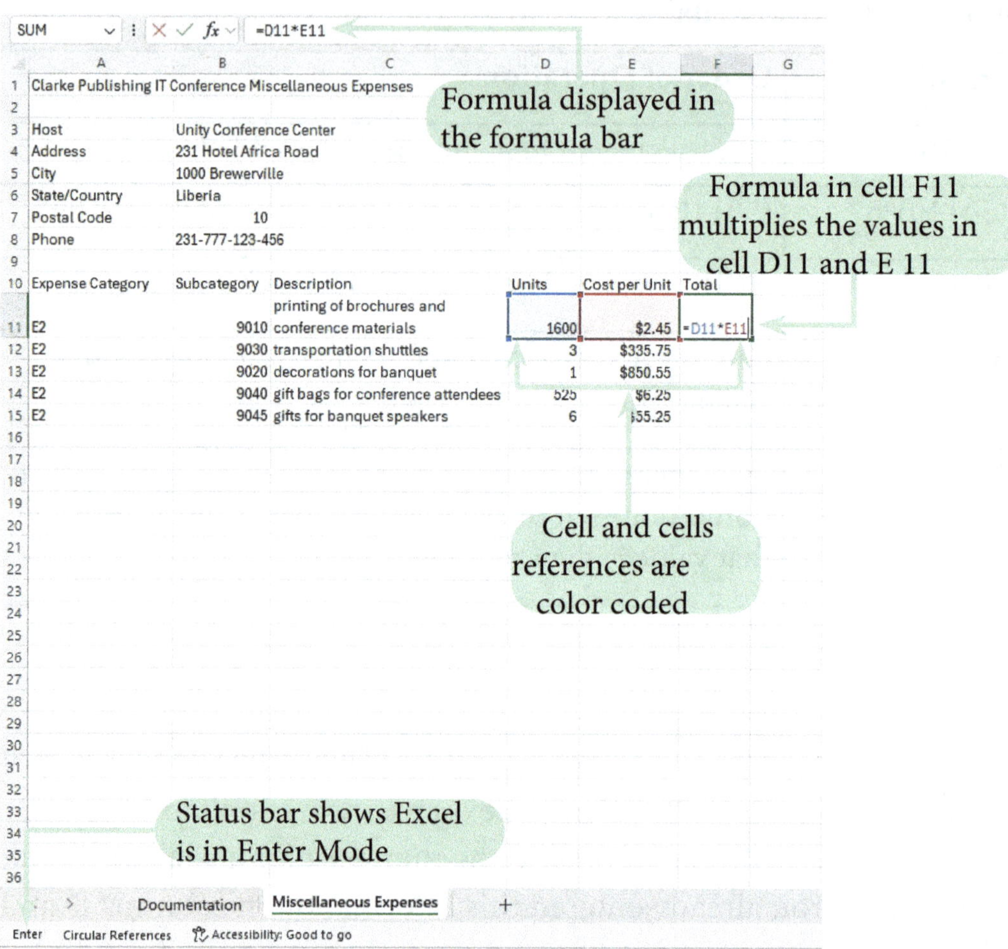

Figure 3-23 Formula being entered in cell F11

(4). Press ENTER. The formula result $3,920.00 appears in cell F11. This value is the total cost of printing 1600 brochures and other conference materials. The result is displayed as currency because cell D11, which is referenced in the formula, contains a currency value.

(5). Click cell **F11** again to make it the active cell. The cell displays the result of the formula, but the formula bar displays the formula you entered so that you can see at a glance both the formula and its value.

For the first item, you entered the formula by typing each cell reference in the expression. You can also insert a cell reference by clicking the cell as you type the formula. This technique reduces the possibility of error caused by typing an incorrect cell reference. You will use this method to enter the formula to calculate the charge for renting vans to shuttle conference attendees between the airport and the hotel.

To enter a formula to calculate the cost of shuttle using the mouse:

(1). **Click** cell **F12** to make it the active cell.

(2). **Type** = to indicate that you are entering a formula. Any cell you click from now on inserts the cell reference of the selected cell into the formula until you complete the formula by pressing Enter or Tab.

(3). **Click** cell **D12**. The cell reference is inserted into the formula in the formula bar. At this point, any cell you click changes the cell reference used in the formula. The cell reference isn't locked until you type an operator.

(4). **Type** * to enter the multiplication operator. The cell reference for cell D12 is locked in the formula, and the next cell you click will be inserted after the operator.

(5). **Click** cell **E12** to enter its cell reference in the formula. The formula, **=D12*E12**, is complete.

(6). **Press ENTER**. Cell **F12** displays the value $1,007.25, which is the cost of renting vans to transport attendees to and from the conference.

Next, you will enter formulas to complete the calculations of the remaining miscellaneous expenses.

COPYING AND PASTING FORMULAS

Many worksheets have the same formula repeated across several rows or columns. Rather than retyping the formula, you can copy a formula from one cell and paste it into another cell. When you copy a formula, Excel places the formula onto the **Clipboard**, which is a temporary storage area for selections you copy or cut. When you **Paste**, Excel retrieves the formula from the Clipboard and places it into the selected cell or range.

The cell references in the copied formula change to reflect the formula's new location in the worksheet. For example, consider a formula from a cell in row 12 that adds other values in row 12. When that formula is copied to row 15, the formula changes to add the corresponding values in row 15. By automatically updating the formula based on its new location, Excel makes it easy to quickly enter the same general formula throughout a worksheet.

You will calculate the costs of the remaining expense categories by copying the formula you entered in cell F12 and pasting it to the range F13:F15.

To copy and paste the formula in cell F12:

(1). **Click** cell **F12** to select the cell that contains the formula you want to copy.

(2). On the Home tab, in the Clipboard group, **Click** the **Copy** button (or **press CTRL+C**). Excel copies the formula to the Clipboard. A blinking green box surrounds the cell being copied.

(3). **Select** the range **F13:F15**. You want to paste the formula into these cells.

(4). In the Clipboard group, **Click** the **Paste** button (or press **CTRL+V**). Excel pastes the formula into the selected cells, adjusting each formula so that the total cost of each item is based on the Units and Cost per Units values in that row. A button appears below the selected range, providing options for pasting formulas and values. See **Figure 3-24** on the next page.

In Figure 3-24 on the next page, the Miscellaneous Expenses sheet shows that the formula entered into cell F11 has been copied to cells F12 through F15. The total costs for the other miscellaneous items have been entered into those cells. The total costs are as follows. Transportation shuttles: $1,007.25. Decorations for banquet: $850.55. Gift bags for conference attendees: $3,281.25. Gifts for banquet speakers: $331.50. A button with an arrow appears below cell F15; clicking the button displays more options for pasting formulas and values. Cell F12 has a dashed box around it to show that its contents have been copied.

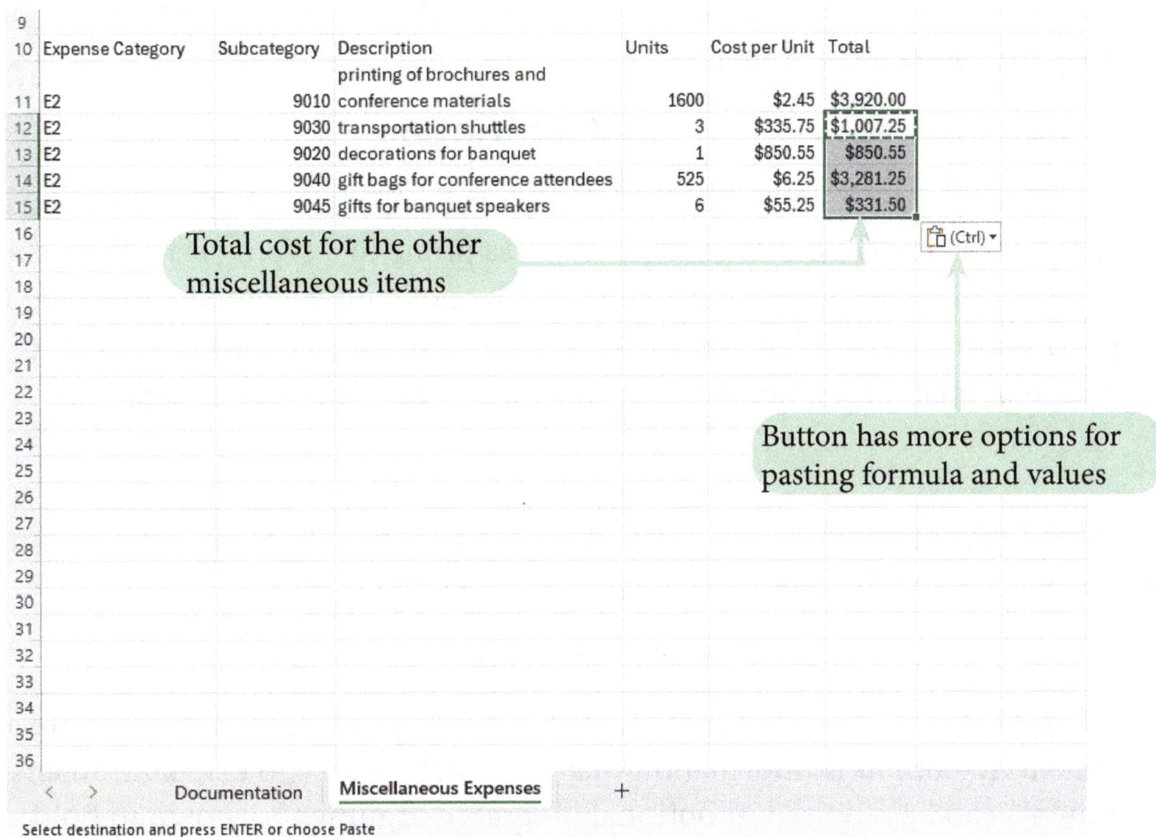

Figure 3-24 Copied and pasted formulas

(5). If necessary, **Click** cell **F13** to make it the active cell. The **Formula =D13*E13** appears in the formula bar. Notice that the cell references in the formula were updated to reflect its current position of the cell in the worksheet.

(6). **Click** cell **F14** to verify that the **Formula =D14*E14** appears in the formula bar, and then click cell F15 to verify that the **Formula =D15*E15** appears in the formula bar.

Another way of performing calculations is to use functions.

Calculating with Functions

A **Function** is a named operation that replaces the arithmetic expression in a formula. Functions are used to simplify long or complex formulas. For example, to add the values from cells A1 through A10, you could enter the following long formula:

=A1+A2+A3+A4+A5+A6+A7+A8+A9+A10

Or, you could use the SUM function to calculate the sum of those cell values by entering the following formula:

=SUM(A1:A10)

In both instances, Excel adds the values in cells A1 through A10, but the SUM function is faster and simpler to enter and less prone to a typing error. You should always use a function, if one is available, in place of a long, complex formula. Excel supports more than 300 functions from the fields of finance, business, science, and engineering, including functions that work with numbers, text, and dates.

UNDERSTANDING FUNCTION SYNTAX

Every function follows a set of rules, or syntax, which specifies how the function should be written. The general syntax of all Excel functions is:

FUNCTION(arg1, arg2, [arg3], [arg4], ...)

where *FUNCTION* is the function name, and ***arg1, arg2,*** and so forth are arguments. An ***Argument*** is information that the function uses to calculate an answer. Arguments can be required or optional. Required arguments, shown in bold, are needed by the function to operate. Optional arguments, enclosed in square brackets, are not required but may be used by the function. Optional arguments are always placed at the end of the argument list. In this case, ***arg1, arg2,*** are required arguments and *arg3, arg4* are optional arguments.

The SUM function shown earlier has the syntax

SUM(number1, [number2], [number], ...)

where ***number1*** is a required argument that indicates the range of values to sum and number2, number3, and so on are optional arguments used for nonadjacent ranges or lists of numbers. For example, the following SUM function calculates the sum of values from the ranges A1:10 and A21:A30:

SUM(A1:A10, A21:A30)

Some functions do not require any arguments and have the syntax FUNCTION( ). Functions without arguments still must include the opening and closing parentheses after the function name. For example, the NOW function does not require any argument values to return the current date and time, as shown in the following formula:

=NOW()

You can learn more about function syntax using Excel Help.

INSERTING FUNCTIONS WITH AUTOSUM

A fast and convenient way to enter commonly used functions is with **AutoSum**. The AutoSum button, located on the Home tab of the ribbon, includes options to insert the following functions into a selected cell or cell range:

- SUM—Sum of the values in the specified range
- AVERAGE—Average value in the specified range
- COUNT—Total count of numeric values in the specified range
- MAX—Maximum value in the specified range
- MIN—Minimum value in the specified range

After you select one of the AutoSum options, Excel determines the most appropriate range from the available data and enters it as the function's argument. You should always verify that the range included in the AutoSum function matches the range that you want to use.

You will use AutoSum to enter the SUM function to add the total cost from all miscellaneous expense categories.

To use AutoSum to sum the miscellaneous expense values:

(1). **Click** cell **E16** to make it the active cell, **Type Subtotal** as the label, and then **Press TAB** to make cell **F16** the active cell.

(2). On the **Home Tab**, in the Editing group, **Click** the **AutoSum Arrow**. The button's menu opens and displays five common functions: Sum, Average, Count Numbers, Max (for maximum), and Min (for minimum).

(3). **Click Sum** to enter the SUM function. The **Formula =SUM(F11:F15)** is entered in cell **F16**. The cells being summed are selected and highlighted on the worksheet so you can quickly confirm that Excel selected the appropriate range from the available data. A ScreenTip appears below the formula describing the function's syntax. See **Figure 3-25** on the next page.

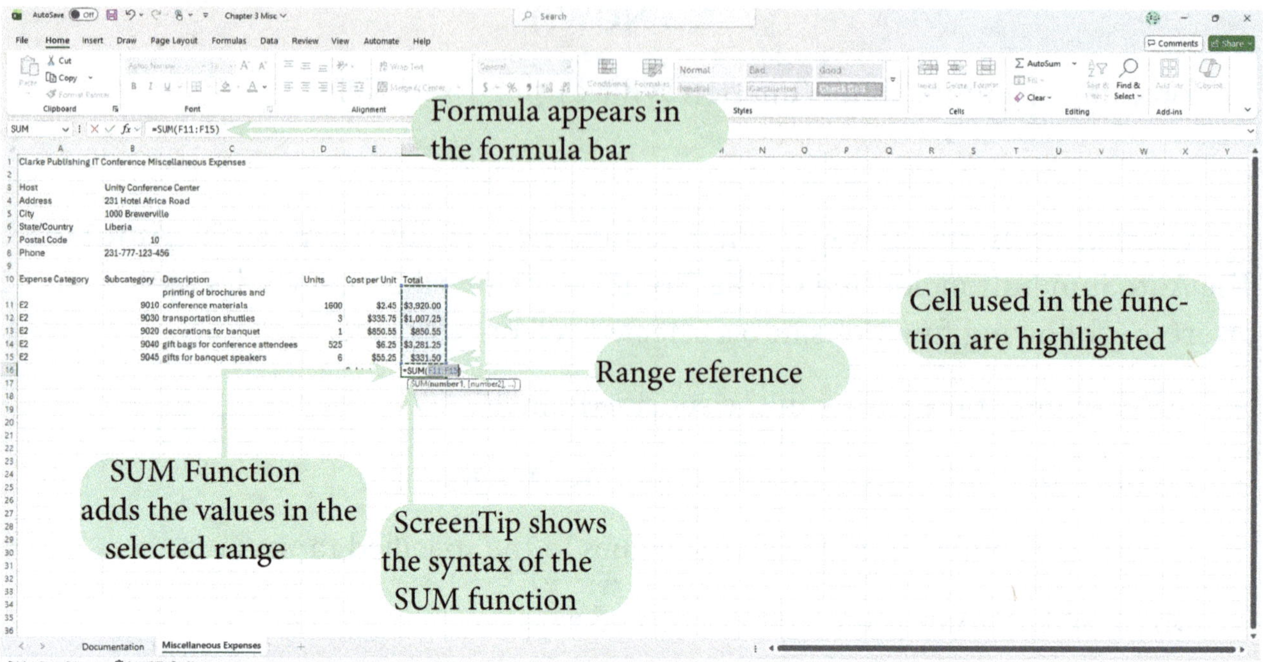

Figure 3-25 SUM function entered using the AutoSum button

(4). **Press ENTER** to accept the formula. The sum of the miscellaneous expenses is $9,390.55.

Mardia wants you to include a 5% sales tax on the miscellaneous expenses. You will calculate the tax, and then add the subtotal value to the tax.

To calculate the sales tax and total expenses:

(1). **Click** cell **E9**, **Type Tax Rate** as the label and then **Press TAB** to make cell F9 the active cell.

(2). **Type 5%** in cell **F9**, and then **Press ENTER**. The 5% value is displayed in cell F9, but the stored value is 0.05. Percentages are displayed with the % symbol but stored as the decimal value.

(3). **Click** cell **E17** to make it the active cell, **Type Est**. Tax as the label, and then **Press TAB** to make F17 the active cell.

(4). **Type** the **Formula =F9*F16** in cell **F17** to calculate the sales tax on all of the miscellaneous expenditures, and then **Press ENTER**. The formula multiplies the sales tax in cell F9 by the order subtotal in cell F16. The estimated taxes of $469.53, which is 5% of the subtotal value of $9,390.55, is displayed in cell F17.

(5). In cell **E18**, **Type TOTAL** as the label, and then **Press TAB** to make cell F18 the active cell.

(6). **Type** the **formula =SUM(F16:F17)** in cell **F18** to calculate the total cost of the miscellaneous expenditures plus tax, and then **Press ENTER**. The overall total is $9,860.08.

(7). **Click** cell **F18** to view the formula in the cell. See **Figure 3-26**.

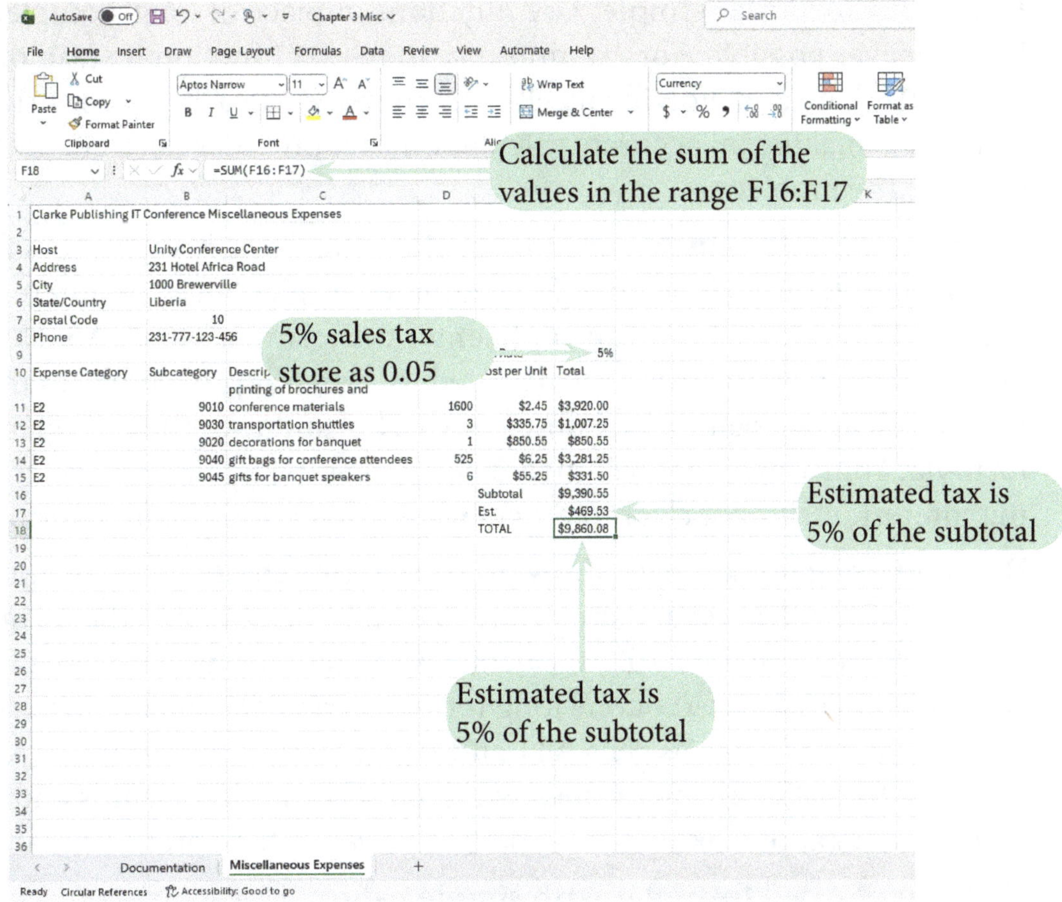

Figure 3-26 Total miscellaneous expenses

If you want to add all of the numbers in a column or row, you need to reference the entire column or row in the SUM function. For example, SUM(E:E) will return the sum of all numeric values in column E and SUM(5:5) will return the sum of all numeric values in row 5.

SMARTSKILLS Setting Column Widths

On You can use formulas to quickly perform calculations and solve problems. First, identify the problem you need to solve. Then, gather the data needed to solve the problem. Finally, create accurate and effective formulas that use the data to answer or resolve the problem. To write effective and useful formulas, consider these guidelines:

- **Keep your formulas simple.** Use functions in place of long, complex formulas whenever possible. For example, use the SUM function instead of entering a formula that adds individual cells, which makes it easier to confirm that the formula is making an accurate calculation as it provides answers needed to evaluate the problem.
- **Do not hide data values within formulas.** The worksheet displays formula results, not the actual formula. For example, to calculate a 5% interest rate on a currency value in cell A5, you could enter the formula =0.05*A5. However, this doesn't show how the value is calculated. A better approach places the 5% value in a cell accompanied by a descriptive label and uses the cell reference in the formula. Your worksheet will then display the interest rate as well as the resulting interest, making it clear to others what calculation is being performed.
- **Break up long formulas to show intermediate results.** Long formulas can be difficult to interpret and are prone to error. For example, the formula =SUM(A1:A10)/SUM(B1:B10) calculates the ratio of two sums but hides the two sum values. Instead of one long formula, consider calculating each sum in a separate cell, such as cells A11 and B11, and use the formula =A11/B11 to calculate the ratio. The worksheet will then show both the sums and the calculation of the ratio, making the workbook easier to interpret and manage.
- **Test complicated formulas with simple values.** Use values you can calculate in your head to confirm that your formula works as intended. For example, using 1s or 10s as the input values makes it easier to verify that your formula is working as intended.

On Finding a solution to a problem requires accurate data and analysis. With workbooks, this means using formulas that are easy to understand, clearly showing the data being used in the calculations and demonstrating how the results are calculated. Only then can you be confident that you are choosing the best problem resolution.

MODIFYING A WORKSHEET

As you develop a worksheet, you will often need to modify its content and structure to create a cleaner and more readable document. You might need to move cells and ranges of cells or you may want to delete rows and columns from the worksheet. You can modify the worksheet's layout without affecting any data or calculations.

MOVING AND COPYING A CELL OR RANGE

One way to move a cell or range is to select it, position the pointer over the bottom border of the selection, drag the selection to a new location, and then release the mouse button. This technique is called **Drag and Drop** because you are dragging the range and dropping it in a new location. If the drop location is not visible, drag the selection to the edge of the workbook window to scroll the worksheet, and then drop the selection.

You can also use the drag-and-drop technique to copy cells by pressing CTRL as you drag the selected range to its new location. A copy of the original range is placed in the new location without removing the original range from the worksheet.

Smart Tips

Moving or Copying a Cell Range

- Select the cell range to move or copy.
- Move the pointer over the border of the selection until the pointer changes shape.
- To move the range, click the border and drag the selection to a new location. To copy the range, hold down CTRL and drag the selection to a new location.

or

- Select the cell range to move or copy.
- On the Home tab, in the Clipboard group, click the Cut or Copy button; or right-click the selection, and then click Cut or Copy on the shortcut menu; or press CTRL+X or CTRL+C.
- Select the cell or the upper-left cell of the range where you want to paste the copied content.
- In the Clipboard group, click the Paste button; or right-click the selection and then click Paste on the shortcut menu; or press CTRL+V.

Mardia wants the labels and value in the range E16:F18 moved down one row to the range E17:F19 to set those calculations off from the list of miscellaneous expenses. You will use the drag-and-drop method to move the range.

To drag and drop the range E16:F18:

(1). **Select** the range **E16:F18**. This is the range you want to move.

(2). **Point** to the **bottom border** of the selected range so that the pointer changes to the **move pointer** .

(3). **Press** and **hold** the **mouse** button to change the pointer to the **Arrow Pointer** , and then drag the selection down one row. Do not release the mouse button. A ScreenTip appears, indicating that the new range of the selected cells will be E17:F19. A dark green border also appears around the new range. See **Figure 3-27**.

(4). Make sure the ScreenTip displays the range **E17:F19**, and then release the mouse button. The selected cells move to their new location.

Some people find dragging and dropping a range difficult and awkward, particularly if the selected range is large or needs to move a long distance in the worksheet. In those situations, it is often more efficient to cut or copy and paste the cell contents. Cutting moves the selected content. Copying duplicates the selected content in the new location.

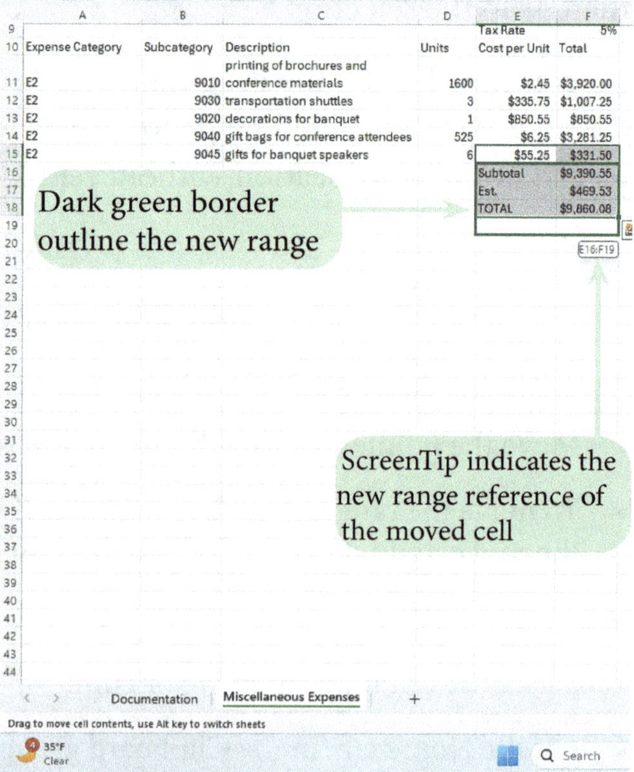

Figure 3-27 Range being dragged

Mardia wants the worksheet to include a summary of the miscellaneous expenses at the top of the worksheet. To free up space for this summary, you'll cut the contents of the range A3:F19 and paste them into the range A7:F23.

To cut and paste the range A3:F19:

(1). **Click** the **Name box** to the left of the formula bar, **Type A3:F19** as the range to select, and then **Press ENTER**. The range A3:F19 is selected.

(2). On the **Home Tab**, in the **Clipboard** group, **Click** the **Cut** button (or **Press CTRL+X**). The range is surrounded by a moving border, indicating that it has been cut.

(3). **Click** cell **A7** to select it. This is the upper-left corner of the range where you want to paste the range that you cut.

(4). In the **Clipboard** group, **Click** the **Paste** button (or **Press CTRL+V**). The range A3:F19 is pasted into the range A7:F23. Note that the cell references in the formulas were automatically updated to reflect the new location of those cells in the worksheet.

Using the COUNT Function

Many financial workbooks need to report the number of entries, such as the number of products in an order or the number of items in an expense or revenue category. To calculate the total number of items, you can use the COUNT function. The COUNT function has the syntax

COUNT(value1, [value2], [value3], ...)

where **value1** is the range of numeric values to count and value2, value3, and so forth specify other ranges.

The **COUNT Function** counts only numeric values. Any cells containing text are not included in the tally. To include cells containing non-numeric data such as text strings, you need to use the COUNTA function. The COUNTA function has the syntax

COUNTA(value1, [value2], [value3], ...)

where **value1** is the range containing numeric or non-numeric values and value2, value3, and so forth specify other ranges to be included in the tally.

Next, you will enter a summary of the miscellaneous expenses by displaying the number of miscellaneous expense categories and the total cost of all the expenses. Because you are interested only in numeric values, you will use the COUNT function to count the number of miscellaneous expense values in the worksheet.

To use the COUNT and SUM functions to create an expense summary:

(1). **Scroll Up** the worksheet, **click** cell **A3** to make it the active cell, **Type Summary** as the label, and then **press ENTER** to make cell A4 the active cell.

(2). In cell **A4**, **Type Expense Categories** as the label, and then **Press TAB** to make cell B4 the active cell.

(3). In cell **B4**, **Type =COUNT**(to begin the COUNT function.

(4). **Select** the range **F15:F19**. The range reference F15:F19 is entered into the COUNT function.

(5). **Type)** to complete the function, and then **Press ENTER** to make cell A5 the active cell. Cell B4 displays 5, indicating that the report includes five types of miscellaneous expenses.

(6). In cell A5, **Type Total Expenses** as the label, and then **Press TAB** to make cell B5 the active cell.

(7). In cell **B5**, **Type =F23** as the formula, and then **Press ENTER**. This formula displays contents of cell F23, which is the cell where you added the subtotal and taxes. See **Figure 3-28**.

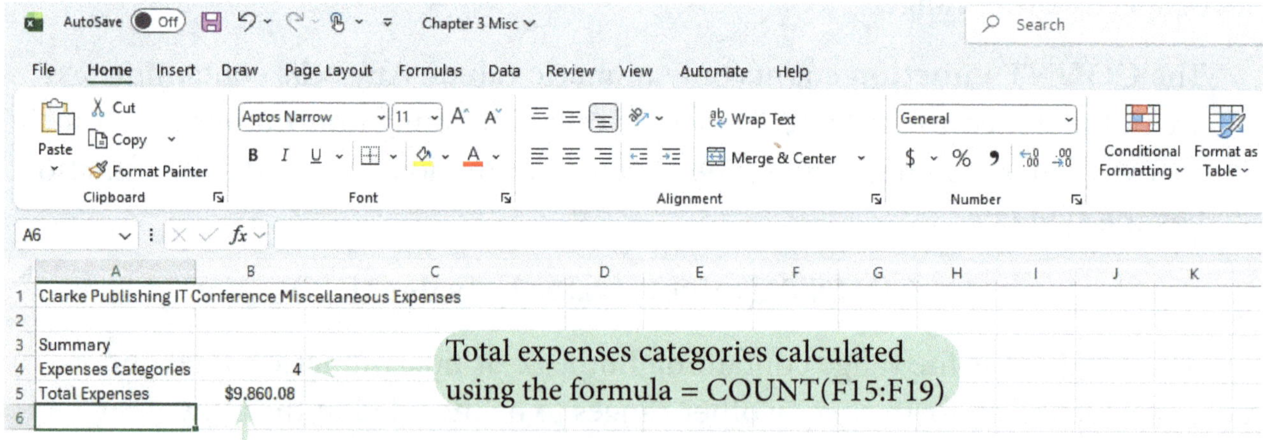

Figure 3-28 Miscellaneous expenses summary

The formula in cell B5 that displays the sum of miscellaneous expenses calculated in cell F23 illustrates an important practice: Don't repeat the same calculation multiple times. Instead, use a formula that references the cell containing the formula results you wish to repeat. This way, if you must later change the formula, you need to edit only that one cell.

Modifying Rows and Columns

Another way to modify the structure of a workbook is by inserting or removing whole rows and columns from a worksheet.

Inserting Rows and Columns

When you insert a new column, the existing columns are shifted to the right, and the new column has the same width as the column directly to its left. When you insert a new row, the existing rows are shifted down, and the new row has the same height as the row above it. Because inserting a new row or column moves the location of the other cells in the worksheet, any cell references in a formula or function are updated to reflect the new layout.

Smart Tips

Inserting and Deleting Rows or Columns

To insert rows and columns into a worksheet:

- Select the row or column headings where you want to insert new content.
- On the Home tab, in the Cells group, click the Insert button; or right-click the selected headings and click Insert on the shortcut menu; or press CTRL+SHIFT+=.

To delete rows or columns:

- Select the row or column headings for the content you want to delete.
- On the Home tab, in the Cells group, click the Delete button; or right-click the selecting headings, and then click Delete on the shortcut menu; or press CTRL+-.

Clarke Publishing is providing the flower arrangements for the tables at the closing banquet of the conference. Mardia asks you to add that expense category to the worksheet. You will insert a new row and enter that expense.

To insert the flower expense category:

(1). **Scroll Down** and **Click** the row **18** heading to select the entire row. You want to add the new expense category as row 18.

(2). On the **Home Tab**, in the Cells group, **Click** the **Insert** button (or **Press CTRL+SHIFT+=**). A new row 18 is inserted in the worksheet, and all the rows below the new row are shifted down.

(3). **Enter E2** in cell **A18**, **Enter 9025** in cell **B18**, **enter flowers for the banquet tables** in cell **C18**, **Enter 20** in cell **D18**, and **Enter $12.50** in cell **E18**.

(4). **Copy** the **Formula** from cell **F17** and **Paste** it into cell **F18**. The formula =D18*E18 entered in cell F18, displaying $250.00 as the category expense total. The formula calculating the overall total of miscellaneous expenses in cell F22 is updated to include the new row. The subtotal, estimated taxes, and grand total are recalculated to include the new category. Cell F24 displays the grand total as ######## because the column is too narrow to display the entire value.

(5). **Increase** the **Width** of **Column F** to **12 characters** using whatever method you choose.

(6). **Click** cell **F22**. See **Figure 3-29**.

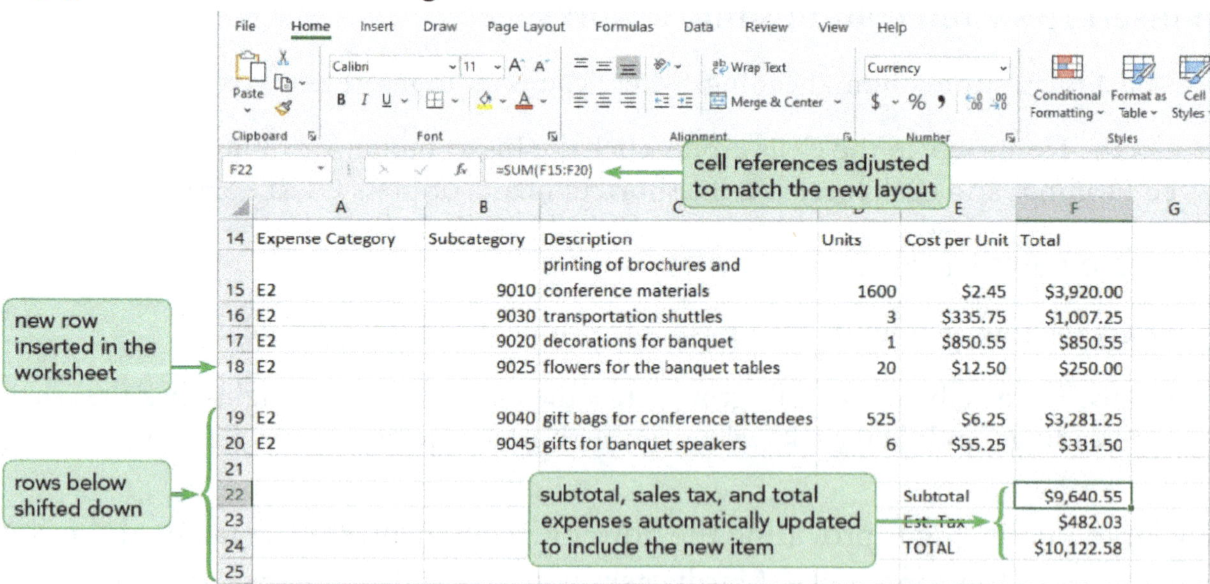

	A	B	C	D	E	F
14	Expense Category	Subcategory	Description	Units	Cost per Unit	Total
15	E2	9010	printing of brochures and conference materials	1600	$2.45	$3,920.00
16	E2	9030	transportation shuttles	3	$335.75	$1,007.25
17	E2	9020	decorations for banquet	1	$850.55	$850.55
18	E2	9025	flowers for the banquet tables	20	$12.50	$250.00
19	E2	9040	gift bags for conference attendees	525	$6.25	$3,281.25
20	E2	9045	gifts for banquet speakers	6	$55.25	$331.50
21						
22					Subtotal	$9,640.55
23					Est. Tax	$482.03
24					TOTAL	$10,122.58
25						

Figure 3-29 New row inserted into the worksheet

Notice that the formula in cell F22 is now =SUM(F15:F20). The range reference was updated to reflect the inserted row. The tax amount increased to $482.03 based on the new subtotal value of $9,640.55, and the total charge increased to $10,122.58 because of the added item. Also, the result of the COUNT function in cell B4 increased to 6 to reflect the added expense.

DELETING ROWS AND COLUMNS

There are two ways of removing content from a worksheet: deleting and clearing. **Deleting** removes both the data and the selected cells from the worksheet. The rows below the deleted row shift up to fill the vacated space. Likewise, the columns to the right of the deleted column shift left to fill the vacated space. Also, all cell references in worksheet formulas are adjusted to reflect the change that removing the row or column makes to the worksheet structure. You click the Delete button in the Cells group on the Home tab to delete selected rows or columns.

Clearing removes the data from the selected cells, leaving those cells blank but preserving the current worksheet structure. You clear data from the selected cells by pressing DELETE.

The IT Conference isn't intended to make money for Clarke Publishing, but the company doesn't want to lose money either. Mardia needs to watch expenses and keep the total miscellaneous costs under $9,000. Mardia negotiated with the management at Hotel Africa own of the Unity Conference Center to provide the transportation shuttles for free and asks you to remove that expense from the worksheet.

To delete the transportation shuttles row from the worksheet:

(1). **Click** the **Row 16** heading to select the entire row.

(2). On the **Home Tab**, in the Cells group, **Click** the **Delete** button (or **Press CTRL+-**). Row 16 is deleted, and the rows below it shift up to fill the space.

All the cell references from the formulas in the worksheet are again updated automatically to reflect the impact of deleting row 16. The subtotal value in cell F21 is now $8,633.30, which is the sum of the range F15:F19. The estimated tax in F22 decreases to $431.67. The total miscellaneous expenses are now $9,064.97, which is closer to the budget that Mardia must meet. Also, the result of the COUNT function in cell B4 returns to 5, reflecting the deleted expense category. As you can see, one of the great advantages of using Excel is that it modifies cell references within the formulas to reflect the additions and deletions made in the worksheet.

INSERTING AND DELETING A RANGE

You can also insert or delete cell ranges within a worksheet. When you use the Insert button to insert a range of cells, the existing cells shift down when the selected range is wider than it is long, and they shift right when the selected range is longer than it is wide, as shown in **Figure 3-30**. When you use the Insert Cells command, you specify whether the existing cells shift right or down, or whether to insert an entire row or column into the new range.

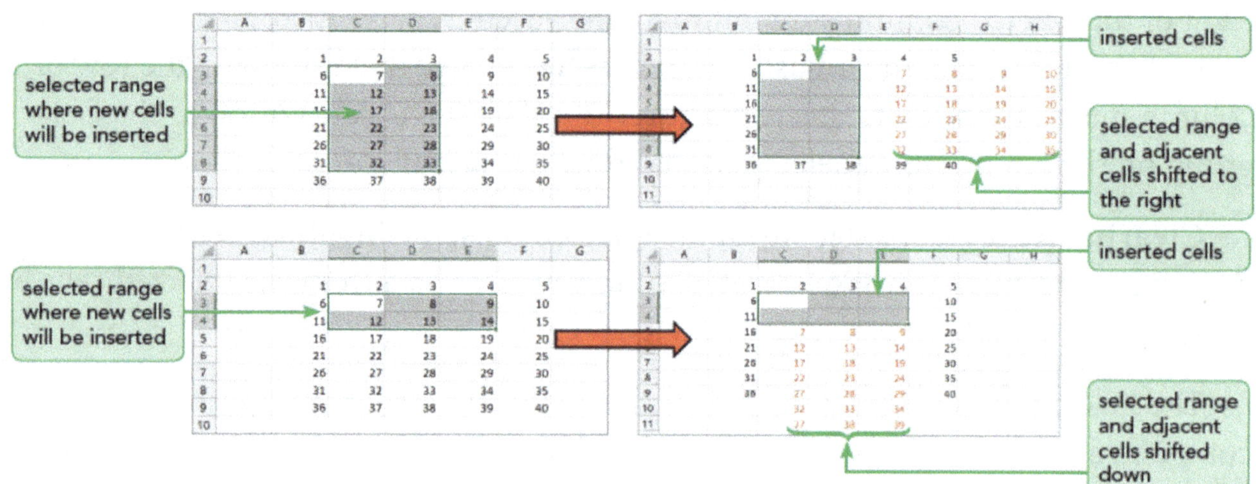

Figure 3-30 Cells inserted into a worksheet

In Figure 3-30 four worksheets are shown. (1) A worksheet with entered into the cell range B2 through F9; the cell range C3 through D8 is selected, and is where new cells will be inserted. (2) the worksheet with the new inserted cells; the selected range and adjacent cells shifted to the right so that there is now a cell range G3 through H8. (3) the same numbers as the first worksheet entered into the cell range B2 through F9.; the cell range C3 through E4 is selected and is where new cells will be inserted. (4) the worksheet shows the new inserted cells.; the selected range and adjacent cells shifted down so that there is now a cell range C10 through E11.

The process works in reverse when you delete a range. As with deleting a row or column, the cells adjacent to the deleted range either move up or left to fill in the space vacated by the deleted cells. The Delete Cells command lets you specify whether you want to shift the adjacent cells left or up or whether you want to delete the entire column or row.

When you insert or delete a range, cells that shift to a new location adopt the width of the columns they move into. As a result, you might need to resize columns and rows in the worksheet.

Smart Tips

Inserting and Deleting a Range

To insert rows and columns into a worksheet:

- Select a range that matches the area you want to insert or delete.
- On the Home tab, in the Cells group, click the Insert button or the Delete button.

or

- Select the range that matches the range you want to insert or delete.
- On the Home tab, in the Cells group, click the Insert arrow and then click Insert Cells or click the Delete arrow and then click Delete Cells; or right-click the selected range, and then click Insert or Delete on the shortcut menu.
- Click the option button for the direction to shift the cells, columns, or rows.
- Click OK.

Clarke Publishing assigns an account ID for each type of revenue and expense item. Mardia asks you to insert this information into the worksheet. You will insert these new cells into the range A13:A23, shifting the adjacent cells to the right.

To insert a range to enter the account IDs:

(1). **Select** the range **A13:A23** using any method you choose.

(2). On the **Home tab**, in the Cells group, **Click** the **Insert Arrow**. A menu of insert options appears.

(3). **Click Insert Cells**. The Insert dialog box opens.

(4). **Verify** that the **Shift cells right** option button is selected.

(5). **Click OK**. New cells are inserted into the selected range, and the adjacent cells move to the right. The shifted content does not fit well in the adjacent columns. You'll resize the columns and rows to fit their data.

(6). **Change** the **Width** of column **B** to **18** characters, the **Width** of column **C** to **12** characters, the width of column **D** to **36** characters, and the widths of columns **F** and G to **14** characters.

(7). **Select rows 15** through **19**.

(8). In the Cells group, **Click** the **Format** button, and then **Click AutoFit Row Height**. The selected rows autofit to their contents.

(9). **Resize** the **height** of row **13** to **42** points, creating additional space between the summary information and the miscellaneous expenses data.

(10). **Click** cell **A14**. See **Figure 3-31**.

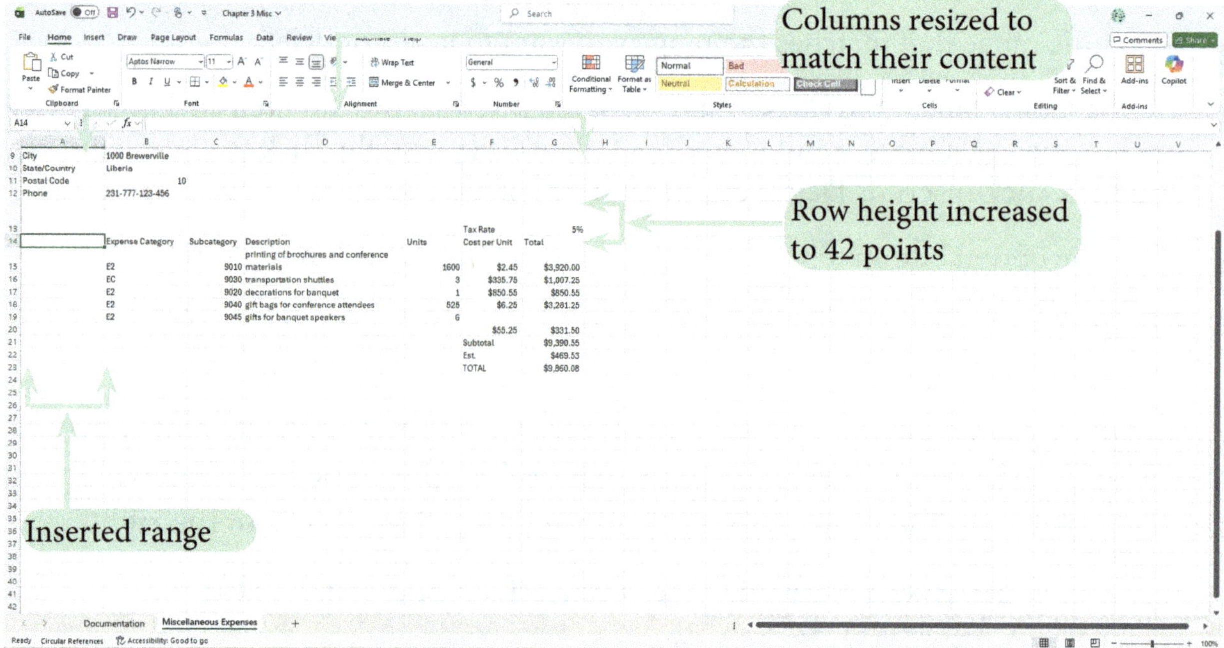

Figure 3-31 Cell range inserted into the worksheet

In Figure 3-31 above, the Miscellaneous Expenses sheet shows that the new cell range A13 through A23 has been added and that all the adjacent cells have shifted to the right. The columns have been resized to match their content. The row height of row 13 has increased to 42 points. Cell A14 is the active cell.

Notice that even though the account IDs will be entered in the range A14:A19, you inserted new cells into the A13:A23 range to retain the layout of the worksheet contents. Selecting the additional rows ensures that the tax rate and summary values still line up with the Cost per Unit and Total columns. Whenever you insert a new range, be sure to consider its impact on the layout of the entire worksheet.

SMARTSKILLS Hiding and Unhiding Rows, Columns, and Worksheets

On Workbooks can become long and complicated, filled with formulas and data that are important for performing calculations but are of little interest to readers. In those situations, you can simplify these workbooks by hiding rows, columns, and even worksheets. Although the contents of hidden cells cannot be seen, the data in those cells is still available for use in formulas and functions throughout the workbook.

Hiding removes a row or column from view while keeping it part of the worksheet. To hide a row or column, select the row or column heading, click the Format button in the Cells group on the Home tab, point to Hide & Unhide on the menu that appears, and then click Hide Rows or Hide Columns. The border of the row or column heading is doubled to mark the location of hidden rows or columns.

A worksheet often is hidden when the entire worksheet contains data that is not of interest to the reader and is better summarised elsewhere in the document. To hide a worksheet, make that worksheet active, click the Format button in the Cells group on the Home tab, point to Hide & Unhide, and then click Hide Sheet.

Unhiding redisplays the hidden content in the workbook. To unhide a row or column, click in a cell below the hidden row or to the right of the hidden column, click the Format button, point to Hide & Unhide, and then click Unhide Rows or Unhide Columns. To unhide a worksheet, click the Format button, point to Hide & Unhide, and then click Unhide Sheet. The Unhide dialog box opens. Click the sheet you want to unhide, and then click OK. The hidden content is redisplayed in the workbook.

Although hiding data can make a worksheet and workbook easier to read, be sure never to hide information that is important to the reader.

You will complete the miscellaneous expenses table by adding the account IDs for each expense. You can use Flash Fill to automatically create the account IDs.

USING FLASH FILL

Flash Fill enters text based on patterns it finds in the data. As shown in **Figure 3-32**, Flash Fill generates names from the first and last names stored in the adjacent columns in the worksheet. To enter the rest of the names, press ENTER; to continue typing the names yourself, press ESC.

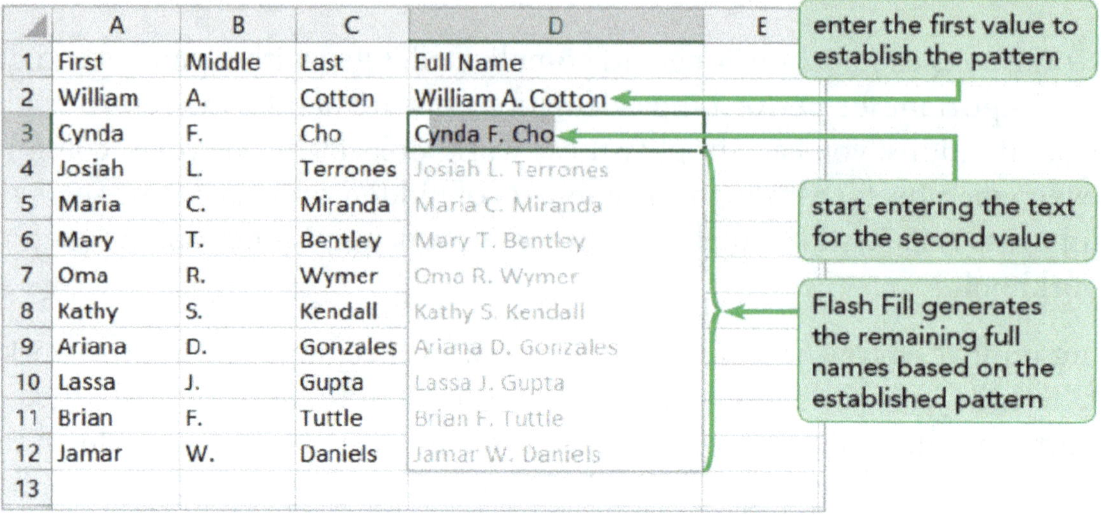

	A	B	C	D	E
1	First	Middle	Last	Full Name	
2	William	A.	Cotton	William A. Cotton	
3	Cynda	F.	Cho	Cynda F. Cho	
4	Josiah	L.	Terrones	Josiah L. Terrones	
5	Maria	C.	Miranda	Maria C. Miranda	
6	Mary	T.	Bentley	Mary T. Bentley	
7	Oma	R.	Wymer	Oma R. Wymer	
8	Kathy	S.	Kendall	Kathy S. Kendall	
9	Ariana	D.	Gonzales	Ariana D. Gonzales	
10	Lassa	J.	Gupta	Lassa J. Gupta	
11	Brian	F.	Tuttle	Brian F. Tuttle	
12	Jamar	W.	Daniels	Jamar W. Daniels	
13					

Figure 3-32 Example of a Flash Fill screen

Flash Fill works best when the pattern is clearly recognized from the values in the data. Be sure to enter the data pattern in the column or row right next to the related data. The data used to generate the pattern must be in a rectangular grid and cannot have blank columns or rows.

Clarke Publishing IT Conference account IDs combine the expense category and subcategory values. For example, an expense from the E2 expense category with a 9010 subcategory has the account ID of E2-9010. Rather than typing this information for every expense item, you will use Flash Fill to complete the data entry.

To enter the account IDs using Flash Fill:

(1). **Type Account ID** in cell **A14**, and then press ENTER. The label is entered in cell A14, and cell A15 is the active cell.

(2). **Type E2-9010** in cell **A15**, and then press ENTER. The first account ID is entered in cell A15, and cell A16 is the active cell.

(3). **Type E2-9020** in cell **A16**. As soon as you complete those characters, Flash Fill generates the remaining entries in the column based on the pattern you entered. See **Figure 3-33** on the next page.

(4). **Press ENTER** to accept the suggested entries.

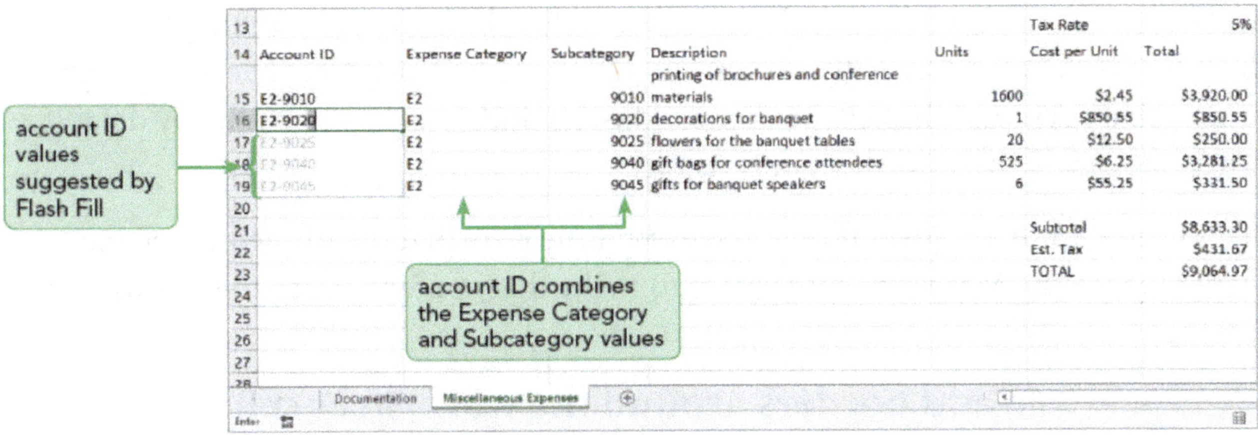

Figure 3-33 Account IDs generated by Flash Fill

Figure 3-33 above of the Miscellaneous Expenses sheet shows that Account IDs are being generated by Flash Fill. The Expense Category in cell B15 is E2. The Subcategory in cell C15 is 9010. The Account ID in cell A15 is E2-9010. The Expense Category in cell B16 is E2. The Subcategory in cell C16 is 9020. The Account ID in cell A16 is E2-9020. The cell range A17 through A19 is generated by Flash Fill to combine the Expense Category and Subcategory values. The remaining Account IDs are the following. Account ID: E2-9025; Expense Category: E2; Subcategory: 9025. Account ID: E2-9040; Expense Category: E2; Subcategory: 9040. Account ID: E2-9045; Expense Category: E2; Subcategory: 9045.

Are You Having Trouble? If you pause for an extended time between entering text to establish the pattern, Flash Fill might not extend the pattern for you.

Note that Flash Fill generates text, not formulas. If you edit or replace the entry originally used to create the Flash Fill pattern, the other entries generated by Flash Fill in the column will not be updated.

Formatting a Worksheet

Formatting enhances the appearance of the worksheet data by changing its font, size, color, or alignment. Two common formatting changes are adding cell borders and changing the font size of text.

Adding Cell Borders

You can make spreadsheet content easier to read by adding borders around the worksheet cells. Borders can be added to the left, top, right, or bottom edge of any cell or range. You can set the color, thickness of and the number of lines in each border. Borders are especially useful when you print a worksheet because the gridlines that surround the cells in the workbook window are not printed by default. They appear on the worksheet only as a guide.

Mardia wants you to add borders around the cells that detail the miscellaneous expenses for the Monrovia conference to make that content easier to read when it's printed.

To add borders around the worksheet cells:

(1). **Select** the range **F13:G13**. You'll add borders to these cells.

(2). On the **Home Tab**, in the Font group, **Click** the **Borders arrow** , and then **click All Borders**. Borders are added around each cell in the selected range. The Borders button changes to reflect the last selected border option, which in this case is All Borders. The name of the selected border option appears in the button's ScreenTip.

(3). **Select** the nonadjacent range **A14:G19,F21:G23**.

(4). On the **Home Tab**, in the Font group, **Click** the **All Borders** button . Borders appear around all the cells in this range as well.

(5). **Click** cell **A24** to deselect the range. See **Figure 3-34**.

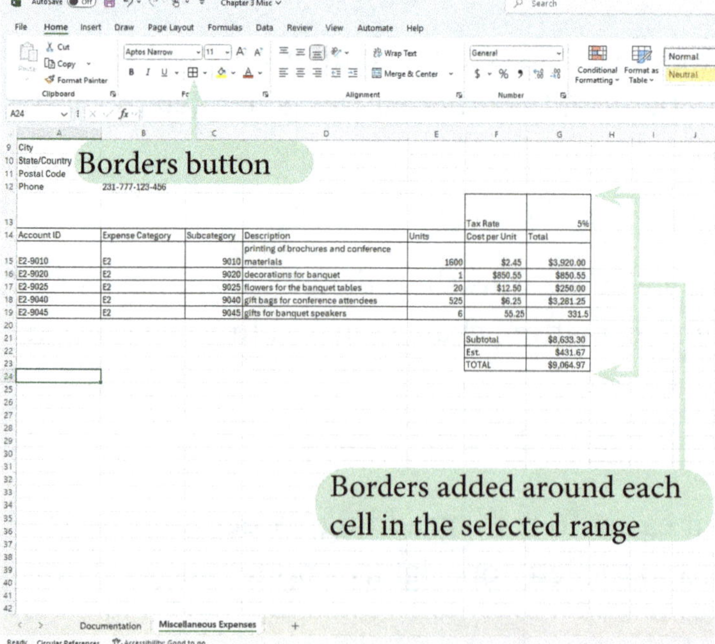

In Figure 3-34, borders have been added around each cell in the cell ranges F13 through G13, A14 through G19, and F21 through G23. The Borders button in the Font group on the Home tab is called out. Cell A24 is the active cell.

Figure 3-34 Borders added to cells

As you prepare for the digital workplace, don't be surprised if you encounter a boss like Mardia who likes perfection.

CHANGING THE FONT SIZE

Changing the size of text in a sheet provides a way to identify different parts of a worksheet, such as distinguishing a title or section heading from data. The size of the text is referred to as the font size and is measured in points. The default font size for worksheets is 11 points, but it can be made larger or smaller as needed. You can resize text in selected cells using the Font Size button in the Font group on the Home tab. You can also use the Increase Font Size and Decrease Font Size buttons to resize cell content to the next higher or lower standard font size.

Mardia wants you to increase the size of the worksheet title to 24 points to make it more visible and stand out from the rest of the worksheet content.

To change the font size of the worksheet title:

(1). **Scroll up** the worksheet and **Click** cell **A1** to select the worksheet title.

(2). On the **Home Tab**, in the Font group, **Click** the **Font Size Arrow** [11] to display a list of font sizes, and then click 24. The worksheet title changes to 24 points. See **Figure 3-35**.

Figure 3-35 Font size increased in cell A1

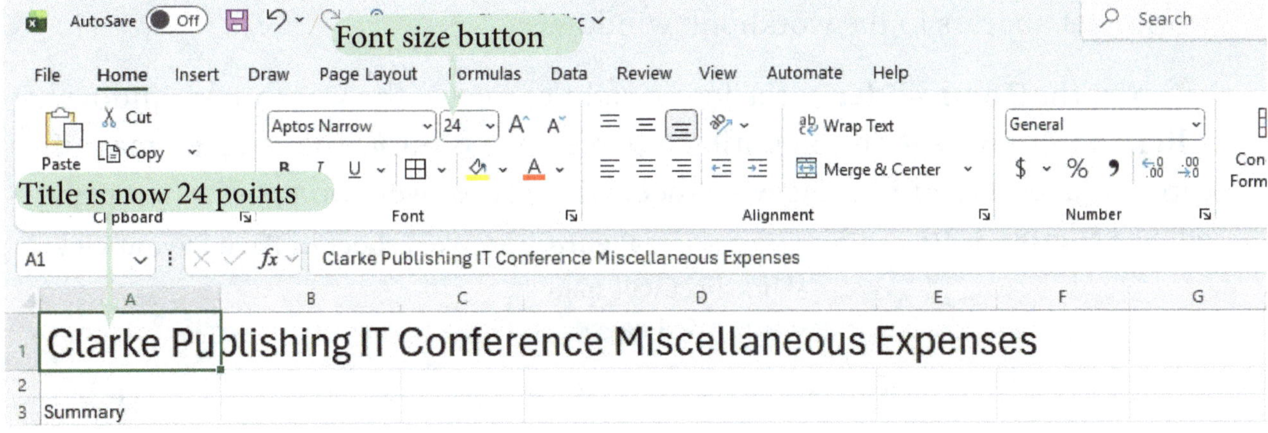

In **Figure 3-35**, the font size of the text in cell A1 was increased to 24 points. The Font Size button in the Font group on the Home tab is called out. Next to this button are buttons that incrementally increase and decrease the font size.

(3). **Press CTRL+S** to save the workbook.

Now that the worksheet content and formatting are final, you can print the worksheet.

PRINTING A WORKBOOK

Excel has many tools to control the print layout and appearance of a workbook. Before printing a worksheet, you will want to preview the printout to make sure that it will print correctly.

CHANGING WORKSHEET VIEWS

You can view a worksheet in three ways. Normal view, which you have been using throughout this module, shows the contents of the worksheet. Page Layout view shows how the worksheet will appear when printed. Page Break Preview displays the location of the different page breaks within the worksheet. This is useful when a worksheet will span several printed pages, and you need to control what content appears on each page.

Mardia wants you to preview the print version of the Miscellaneous Expenses worksheet. You will do this by switching between views.

To switch the Miscellaneous Expenses worksheet between views:

(1). **Click** the **Page Layout** button on the status bar. The page layout of the worksheet appears in the workbook window.

(2). On the **Zoom slider** at the lower-right corner of the workbook window, **Click** the **Zoom Out** button until the percentage is **60%**. The reduced magnification makes it clear that the worksheet will spread over two pages when printed. See **Figure 3-36**.

Figure 3-36 Worksheet in Page Layout view

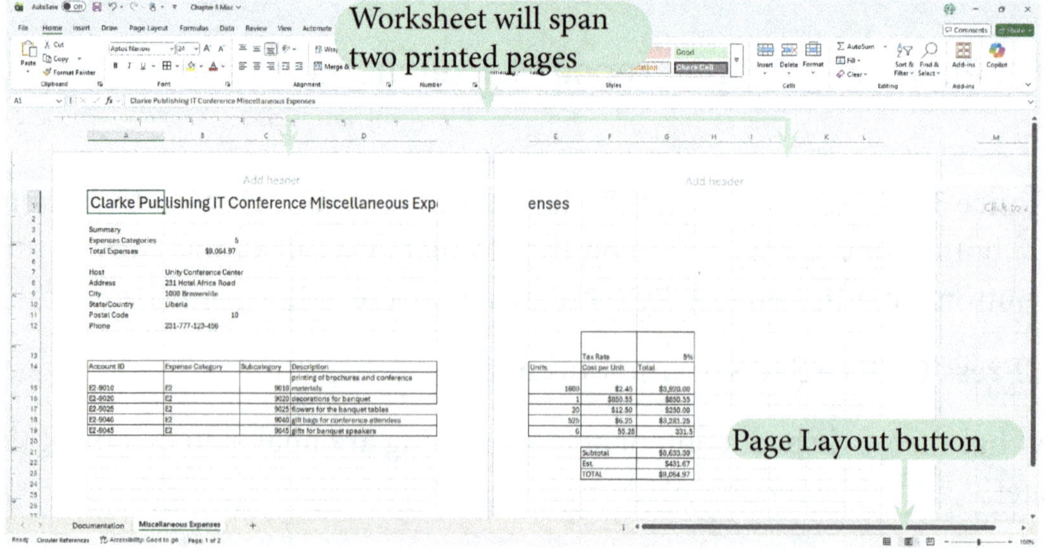

(3). **Click** the **Page Break Preview** button on the status bar. The view switches to Page Break Preview, which shows only those parts of the current worksheet that will print. A dotted blue border separates one page from another.

(4). On the **Zoom slider**, **Drag** the **Slider** button to the **right** until the **percentage** is **80%**. You can now more easily read the contents of the worksheet. See **Figure 3-37**.

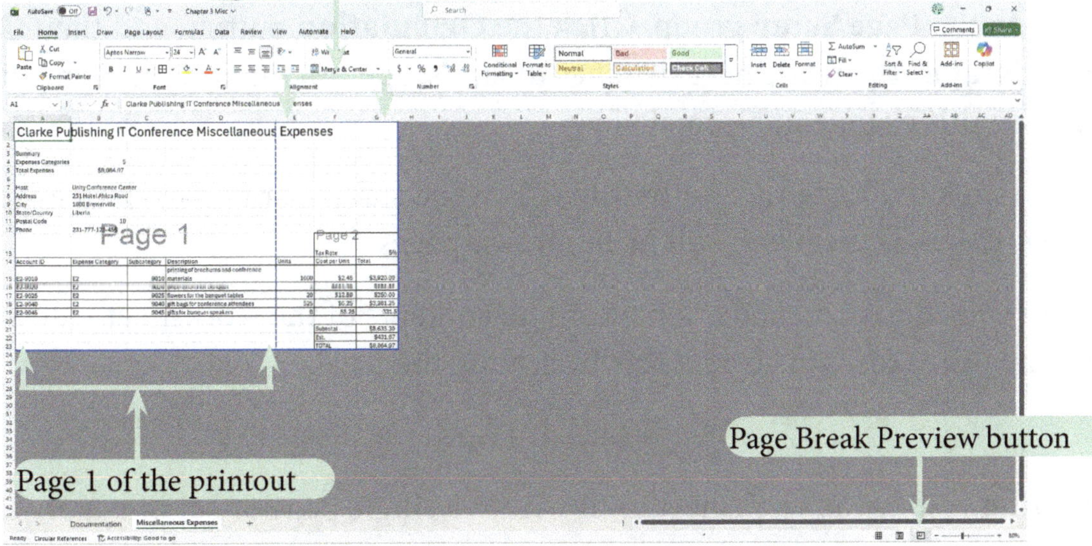

Figure 3-37 Worksheet in Page Break Preview view

(5). **Click** the **Normal** button on the status bar. The worksheet returns to Normal view. Notice that after viewing the worksheet in Page Layout or Page Break Preview, a dotted black line appears between columns D and E in Normal view to show where the page breaks occur.

CHANGING THE PAGE ORIENTATION

Page Orientation specifies in which direction content is printed. In **Portrait Orientation**, the page is taller than it is wide. In **landscape Orientation**, the page is wider than it is tall. By default, Excel displays pages in portrait orientation. Changing the page orientation affects only the active sheet in the workbook and not the other, unselected, sheets.

As you saw in Page Layout view and Page Break Preview, the Miscellaneous Expenses worksheet will print on two pages—columns A through D will print on the first page, and columns E through G will print on the second page. Keep in mind that the columns that print on each page may differ slightly depending on the printer. Mardia wants the entire worksheet to print on a single page, so you'll change the page orientation from portrait to landscape.

To change the page orientation of the worksheet:

(1). On the **Ribbon**, **Click** the **Page Layout Tab**. The tab includes options for changing how the worksheet is arranged.

(2). In the **Page Setup** group, **Click** the **Orientation** button, and then **Click** **Landscape**. The worksheet switches to landscape orientation, though you cannot see this change in Normal view.

(3). **Click** the **Page Layout** button on the status bar to switch to Page Layout view. The worksheet will still print on two pages.

Even with the landscape orientation the contents of the worksheet will still not fit on one page. You can correct this by scaling the page.

SETTING THE SCALING OPTIONS

Scaling resizes the worksheet to fit within a single page or set of pages. There are several options for scaling a printout of a worksheet. You can scale the width or the height of the printout so that all the columns or all of the rows fit on a single page. You can also scale the printout to fit the entire worksheet (both columns and rows) on a single page. If the worksheet is too large to fit on one page, you can scale the print to fit on the number of pages you select. You can also scale the worksheet to a percentage of its size. For example, scaling a worksheet to 50% reduces the size of the sheet by half when it is sent to the printer. When scaling a printout, make sure that the worksheet is still readable after it is resized. Scaling affects only the active worksheet, so you can scale each worksheet separately to best fit its contents.

Mardia asks you to scale the printout so that the Miscellaneous Expenses worksheet fits on one page in landscape orientation.

To scale the printout of the worksheet:

(1). On the **Page Layout tab**, in the **Scale to Fit** group, **Click** the **Width Arrow**, and then **Click 1 page** on the menu that appears. All the columns in the worksheet now fit on one page.

If more rows are added to the worksheet, Mardia wants to ensure that they still fit within a single sheet.

(2). In the **Scale to Fit** group, **Click** the **Height arrow**, and then **Click 1 page**. All the rows in the worksheet will now always fit on one page. See **Figure 3-38.**

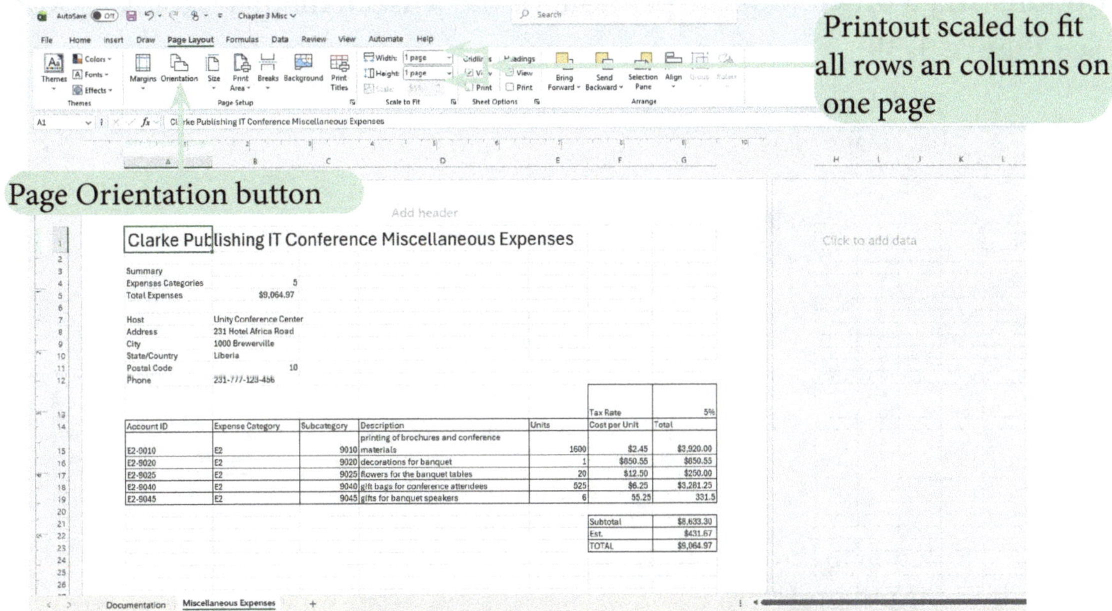

Figure 3-38 Printout scaled to fit on one page

SETTING THE PRINT OPTIONS

You can print the contents of a workbook by using the Print screen in Backstage view. The Print screen provides options for choosing where to print, what to print, and how to print. For example, you can specify the number of copies to print, which printer to use, and what to print. You can choose to print only the selected cells, only the active sheets, or all the worksheets in the workbook that contain data. The printout will include only the data in the worksheet. The other elements in the worksheet, such as the row and column headings and the gridlines around the worksheet cells, will not print by default. The preview shows you exactly how the printed pages will look with the current settings. You should always preview before printing to ensure that the printout looks exactly as you intended and avoid unnecessary reprinting. To print the gridlines or the column and row headings, click the corresponding Print check box in the Sheet Options group on the Page Layout tab.

Mardia asks you to preview and print the workbook containing the estimates of miscellaneous expenses for the conference.

To preview and print the workbook:

(1). On the **ribbon**, **Click** the **File Tab** to display Backstage view.

(2). **Click Print** in the navigation bar. The Print screen appears with the print options and a preview of the printout of the Miscellaneous Expenses worksheet. See **Figure 3-39**.

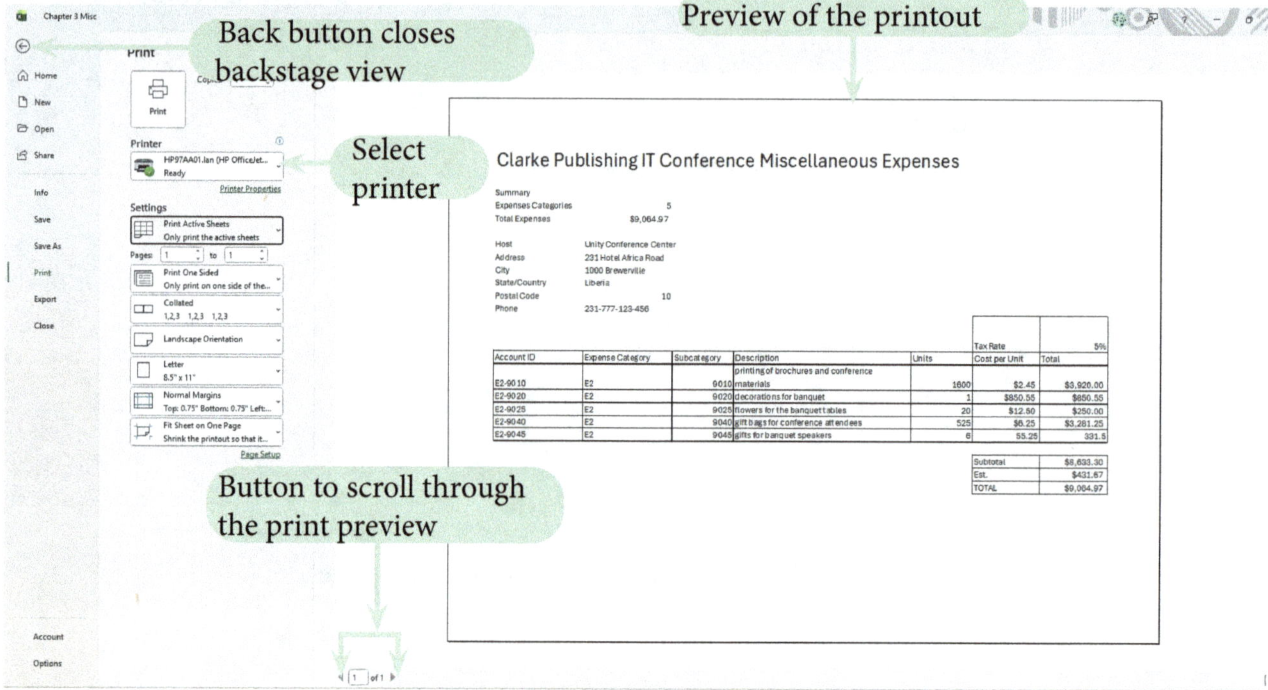

Figure 3-39 Print screen in Backstage view

(3). **Click** the **Printer** button, and then click the printer you want to print to, if it is not already selected. By default, Excel will print only the active sheet.

(4). In the Settings options, click the top button, and then **Click Print Entire Workbook** to print all of the sheets in the workbook—in this case, both the Documentation and the Miscellaneous Expenses worksheets. The preview shows the first sheet in the workbook—the Documentation worksheet. Note that this sheet is still in portrait orientation.

(5). Below the preview, **Click** the **Next Page** button ▶ to view the print preview for the Miscellaneous Expenses worksheet, which will print on a single page in landscape orientation.

(6). If you are instructed to print, **Click** the **Print** button to send the contents of the workbook to the specified printer. If you are not instructed to print, **Click** the **Back** button in the navigation bar to exit Backstage view.

VIEWING WORKSHEET FORMULAS

Most of the time, you will be interested in only the final results of a worksheet, not the formulas used to calculate those results. However, in some cases, you might want to view the formulas used to develop the workbook. This is particularly useful when you encounter unexpected results and want to examine the underlying formulas, or you want to discuss the formulas with a colleague. You can display the formulas instead of the resulting values in cells.

If you print the worksheet while the formulas are displayed, the printout shows the formulas instead of the values. To make the printout easier to read, you should print the worksheet gridlines as well as the row and column headings so that cell references in the formulas are easy to find in the printed version of the worksheet.

You'll look at the formulas in the Miscellaneous Expenses worksheet.

To look at the formulas in the worksheet:

(1). Make sure the **Miscellaneous Expenses** worksheet is displayed in **Page Layout View**.

(2). On the **Ribbon**, **Click** the **Formulas Tab**.

(3). In the Formula Auditing group, **Click** the **Show Formulas** button. The worksheet changes to display formulas instead of the values. Notice that the columns widen to display the complete formula text within each cell.

(4). When you are done reviewing the formulas, **Click** the **Show Formulas** button again to hide the formulas and display the resulting values.

(5). **Click** the **Normal** button on the status bar to return the workbook to Normal view.

(6). On the **Zoom slider**, **drag** the **slider** button to the **right** until the percentage is **120%** (or the magnification you want to use). See **Figure 3-40** on the next page.

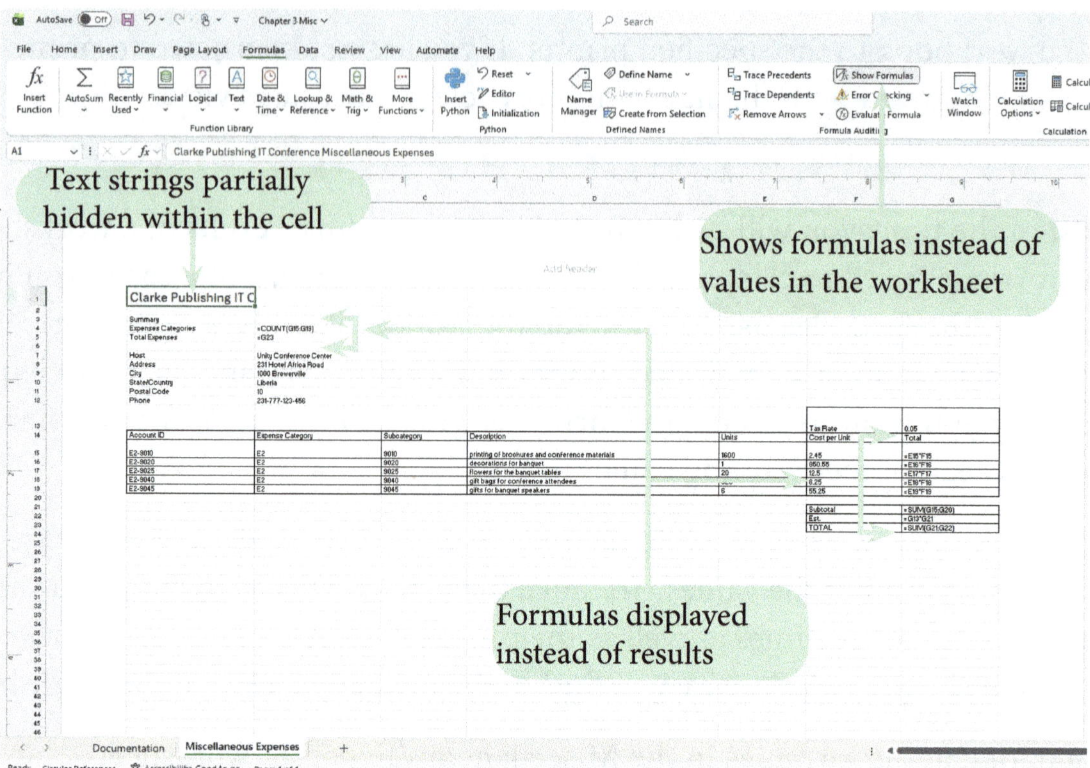

Figure 3-40 Worksheet with formulas displayed

(7). Save the workbook, and then close it.

At this point, Mardia is well pleased with the workbook you created and will use your estimates in the workbook she is using to track all of the cost estimates for the Clarke Publishing IT Conference. Mardia will continue to adjust the revenue and expense projections as the conference data approaches and will get back to you if she needs you to do additional analysis.

Quick Review

1. What is the formula to add values in cells A1, B1, and C1? What function will achieve the same result?
2. What is the formula to count the number of numeric values in the range A1:A30?
3. If you insert cells into the range C1:D10, shifting the cells to the right, what is the new location of the data that was previously in cell F4?
4. Cell E11 contains the formula =SUM(E1:E10). How is the formula adjusted when a new row is inserted above row 5?
5. In the following function, which arguments are required and which arguments are optional:

Quick Review

AVERAGE(number1, [number2], [number3], ...)

6. What is the formula to sum all the numeric values in column C.
7. Describe the four ways of viewing worksheet contents in Excel.
8. How are page breaks indicated in Page Break Preview?

CHAPTER REVIEW

Lab Project 1

CHAPTER REVIEW LABS

There are no Data Files needed for the Review Assignments.

Mardia needs to estimate the total costs for supplying computers, audio/video equipment, and Internet access to the participants at the Clarke Publishing IT Conference. She has some documentation on the cost of different expense items. Mardia asks you to enter that information into a workbook and calculate the total cost of items in this expense category. Complete the following:

(1). **Create** a **new**, **blank workbook**, and then **save** it as **N Chapter 3 Equipment** in the location specified by your instructor.

(2). **Rename** the **Sheet1 worksheet** using **Documentation** as the new name.

(3). **Enter** the data shown in **Figure 3-41** in the specified cells.

Figure 3-41 Documentation sheet data

Cell	Text
A1	Clarke Publishing Communications & BPO, Inc.
A3	Author
A4	Date
A5	Purpose
B3	your name
B4	current date
B5	To estimate the cost of renting computers, audio/video equipment, Internet access, and hiring technical support at the Clarke Publishing Conference in Monrovia at the Unity Conference Center

CHAPTER REVIEW

Lab Project 1

CHAPTER REVIEW LABS

(4). **Set** the **font size** of the title text in cell **A1** to **24** points.

(5). **Set** the **width** of column **B** to **32** characters, and then **wrap** the contents of cell **B5** within the cell.

(6). **Add borders** around all of the cells in the range **A3:B5**.

(7). **Add** a **new worksheet** after the **Documentation** sheet, and then **rename** the worksheet using **Equipment Expenses** as the new name.

(8). In cell **A1**, **Enter Clarke Publishing Conference Equipment Expenses** as the worksheet title. **Set** the **font size** of the title text in cell **A1** to **20** points.

(9). **Enter** the data summarizing the conference in the specified cells as shown in **Figure 3-42**. Make sure that the postal code value is treated as text rather than a number.

Figure 3-42 Conference Summary data

Cell	Text
A4	Equipment Categories
A5	Total Expenses
A7	Vendor
A8	Street Address
A9	City
A10	State
A11	Postal Code
A12	Phone
B7	Conference Connections
B8	33 Parker Corner
B9	Brewerville
B10	Liberia
B11	1000
B12	231-888-123-456

CHAPTER REVIEW

Lab Project 1

CHAPTER REVIEW LABS

(10). **Enter** the column titles and expenses for various equipment that will be used at the conference in the range **A14:E20** as shown in **Figure 3-43**.

Expense Category	Subcategory	Description	Units	Cost per Unit
E2	5010	computer workstation rental	25	$105.00
E2	5020	audio/video rental	10	$85.00
E2	5030	screen projector rentals	10	$75.00
E2	5040	high-speed Internet access	1	$450.00
E2	5050	onsite wiring	1	$500.00
E2	5060	web hosting	1	$700.00

Figure 3-43 Equipment expenses

(11). In cell **F14**, **Enter Total** as the label. In the range F15:F20, calculate the total cost of each equipment item by entering formulas that return the value of the number of units ordered multiplied by the cost per unit.

(12). In cell **E22**, **Enter Subtotal** as the label, and then in cell **F22**, use the **SUM Function** to calculate the sum of the values in the range F15:F20.

(13). In cell **E13**, **Enter Tax Rate** as the label, and then in cell **F13**, **Enter 3%** as the value.

(14). In cell **E23**, **enter Est. Tax** as the label, and then in cell **F23**, calculate the estimated tax by **entering a formula that multiplies the subtotal value** in cell **F22** by the **tax rate** in cell **F13**.

CHAPTER REVIEW

Lab Project 1

CHAPTER REVIEW LABS

(15). In cell **E24**, **Enter TOTAL** as the label, and then in cell **F24**, use the **SUM Function** to calculate the sum of the subtotal value in cell F22 and the estimated tax in cell **F23**.

(16). **Insert new cells** in the range **A13:A24**, shifting the other cells to the right.

(17). In cell **A14**, **enter Account ID** as the label. In cell **A15**, **enter E2-5010** as the first ID. In cell **A16**, **enter E2-5020** as the second ID, and allow Flash Fill to enter the remaining IDs.

(18). **Add Borders** around all of the cells in the range **F13:G13,A14:G20,F22:G24**.

(19). **Set** the **Width** of columns **A** and **B** to **22** characters. **Set** the **Width** of columns **C, E, F,** and **G** to **13** characters. **Set** the **Width** of column **D** to **24** characters. **Set** the **Height** of row **13** to **30** points.

(20). **Wrap** the text in the range **D15:D20** so all of the content is visible.

(21). In cell **B4**, use the **COUNT Function** to count the number of numeric values in the range **E15:E20**. In cell **B5**, display the value of the total expenses that was calculated in cell **G24**.

(22). Mardia wants to keep the equipment budget under **$6,000**. If the total cost of the equipment is less than **$6,000**, **Enter within budget** in cell **B3**, otherwise **Enter over budget** in the cell.

(23). **Change** the page orientation of the **Equipment Expenses** worksheet to **landscape orientation**, and then scale the **width** and **height** of the **Equipment Expenses worksheet** to print on a **single page**.

(24). **Save** the workbook. If you are instructed to print, then print the entire workbook.

CHAPTER REVIEW

Lab Project 1

CHAPTER REVIEW LABS

(25). **Display the formulas** in the **Equipment Expenses worksheet**. If you are instructed to print, then print the worksheet. **Remove** the worksheet from **formula view**.

(26). **Save** the workbook, and then **close** it.

Lab Project 2

Self Challenge Lab

Data File needed for this Self Driven Problem: Chapter 5 Tracking Log.xlsx

Kaduna State Trucking Zubby Talaye is a dispatch manager at Kaduna State Trucking, a major trucking company based in Kaduna City, Nigeria. Zubby needs to develop a workbook that will summarise the driving log of Kaduna State Trucking drivers. Zubby has a month of travel data from one of the company's drivers and wants you to create a workbook that he can use to analyse this data. Complete the following:

(1). **Open** the **Chapter 3 Tracking Log** workbook located in the **Chapter 3 > Labs** folder included with your Data Files. Save the workbook as **Drivers Tracking** in the location specified by your instructor.

(2). **Change** the **name** of the **Sheet1** worksheet using **Travel Log** as the name, and then **move** it to the **end** of the workbook. **Rename** the **Sheet2** worksheet using **Documentation** as the name.

(3). In the **Documentation** sheet, **enter your name** in cell **B3** and the **current date** in cell **B4**.

(4). **Go to** the **Travel Log** worksheet. **Resize** the **columns** so that all the data is visible in the cells.

(5). **Between** the **Odometer Ending** column and the **Hours column**, **insert** a **new column**. **Enter Miles** as the column label in cell **I4**.

CHAPTER REVIEW

Lab Project 2

CHAPTER REVIEW LABS

(6). In the **Miles** column, in the range **I5:I25**, **enter formulas** to calculate the number of miles driven each day by subtracting the odometer beginning value from the odometer ending value.

(7). **Between** the **Price per Gallon** column and the **Seller** column, **insert** a **new column**. **Enter Total Fuel Purchase** as the label in cell **N4**.

(8). In the **Total Fuel Purchase** column, in the range **N5:N25**, enter formulas to calculate the total amount spent on fuel each day by multiplying the gallons value by the price per gallon value.

(9). In cell **B5**, use the **COUNT Function** to calculate the total number of driving days using the values in the **Date** column in the range **D5:D25**.

(10). In cell **B6**, use the **SUM Function** to calculate the total number of driving hours using the values in the **Hours** column in the range **J5:J25**.

(11). In cell **B7**, **enter** a **formula** that divides the hours of driving value (cell **B6**) by the days of driving value (cell **B5**) to calculate the average hours of driving per day.

(12). In cell **B9**, use the **SUM Function** to calculate the driver's total expenses for the month for fuel, tolls, and miscellaneous expenditures using the nonadjacent range **N5:N25,P5:Q25**.

(13). In cell **B10**, use the **SUM** function to calculate the total amount the driver spent on fuel using the range **N5:N25**. In cell **B11**, use the **SUM** function to calculate the total amount spent on tolls and miscellaneous expenses using the range **P5:Q25**.

(14). In cell **B13**, use the **SUM Function** to calculate the total gallons of gas the driver used during the month using the range **L5:L25**. In cell **B14**, use the

CHAPTER REVIEW

Lab Project 2

CHAPTER REVIEW LABS

SUM function to calculate the total number of miles driven that month using the range **I5:I25**.

(15). In cell **B15**, **Enter** a **formula** that **divides** the **total miles** (cell **B14**) by the total gallons (cell **B13**) to calculate the miles per gallon. In cell **B16**, **enter** a **formula** that divides the total expenses (cell **B9**) by the total miles driven (cell **B14**) to calculate the cost per mile.

(16). **Kaduna State Trucking** wants to keep the cost of driving and delivering goods to less than 55 US cents or its Naira equivalent per mile. If the cost per mile is greater than 55 US cents or its Naira equivalent, **enter over budget** in cell **B4**, otherwise **enter on budget** in cell **B4**.

(17). In cell **A1**, increase the font size of the text to **28 points**. In cell **D3** and cell **L3**, increase the font size of the Mileage Table and Expense Table labels to **18 points**.

(18). **Change** the worksheet to **landscape orientation**. Scale the printout so that the worksheet is scaled to **1 page wide** by **1 page tall**. If you are instructed to print, print the entire workbook.

(19). **Save** the workbook, and then **Close** it.

CHAPTER 4 HIGHLIGHT

Microsoft PowerPoint: Creating a Presentation

After completing this chapter, you will be able to:

1. Plan and create a new presentation
2. Create a title slide and slides with lists
3. Edit and format text
4. Move and copy text
5. Duplicate, rearrange, and delete slides
6. Change the theme and theme variant
7. Close a presentation
8. Open an existing presentation
9. Insert and crop photos
10. Resize and move objects
11. Modify photo compression options
12. Convert a list to a SmartArt diagram
13. Create speaker notes
14. Check the spelling
15. Run a slide show
16. Print slides, handouts, speaker notes, and the outline

CHAPTER 4 | Microsoft PowerPoint

AIICO Insurance Plc. is an insurance company headquartered at Plot PC 12, Churchgate Street, Victoria Island, Lagos, Nigeria. The company was founded in 1963 and has offices all over the Nigeria. Oladele Kanu (combination of two employees' name), a sales manager in the Lagos office, recently hired you as his assistant. Oladele frequently visits companies to try to convince them to offer insurance plans from AIICO Insurance Plc. to their employees. Many businesses have opened offices in the Abuja and Lagos areas over the past several years. Oladele wants to use a PowerPoint presentation when he visits these businesses. He asks you to prepare a presentation to which he will later add data and cost information.

Microsoft PowerPoint (or simply PowerPoint) is a complete presentation application or app that lets you produce professional-looking presentation files and then deliver them to an audience. In this chapter, you'll use PowerPoint to create a file that includes text, graphics, and speaker notes. Oladele can use the presentation as a starting point for his more comprehensive sales pitch. Before you give the presentation to Oladele, you'll check the spelling, run the slide show to evaluate it, and print the file.

Planning a Presentation

A **Presentation** is a talk, formal lecture, or prepared file in which the person speaking or the person who prepared the file wants to communicate with an audience to explain new concepts or ideas, sell a product or service, entertain, or train the audience in a new skill or technique, or any of a wide variety of other topics.

Most people find it helpful to use **Presentation Media**—visual and audio aids that support key points and engage the audience's attention. PowerPoint is one of the most commonly used tools for creating effective presentation media. The features of PowerPoint make it easy to incorporate text with photos, drawings, music, and video to illustrate key points of a presentation.

SmartSkills **Verbal Communication: Planning a Presentation**

Answering a few key questions will help you create a presentation using appropriate presentation media that successfully delivers its message or motivates the audience to take an action.

- What is the purpose of your presentation? Consider the action or response you want your

SMARTSKILLS Verbal Communication: Planning a Presentation

audience to have. Do you want them to buy something, follow instructions, or make a decision?

- Who is your audience? Think about the needs and interests of your audience as well as any decisions they will make because of what you have to say. What you choose to say to your audience must be relevant to their needs, interests, and decisions.
- How will you ensure members of your audience with visual or hearing impairments will be able to experience your presentation?
- What are the main points of your presentation? Identify the information your audience will find most relevant.
- What presentation media will help your audience absorb the information and remember it later? Do you need lists, photos, charts, and/or tables?
- What is the format for your presentation? Will you deliver the presentation orally or will you create a presentation file for people to view on their own?
- How much time do you have for the presentation? Keep that in mind as you prepare the presentation content so that you have enough time to present all of your key points. Practicing your presentation out loud will help you determine the timing.
- Consider whether distributing handouts will help your audience follow along with your presentation or steal their attention when you want them to be focused on you during the presentation.

Before you create a presentation, you should spend some time planning its content. The purpose of Oladele's presentation is to convince businesses to offer AIICO Insurance Plc. plans to their employees. His audience will be members of Human Resource departments or Boards of Directors. Oladele will use PowerPoint to display lists and graphics to help make his message clear. He plans to deliver his presentation orally to small groups of people in conference rooms at each business, and his presentation will be about 10 minutes long. He will not distribute anything before speaking because he wants the audience's full attention to be on him at the beginning of his presentation. He plans to distribute informational handouts with specific details about the insurance plans available at an appropriate point during the presentation. After the presentation is over and he has answered all of his audience's questions, he will distribute business cards with his contact information.

Once you know what you want to say, you can prepare the presentation media to help communicate your ideas.

STARTING POWERPOINT AND CREATING A NEW PRESENTATION

PowerPoint is a tool you can use to create and display visual and audio aids on slides to help clarify the points you want to make in your presentation. You also can use PowerPoint to create a presentation that people view on their own without you.

When PowerPoint starts, Backstage view appears, showing the Home screen. Backstage view is the view that contains commands that allow you to manage the file and program settings. When you first start PowerPoint, the actions available to you in Backstage view are to create a new PowerPoint file, open an existing PowerPoint file, view your Account settings, submit feedback to Microsoft, and open the PowerPoint Options dialog box to change app settings.

Now, get yourself ready to start PowerPoint and create Oladele's presentation.

To start PowerPoint:

(1). On the Windows taskbar, **Click** the **Start** button ⊞. The Start menu opens.

(1). On the **Start Menu**, scroll the list of applications on the left, and then **Click** **PowerPoint**. PowerPoint starts and displays the Home screen in Backstage view. Options for creating new presentations appear in a row at the top of the screen, and if you have recently viewed PowerPoint files, they appear below this row. See **Figure 4-1** on the next page.

In Figure 4-1 displays Microsoft PowerPoint window with the Home screen in Backstage view. The Home, New, and Open buttons are on the left pane, and the Home button is selected. The Recent list might show a list of recently opened presentations. A list of templates displays in the window, and includes an option for creating a new, blank presentation. The template choices might differ.

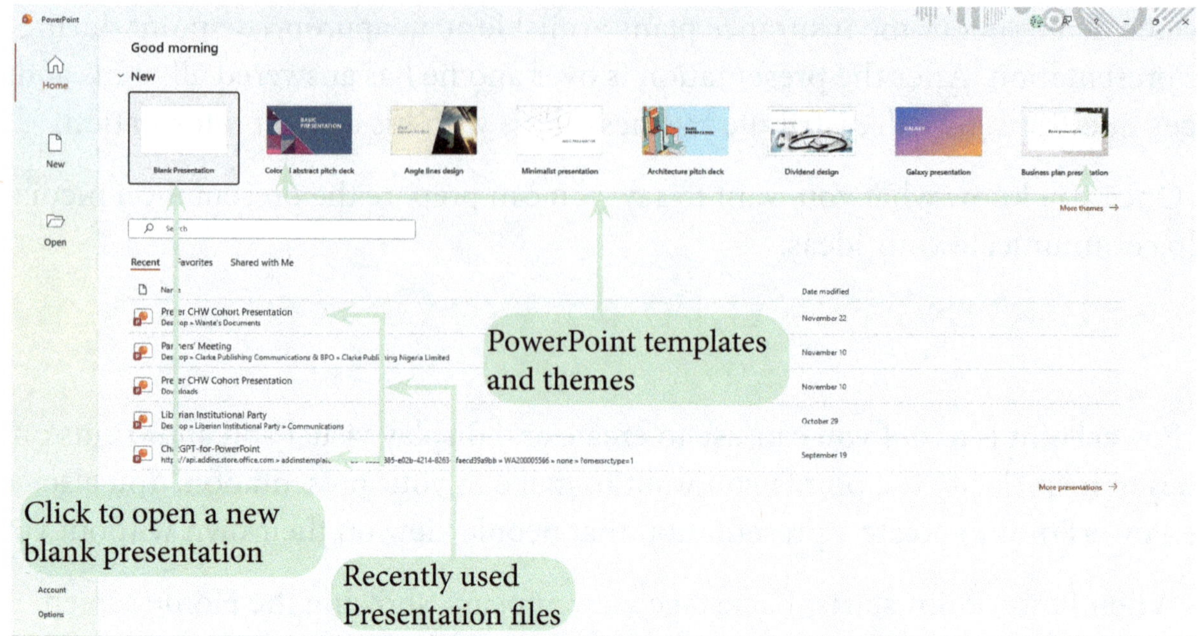

Figure 4-1 Home screen in Backstage view

(3). **Click Blank Presentation.** Backstage view closes and a new presentation window appears. The temporary filename "Presentation1" appears in the title bar. There is only one slide in the new presentation—Slide 1.

To create a new blank presentation when PowerPoint is already running, click the File tab on the ribbon, click New in the navigation pane, and then click Blank Presentation.

Are You Having Trouble? If you do not see the area on the ribbon that contains buttons and you see only the ribbon tab names, click the Home tab to expand the ribbon and display the commands, and then in the bottom-right corner of the ribbon, click the Pin the ribbon button .

Are You Having Trouble? If the window does not appear maximized, click the Maximize button in the upper-right corner.

Because you just started PowerPoint, you clicked Blank presentation on the Home screen. If PowerPoint was already running and you wanted to create a new, blank presentation, you would click the File tab, click New in the navigation pane, and then click Blank Presentation on the New screen in Backstage view.

SMARTSKILLS Using QuickStarter

QuickStarter is a feature in PowerPoint that creates slide titles based on a topic you enter. To use QuickStarter when you first start PowerPoint, click QuickStarter on the Home screen. If you already are using PowerPoint, click the File tab, click New, and then click QuickStarter on the New screen. The Search here to get started window opens. (The first time you use this feature, the Welcome to PowerPoint QuickStarter window appears. Click Get started to open the Search here to get started window. Also, if the Intelligent Services for Your Work window opens, click Turn On to start using the feature.) Type a topic in the Search box, and then click Search. Suggested presentation ideas appear in the window. Click the one you want to use to display starter slides in the window. If you do not want to include one of the starter slides, click it to deselect it. Click Next to open the Pick a look window, in which you select a theme. Finally, click Create to generate a presentation containing the starter slides. Some presentations created using QuickStarter will also include slides with additional information based on your search topic or provide a list of suggested related topics that you can use to search for more information.

When you create a new presentation, it appears in **Normal View**. Normal view is the view in which the selected slide appears enlarged so you can add and manipulate objects on the slide, and thumbnails of all the slides in the presentation appear in the pane on the left. A **Thumbnail** is a reduced-size version of a larger graphic image. In this case, each thumbnail represents a slide in the presentation. The Home tab on the ribbon is selected when you first open or create a presentation.

WORKING IN TOUCH MODE

In Office 2024, you can work with a mouse or, if you have a touch screen, you can work in Touch Mode. In **Touch Mode**, the ribbon increases in height, the buttons are larger, and more space appears around buttons so you can more easily use your finger or stylus to tap screen elements. Also, in the placeholders on the slide, "Double tap" replaces the instruction telling you to "Click." Note that the figures in this text show the screen with Mouse Mode on. You'll switch to Touch Mode and then back to Mouse Mode now.

To switch between Touch Mode and Mouse Mode:

(1). On the Quick Access Toolbar, **Click** the **Customize Quick Access Toolbar** button . A menu opens. The Touch/Mouse Mode command near the bottom of the menu does not have a checkmark next to it.

Are You Having Trouble? If the Touch/Mouse Mode command has a checkmark next to it, press the Esc key to close the menu, and then skip Step 2.

(2). On the menu, **Click Touch/Mouse Mode**. The menu closes, and the Touch/Mouse Mode button appears on the Quick Access Toolbar.

(3). On the Quick Access Toolbar, **Click** the **Touch/Mouse Mode** button . A menu opens listing Mouse and Touch. The icon next to Mouse is shaded to indicate it is selected.

Are You Having Trouble? If the icon next to Touch is shaded, press ESC to close the menu and skip **Step 4**.

(4). On the menu, **Click Touch**. The menu closes, and the ribbon increases in height so that there is more space around each button on the ribbon. Notice that the instructions in the placeholders on the slide changed by replacing the instruction to "**Click**" with the instruction to "Double tap." See **Figure 4-2**. Now you'll change back to Mouse Mode.

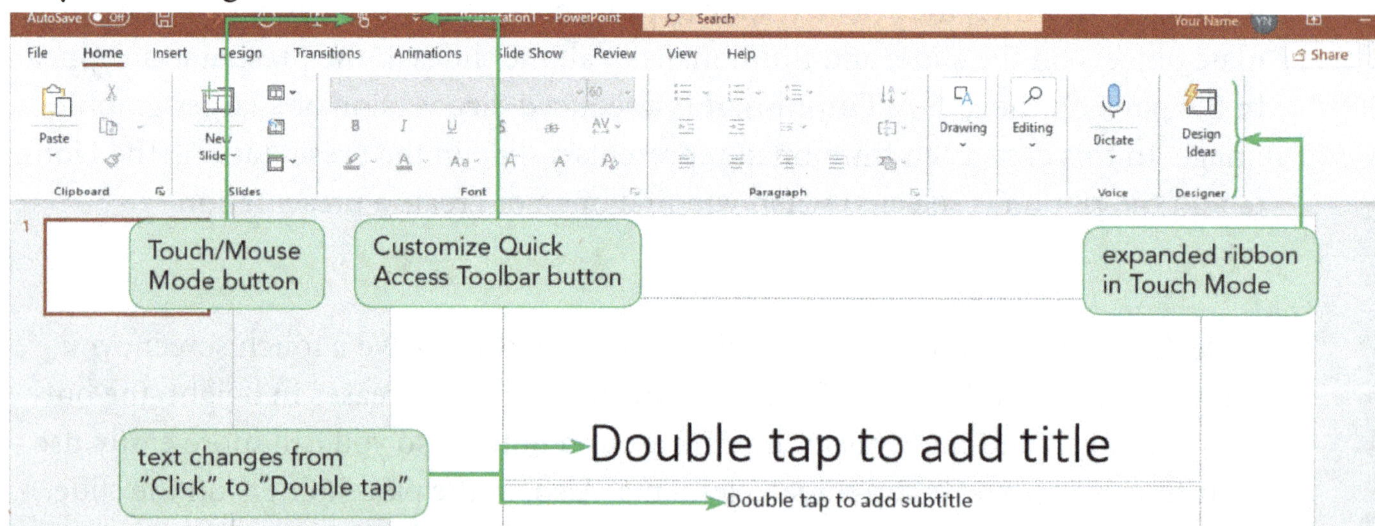

Figure 4-2 PowerPoint window with Touch Mode active

Figure 4-2 shows the PowerPoint window with a blank presentation. The Touch/Mouse Mode button was clicked and Touch selected, displaying the PowerPoint window in Touch Mode. The placeholder instructions have changed from "Click" to "Double tap." The ribbon has expanded vertically.

Are You Having Trouble? If you are working with a touch screen and want to use Touch Mode, skip Steps 5 and 6.

(5). **Click** the **Touch/Mouse Mode** button, and then **click Mouse**. The ribbon and the instructions change back to Mouse Mode defaults.

(6). **Click** the **Customize Quick Access Toolbar** button, and then **Click Touch/Mouse Mode** to deselect this option and remove the checkmark. The Touch/Mouse Mode button disappears from the Quick Access Toolbar.

CREATING A TITLE SLIDE

The **Title Slide** is the first slide in a presentation. It usually contains the presentation title and other identifying information, such as a company name or logo, a company's slogan, or the presenter's name. The **Font**—a set of letters, numbers, and symbols that all have the same style and appearance—used in the title and subtitle may be the same or may be different fonts that complement each other.

The title slide contains two objects called text placeholders. A **Text Placeholder** is a placeholder designed to contain text and that contains a prompt that instructs you to click to add text and might describe the purpose of the placeholder. The large placeholder on the title slide is for the presentation title. The small placeholder is for a subtitle. Once you enter text into a text placeholder, the instructional text disappears and it becomes an object called a text box. A **Text Box** is an object that contains text.

When you click in the placeholder, the insertion point appears. The **Insertion Point** is a blinking vertical line that indicates where new text will be inserted. Also, a new tab, the Shape Format tab, appears on the ribbon. This tab is a contextual tab. A **Contextual Tab** appears only in context—that is, when a particular type of object is selected or is active—and contains commands for modifying that object.

You'll add a title and subtitle for Oladele's presentation now. Oladele wants the title slide to contain the company name and slogan.

To add the company name and slogan to the title slide:

(1). On **Slide 1**, move the pointer to position it in the title text placeholder (where it says "**Click to add title**") so that the pointer changes to the **I-beam**

Pointer I , and then click. The insertion point replaces the placeholder text, and the Shape Format contextual tab appears as the rightmost tab on the ribbon. Note that in the Font group on the Home tab, the Font box identifies the title font as Calibri Light (Headings). See **Figure 4-3**.

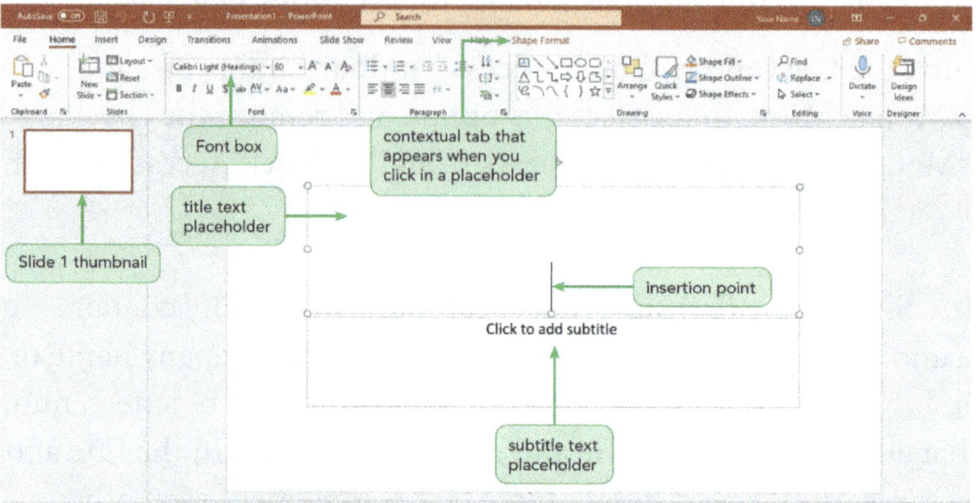

Figure 4-3 Title text placeholder after clicking in it

(2). **Type AIICO Insurance Plc**. in the placeholder. The placeholder is now a text box.

(3). **Click** a blank area of the slide. The border of the text box disappears, and the Shape Format tab no longer appears on the ribbon.

(4). **Click** in the subtitle text placeholder (where it says "Click to add subtitle"), 63and then **Type The Best in Health Care Since 1963** in the placeholder. Notice in the Font group that the subtitle font is Calibri (Body), a font that works well with the Calibri Light font used in the title text.

(5). **Click** a blank area of the slide.

SAVING AND EDITING A PRESENTATION

Once you have created a presentation, you should name and save the presentation file. You can save the file on a hard drive or a network drive, on an external drive such as a USB drive, or to your account on OneDrive, Microsoft's free online storage area if you have one setup already.

To save the presentation for the first time:

(1). On the **Quick Access Toolbar**, point to the **Save** button . A box called a ScreenTip appears. A ScreenTip is a label that appears when you point to a button or object, which may include the name, purpose, or keyboard shortcut for the object, and may include a link to associated help topics.

(2). **Click** the **Save** button . The Save this file dialog box opens, which you use to save a file on OneDrive. To save a new presentation on your hard drive or an external drive, you use the Save As screen.

(3). **Click** the **More Options** link. The Save As screen in Backstage view appears. See **Figure 4-4**. The navigation pane is the pane on the left that contains commands for working with the file and program options. Recently used folders on the selected drive appear in a list on the right.

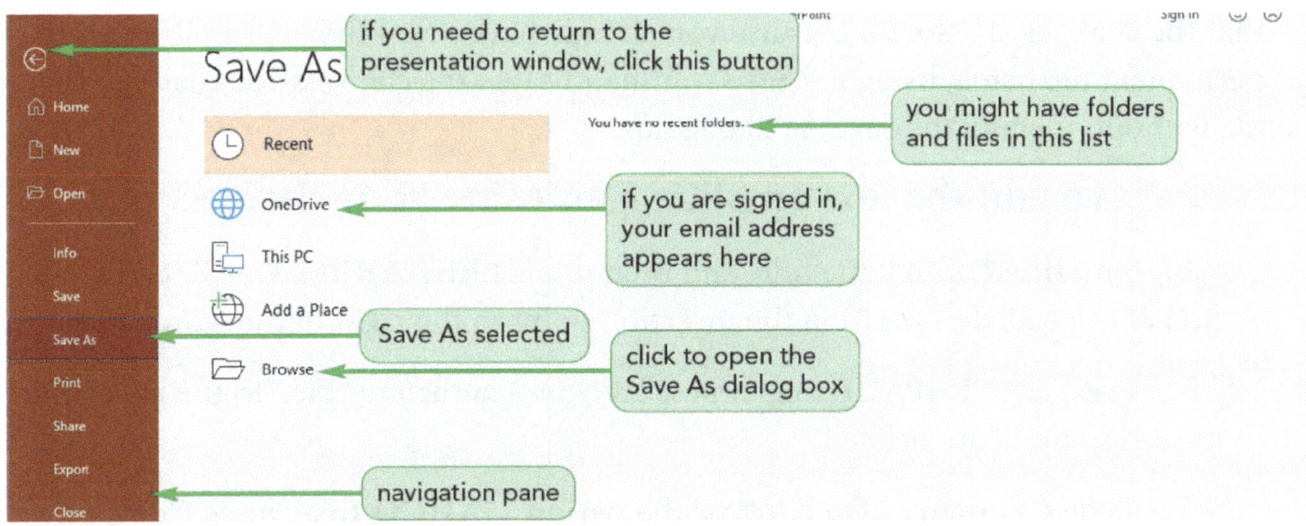

Figure 4-4 Save As screen in Backstage view

(4). **Click Browse**. The **Save As dialog box** opens with folders options.

(5). **Navigate** to the drive and folder where you are storing your **Data Files**, and then **Click** in the **File name box.** The suggested file name, **AIICO Insurance Plc.**, is selected.

(6). **Type Chapter 4 New Business** to replace the selected text in the File name box.

(7). **Click Save**. The file is saved, the dialog box and Backstage view close, and the presentation window appears again with the new file name in the title bar.

Once you have created a presentation, you can make changes to it. For example, if you need to change text in a text box, you can edit it easily. The Backspace key removes characters to the left of the insertion point, and the Delete key removes characters to the right of the insertion point.

If you mistype or misspell a word, you might not need to correct it because the **AutoCorrect Feature** automatically detects and corrects commonly mistyped and misspelled words. For instance, if you type "cna" and then press SPACEBAR, PowerPoint corrects the word to "can." If you want AutoCorrect to stop making a particular change, you can display the AutoCorrect Options menu, and then click Stop Automatically Correcting. (The exact wording will differ depending on the change made.)

After you make changes to a presentation, you will need to save the file again so that the changes are stored. Because you already have saved the presentation with a permanent file name, using the Save command saves the changes you made to the file without opening the Save As dialog box.

To edit the text on Slide 1 and save your changes:

(1). On **Slide 1**, **Click** the title, and then **press LEFT ARROW** or **RIGHT ARROW** as needed to position the insertion point to the right of the word "Plc."

(2). **Press BACKSPACE** four times. The four characters "Plc." to the left of the insertion point are deleted.

(3). **Type Company**. (Do not type the period.) "**AIICO Insurance Company**" now appears as the title.

(4). In the subtitle text box, **Click** to the left of the word "**Best**" to position the insertion point in front of that word, **Type Teh**, and then **press SPACEBAR**. "**Teh**" is corrected to "**The**" after you press SPACEBAR. "**The Best in Health Care Since 1963**" now appears as the subtitle.

(5). **Move** the pointer over the word "**The**." A small, faint rectangle appears below the first letter of the word. This rectangle indicates that an autocorrection was made.

Are You Having Trouble? If you can't see the rectangle, point to the letter "T," and then slowly move the pointer down until it is on top of the rectangle.

(6). **Move** the **pointer** on top of the rectangle so that it changes to the AutoCorrect Options button , and then **click** the **AutoCorrect Options** button . A menu opens, as shown in **Figure 4-5**. You can change the word back to what you originally typed, instruct PowerPoint to stop making this type of correction in this file, or open the AutoCorrect dialog box.

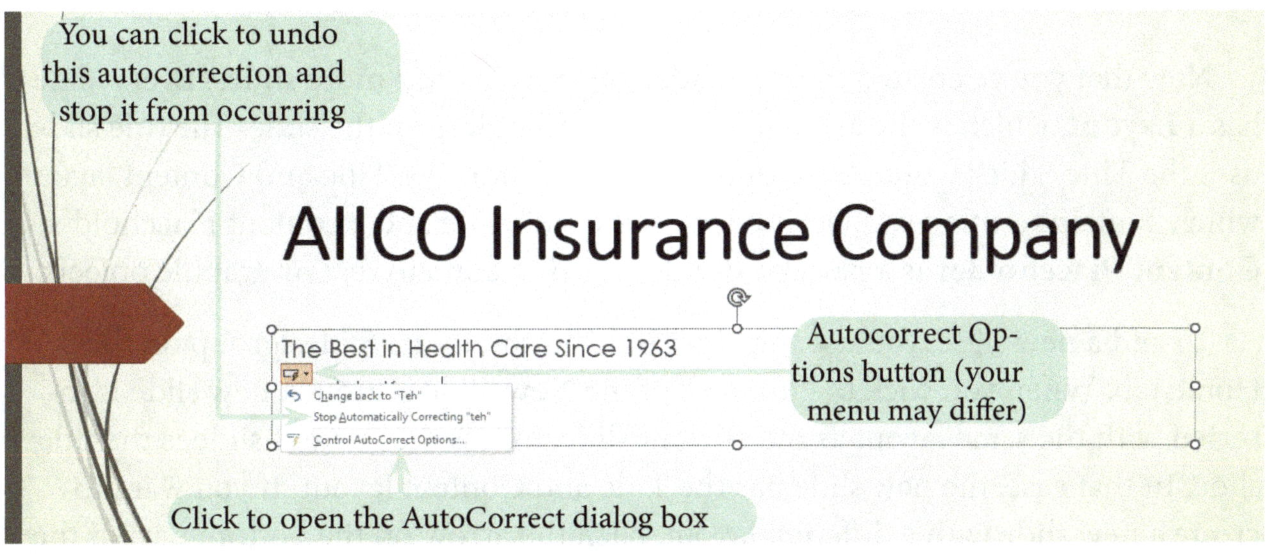

Figure 4-5 AutoCorrect Options button menu

(7). **Click Control AutoCorrect Options**. The AutoCorrect dialog box opens with the AutoCorrect tab selected. See **Figure 4-6**.

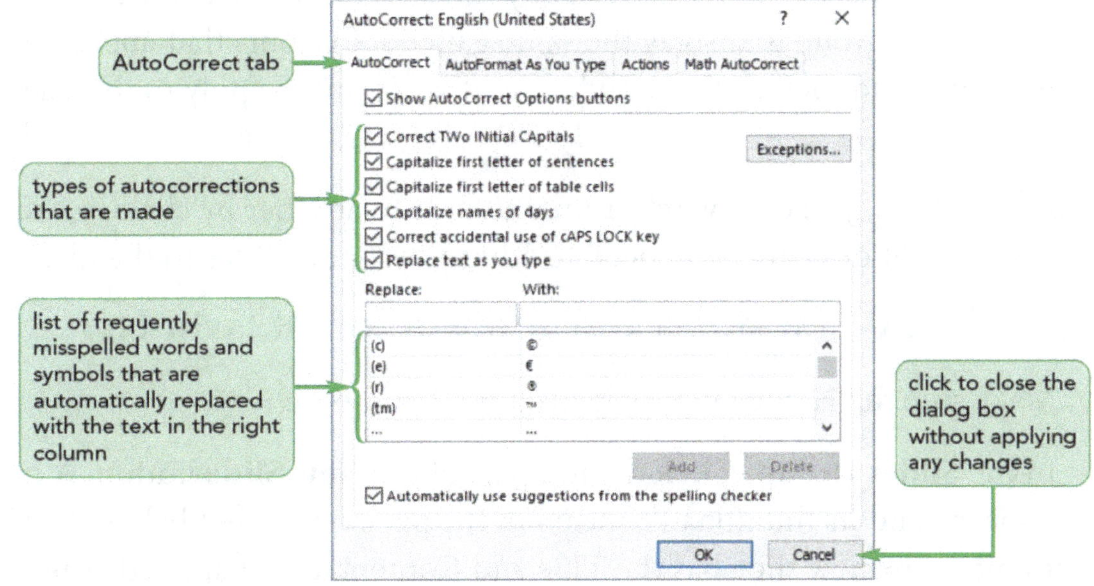

Figure 4-6 AutoCorrect tab in the AutoCorrect dialog box

(8). Examine the types of changes the AutoCorrect feature makes, and then **Click Cancel**.

(9). On the Quick Access Toolbar, **Click** the **Save** button . The saved file now includes the new changes you made.

Are You Having Trouble? If AutoSave is enabled, skip **Step 9**.

ADDING NEW SLIDES

Now that you've created the title slide, you need to add more slides. Every slide has a **Layout**, which is the arrangement of placeholders on the slide. The title slide uses the Title Slide layout. A commonly used layout is the Title and Content layout, which contains a title text placeholder for the slide title and a content placeholder. A **Content Placeholder** is a placeholder designed to contain text or graphic objects.

To add a new slide, you use the New Slide button in the Slides group on the Home tab. When you click the top part of the New Slide button, a new slide is inserted with the same layout as the current slide, unless the current slide is the title slide. In that case, the new slide has the Title and Content layout. If you want to create a new slide with a different layout, click the arrow on the bottom part of the New Slide button to open a gallery of layouts, and then click the layout you want to use.

You can change the layout of a slide at any time. To do this, click the Layout button in the Slides group to display the same gallery of layouts that appears in the New Slide gallery, and then click the slide layout you want to apply to the selected slide.

As you add slides, you can switch from one slide to another by clicking the slide thumbnails in the Slides pane. You need to add several new slides to the file.

To add new slides and apply different layouts:

(1). Make sure the Home tab is displayed on the ribbon.

(2). In the **Slides Group**, **Click** the top part of the **New Slide** button. A new slide appears, and its thumbnail appears in the pane on the left below the Slide 1 thumbnail. The new slide has the Title and Content layout applied. This layout contains a title text placeholder and a content placeholder. An orange border appears around the new Slide 2 thumbnail, indicating that it is the current slide.

(3). In the **Slides Group**, **Click** the **New Slid**e button again. A new Slide 3 is added. Because Slide 2 had the Title and Content layout applied, Slide 3 also has that layout applied.

(4). In the **Slides Group**, **Click** the **New Slide arrow** (the bottom part of the New Slide button). A gallery of the available layouts appears. See **Figure 4-7**.

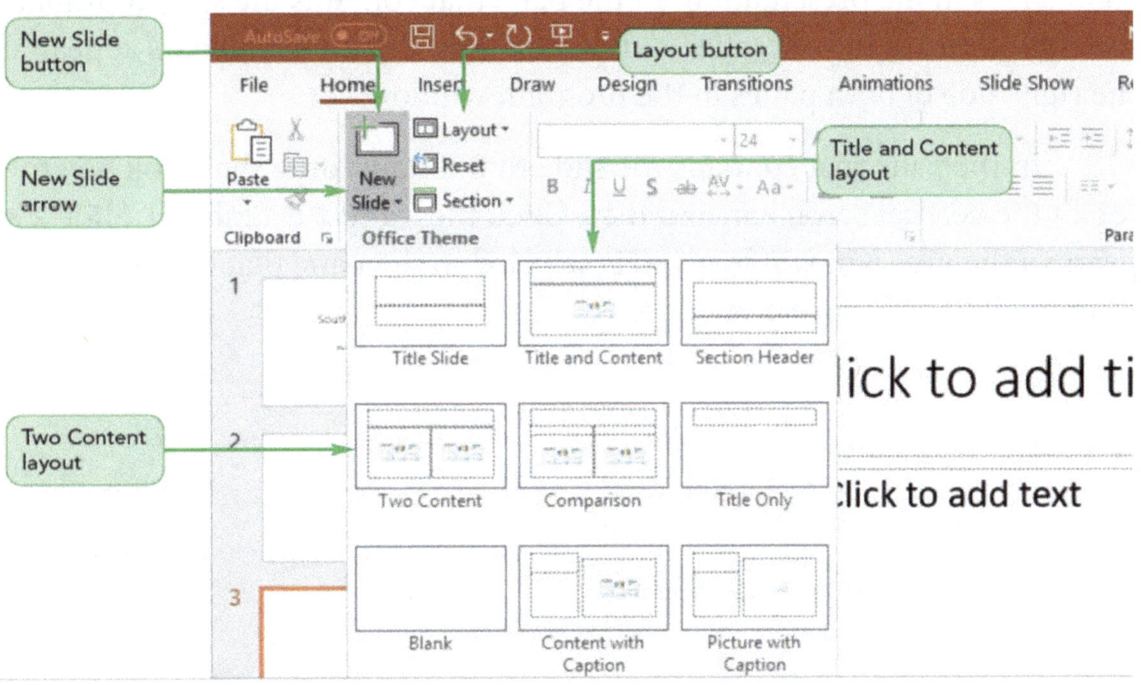

Figure 4-7 Gallery of layouts on the New Slide menu

(5). In the gallery, **Click** the **Two Content Layout**. The gallery closes, and a new Slide 4 is inserted with the Two Content layout applied. This layout includes three objects: a title text placeholder and two content placeholders.

(6). In the **Slides Group**, **Click** the **New Slide** button **Twice**. New Slides 5 and 6 are added to the presentation. Because Slide 4 had the Two Content layout applied, that layout is also applied to the new slides. You need to change the layout of **Slide 6**.

(7). In the **Slides Group**, **Click** the **Layout** button. The same gallery of layouts that appeared when you clicked the New Slide arrow appears. The shading behind the Two Content layout indicates that it is applied to the current slide.

(8). **Click** the **Title** and **Content layout**. The layout of **Slide 6** changes to Title and Content.

Are You Having Trouble? If the Design Ideas pane opens, Click its Close button ☒

(9). In the **Slides Group**, **Click** the **New Slide** button to add **Slide 7** with the Title and Content layout.

(10). Add one more new slide with the Two Content layout. There are now eight slides in the presentation. In the pane that contains the slide thumbnails, some thumbnails have scrolled out of view, and vertical scroll bars appear along the right side of both panes in the program window.

(11). In the pane that contains the slide thumbnails, drag the scroll box to the top of the vertical scroll bar, and then **Click** the **Slide 2** thumbnail. Slide 2 appears in the program window and is selected in the pane that contains the slide thumbnails. See **Figure 4-8**.

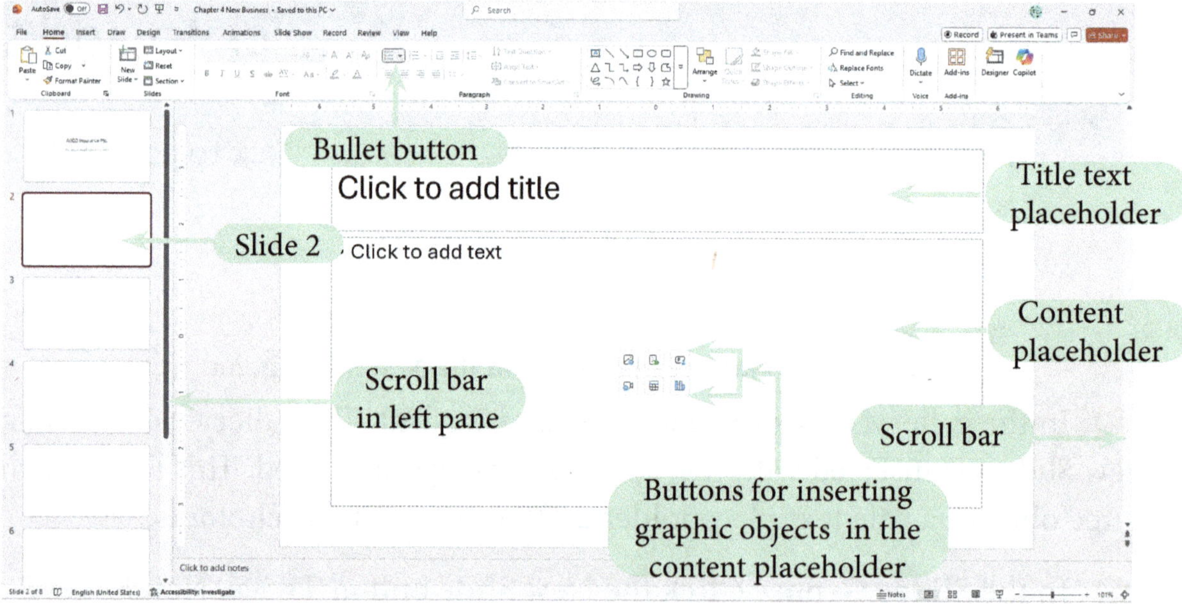

Figure 4-8 Slide 2 with the Title and Content layout

(12). On the Quick Access Toolbar, **Click** the **Save** button . The changes you made are saved in the file.

If you accidentally close a presentation without saving changes and need to recover it, click the File tab, click Open in the navigation bar, and then click the Recover Unsaved Presentations button.

CREATING LISTS

You can use a list to help explain a topic or concept. If you are preparing an oral presentation (one that you give in front of an audience), lists on your slides should enhance the oral presentation, not replace it. If you are preparing a self-running presentation (one that others will view on their own), list items might need to be longer and more descriptive.

Each item in a list is a paragraph. Items in a list can appear at different levels. A first-level item is a main item in a list. A second-level item is an item beneath and indented from a first-level item. A third-level item is an item beneath and indented from a second-level item, and so on. All items below the first level are subitems. A **Subitem** is any item in a list that is beneath and indented from a higher-level item.

Usually, the size of the text in subitems on a slide is smaller than the size of the text in the level above. Text is measured in points. A **Point** is the unit of measurement used for text size. One point is equal to 1/72 of an inch. Text in a book typically is printed in 10- or 12-point type. We wrote this point using 14-poin type using Minion Pro Style to accommodate students with visual problems. Text on a slide in a presentation that will be shown to an audience needs to be much larger so that the audience can easily read it.

CREATING A BULLETED LIST

A **Bulleted List** is a series of paragraphs, each beginning with a bullet character, such as a dot or checkmark. Subitems in a list often begin with a different or smaller bullet symbol. Use bulleted lists when the order of the items is not important.

You need to create a bulleted list that describes the types of insurance plans that AIICO Insurance Plc. offers and one that highlights why it would be the best insurance company for businesses to create a relationship with.

To create a bulleted list on Slides 2 and 3:

(1). On **Slide 2**, **Click** in the title text placeholder (with the placeholder text “Click to add title”), and then **Type Types of Plans**. (*Do not type the period.*)

(2). In the content placeholder, **Click** any area where the pointer is the I-beam pointer (anywhere except on one of the buttons in the center of the placeholder). The placeholder text "Click to add text" disappears, the insertion point appears, and a light gray bullet symbol appears.

(3). **Type Life** in the placeholder. As soon as you type the first character, the icons in the center of the content placeholder disappear, the bullet symbol darkens, and the content placeholder changes to a text box. On the Home tab, in the Paragraph group, the Bullets button is shaded to indicate that it is selected.

(4). **Press ENTER**. The insertion point moves to a new line, and a light gray bullet appears on the new line.

(5). **Type Health**, and then **Press ENTER**. The bulleted list now consists of two first-level items, and the insertion point is next to a light gray bullet on the third line in the text box. On the Home tab, in the Font group, the point size in the Font Size box is 28 points.

(6). **Press TAB**. The bullet symbol and the insertion point indent one-half inch to the right, the bullet symbol changes to a smaller size, and the number in the Font Size box changes to 24. See **Figure 4-9**.

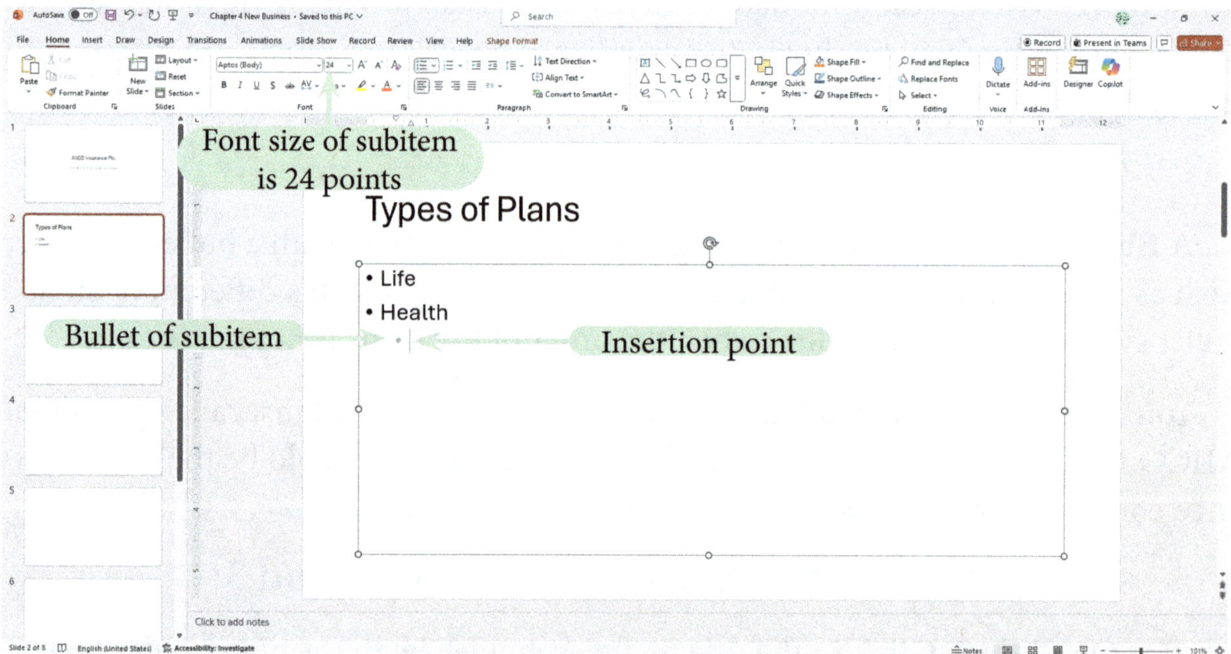

Figure 4-9 Subitem created on Slide 2

(7). **Type HMO** and then **Press ENTER**.

(8). **Type PPO, press ENTER, type POS**, and then **press ENTER**. A fourth subitem is created. You will change it to a first-level item using a key combination. In this lab textbook, when you need to press two keys at the same time, the keys will be separated by a plus sign.

(9). **Press SHIFT+TAB**. The bullet symbol and the insertion point shift back to the left margin of the text box, the bullet symbol changes back to the larger size, and 28 again appears in the Font Size box because this line is now a first-level bulleted item. But you really don't have to press both keys at the same time.

(10). **Type Disability**, and then **Press ENTER**. A fourth first-level item is created. You need to enter subitems for the "Disability" first-level item.

(11). On the **Home Tab**, in the Paragraph group, **Click** the **Increase List Level** button . Clicking the Increase List Level button is an alternative to pressing TAB to create a subitem.

(12). **Type Long-term, Press ENTER, Type Short-term**, and then **Press ENTER**. A third second-level item is created. You need to create a fourth first-level item.

(13). In the Paragraph Group, **Click** the **Decrease List Level** button . **Clicking** the **Decrease List Level** button is an alternative to pressing **SHIFT+TAB** to change a lower-level item to a higher-level item.

(14). **Type Stable**. The list now contains four first-level items.

If you add more text than will fit in the content placeholder, **AutoFit** adjusts the font size and line spacing to make the text fit. When AutoFit is active, the AutoFit Options button appears below the text box. You can click this button and then select from among several options, including turning off AutoFit for this text box and splitting the text between two slides. Although AutoFit can be helpful, be aware that it also enables you to crowd text on a slide, making the slide more difficult to read.

CREATING A NUMBERED LIST

A **Numbered List** is a group of paragraphs in which each one is preceded by a number, with the paragraphs numbered consecutively. The numbers can be followed by a separator character, such as a period or parenthesis. Generally, you use a numbered list when the order of the items is important. For example, you would use a numbered list if you are presenting a list of step-by-step instructions that need to be followed in sequence to complete a task successfully.

You will create a numbered list on Slide 5 to explain why AIICO Insurance Plc. is a good choice for businesses to use.

To create a numbered list on Slide 5:

(1). In the pane containing the thumbnails, **Click** the **Slide 5** thumbnail to display Slide 5, and then **Type Choose AIICO Insurance Plc.** in the title text placeholder.

(2). In the left content placeholder, **Click** the **placeholder text**.

(3). On the **Home Tab**, in the Paragraph group, **Click** the **Numbering** button . The Numbering button is selected, the Bullets button is deselected, and in the content placeholder, the number 1 followed by a period replaces the bullet symbol.

If a menu containing a gallery of numbering styles appears, you clicked the Numbering arrow on the right side of the button. Click the Numbering arrow again to close the menu, and then click the left part of the Numbering button.

(4). **Type Reliable**, and then **Press ENTER**. As soon as you start typing, the number 1 darkens to black. After you press ENTER, the insertion point moves to the next line, next to the light gray number 2.

(5). **Type Customer-focused**, and then **Press ENTER**. The number 3 appears on the next line.

(6). In the Paragraph group, **Click** the **Increase List Level** button . The third line is an indented subitem under the second item, and the number 3 changes to a number 1 in a smaller font size than the first-level items.

(7). **Type Dedicated customer service team**, **Press ENTER**, type **24/7 support**, and then **press ENTER**.

(8). In the Paragraph group, **Click** the **Decrease List Level** button . The fifth line becomes a first-level item, and the number 3 appears next to it.

(9). **Type Dependable.** The list now consists of three first-level numbered items and two subitems under number 2.

(10). In the second item, **Click** before the word "**Customer**," and then **press ENTER**. A blank line is inserted above the second item.

(11). **Press UP ARROW**. A light-gray number 2 appears in the blank line. The item on the third line in the list is still numbered 2.

(12). **Type Trustworthy**. As soon as you start typing, the new number 2 darkens in the second line, and the number of the third item in the list changes to 3. Compare your screen to **Figure 4-10**.

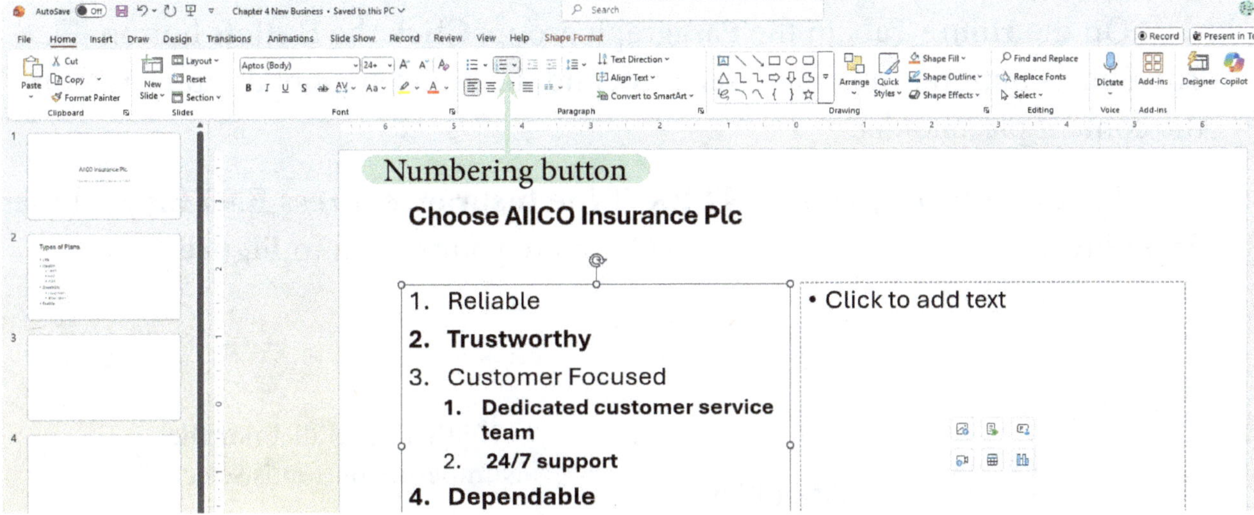

Figure 4-10 Numbered list on Slide 5

CREATING AN UNNUMBERED LIST

An **Unnumbered List** is a list that does not have bullets or numbers preceding each item. Unnumbered lists are useful when you want to present information on multiple lines, but you do not want to start each item with a bullet or number.

Each item in lists is a paragraph. When you press ENTER to create a new item, you create a new paragraph with a little bit of extra space between the new item and the previous item. Sometimes, you don't want to create a new item, or you do not want extra space between lines. In that case, you can create a new line without creating a new paragraph by pressing SHIFT+ENTER. When you do this, the insertion point moves to the next line, but there is no extra space above it. If you do this in a bulleted or numbered list, the new line will not have a bullet or number next to it because it is not a new item.

You need to create a slide that highlights the company's name. Also, Oladele asks you to create a slide containing contact information

To create unnumbered lists on Slides 4 and 7:

(1). In the pane containing the thumbnails, **Click** the **Slide 4** thumbnail to display Slide 4. Slide 4 has the Two Content layout applied.

(2). **Type About Us** in the title text placeholder, and then in the left content placeholder, **Click** the **placeholder text**.

(3). On the **Home Tab**, in the Paragraph group, **Click** the **Bullets** button . The Bullets button is no longer selected, and the bullet symbol disappears from the content placeholder.

(4). **Type Southwest**, **Tress ENTER**, **Type Insurance**, **Press ENTER**, and then **Type Plc**. (*Do not type the period.*) Compare your screen to **Figure 4-11**.

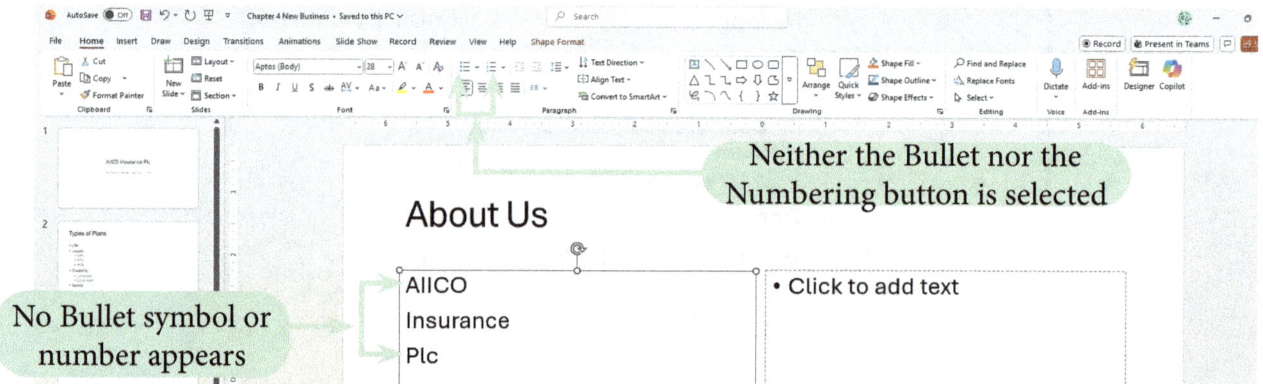

Figure 4-11 Unnumbered list on Slide 4

(5). **Switch** to **Slide 7**, **Type For More Information** in the title text placeholder, and then in the content placeholder, click the placeholder text.

(6). In the Paragraph group, **Click** the **Bullets** button to remove the bullets, **Type AIICO Insurance Plc**, and then **Press ENTER**. A new line is created, but there is extra space above the insertion point. You want the address information to appear on multiple lines, but without the extra spacing between each line.

(7). **Press BACKSPACE** to delete the new line and move the insertion point back to the end of the first line, and then **Press SHIFT+ENTER**. The insertion point moves to the next line. There is no extra space above the line, and the insertion point is aligned below the first character in the first line.

(8). **Type Plot PC 12, Churchgate Street**, **Press SHIFT+ENTER**, and then **Type Victoria Island, Lagos, Nigeria.** (*Do not type the period.*) You need to insert the phone number on the next line, Oladeles email address on the line after that, and the website address on the last line. The extra space above these lines will set this information apart from the address and make it easier to read.

(9). **Press ENTER** to create a new line with extra space above it, **type (234) 700-2442-6682**, **Press ENTER**, **type aiicontact@aiicoplc.com**. (*Do not type the period.*)

Are You Having Trouble? If the first character in the email address changed to an uppercase letter "A," move the pointer on top of the "A" so that the AutoCorrect rectangle appears, move the pointer on top of the AutoCorrect rectangle so that the AutoCorrect Options button appears, click the AutoCorrect Options button, and then click Undo Automatic Capitalization.

(10). **Press ENTER**. The insertion point moves to a new line with extra space above it, and the email address you typed changes color to blue and is underlined.

When you type text that PowerPoint recognizes as an email or website address and then press SPACEBAR or ENTER, the text is automatically formatted as a link that can be clicked during a slide show. Formatted links generally appear in a different color and are underlined.

(11). **Type https://www.aiicoplc.com** , and then press SPACEBAR. The text is formatted as a link. Oladele plans to click this link during his presentation to show the audience the website, so he wants it to stay formatted as a link. However, there is no need to have the email address formatted as a link.

(12). **Right-Click aiicontact@aiicoplc.com**. A shortcut menu opens.

(13). On the shortcut menu, **Click Remove Link**. The email address is no longer formatted as a link. Compare your screen to **Figure 4-12**.

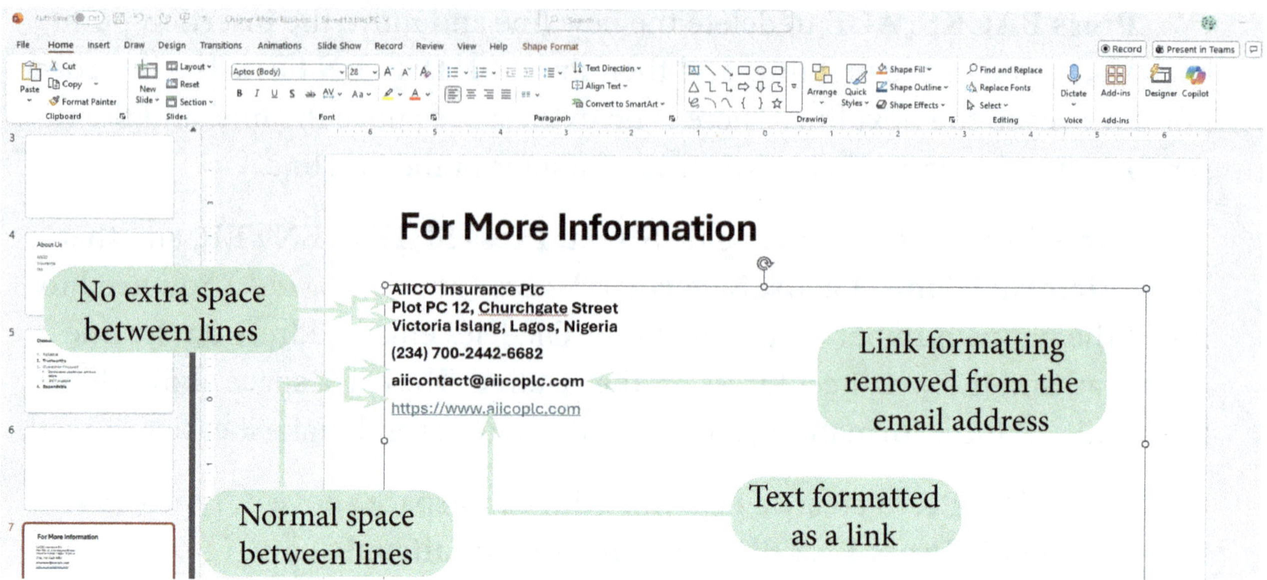

Figure 4-12 List on Slide 7

(14). On the **Quick Access Toolbar**, **click** the **Save** button to save the changes.

FORMATTING TEXT

Slides in a presentation should have a consistent look and feel. For example, the slide titles and the text in content placeholders should be in complementary fonts. There are times, however, when you need to change the format of text. For instance, you might want to make specific words bold to make them stand out more.

The commands in the Font group on the Home tab are used to apply formatting to selected text. **Figure 4-13** on the next page describes the buttons in this group.

To apply formatting to text, you must first select either the text or the text box. If you want to apply the same formatting to all the text in a text box, you can click the border of the text box. When you do this, the dotted line border changes to a solid line to indicate that the contents of the entire text box are selected. After you select the text or the text box, you click the button on the ribbon, or click the arrow, and then click an option in the menu or gallery that opens. For example, if you wanted to change the font, you would click the Font arrow, and then click the font you want to use.

To remove all formatting from selected text, click the Clear All Formatting button in the Font group.

Button	Name	Description
Calibri (Body)	Font	Change the font.
11	Font Size	Change the font size; click a size on the menu or type any value between 1 and 3600 in increments of 0.1 (for example, 42.4).
A^	Increase Font Size	Increase the font size to the next size up listed on the Font Size menu.
A˅	Decrease Font Size	Decrease the font size to the next size down listed on the Font Size menu.
	Clear All Formatting	Remove formatting of selected text.
B	Bold	Format text as bold.
I	Italic	Italicize text.
U	Underline	Underline text.
S	Text Shadow	Apply a shadow to text.
ab	Strikethrough	Add a line through text.
	Character Spacing	Change the spacing between characters.
Aa	Change Case	Change the case of selected text (for example, change to all uppercase).
	Text Highlight Color	Add a highlight color to selected text.
A	Font Color	Change the color of text.

Figure 4-13 Formatting commands in the Font group on the Home tab

Some of the formatting commands are also available on the Mini toolbar, which appears when you select text with the mouse or when you right-click on a slide. The **Mini Toolbar** is a small toolbar that appears next to text you select using the mouse or when you right-click a slide and that contains the most frequently-used text formatting commands, such as bold, italic, font color, and font size. If the Mini toolbar appears, you can use the buttons on it instead of those in the Font group.

Some of the commands in the Font group have menus or galleries that use the Microsoft Office **Live Preview** feature, which shows the results that would occur in your file, such as the effects of formatting options, if you clicked the option you are pointing to.

Oladele wants the contact information on Slide 7 ("For More Information") to be larger. He also wants the first letter of each item in the unnumbered list on Slide 4 ("About Us") formatted so it is more prominent.

To format the text on Slides 7 and 4:

(1). On **Slide 7** ("**For More Information**"), position the pointer on the border of the text box containing the contact information so that it changes to the move pointer ✥ , and then click the border of the text box. The border changes to a solid line to indicate that the entire text box is selected.

(2). On the **Home Tab**, in the Font group, **Click** the **Increase Font Size** button [A^] **twice**. All the text in the text box increases in size with each click and is now **36 points**.

(3). In the pane containing the thumbnails, **Click** the **Slide 4** thumbnail to display that slide.

(4). In the unnumbered list, **Click** to the left of "**AIICO**," **Press** and hold **SHIFT**, **Press RIGHT ARROW**, and then **Release SHIFT**. The letter "A" is selected. See **Figure 4-14**.

(5). In the Font group, **Click** the **Bold** button [B]. The Bold button becomes selected, and the selected text is formatted as bold.

(6). In the Font group, **Click** the **Font Size Arrow** to open the Font Size menu, and then **click 48**. The selected text is now 48 points.

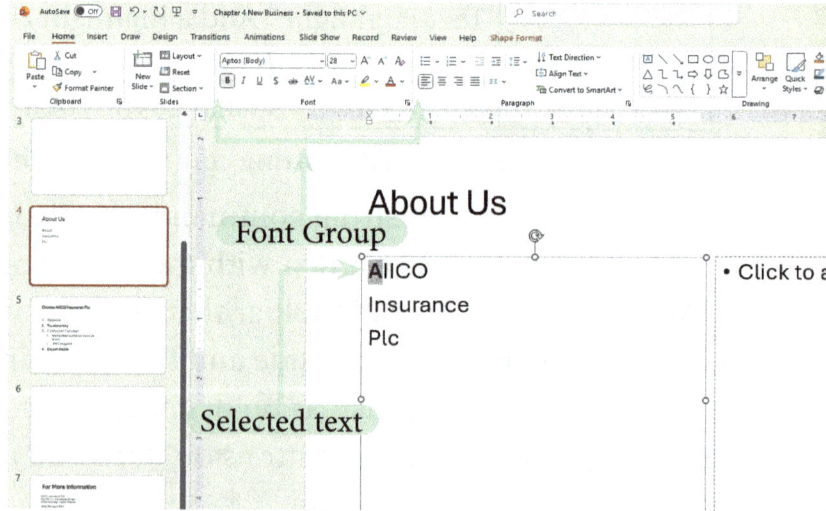

Figure 4-14 Text selected to be formatted

(7). In the Font group, **Click** the **Font Color Arrow** [A ▾] . A menu containing color options opens.

(8). Under Theme Colors, **Move** the **Pointer** over each color, noting the Screen-

Tips that appear and watching as Live Preview changes the color of the selected text as you point to each color. **Figure 4-15** shows the pointer pointing to the **Green, Accent 6, Darker 25%** color.

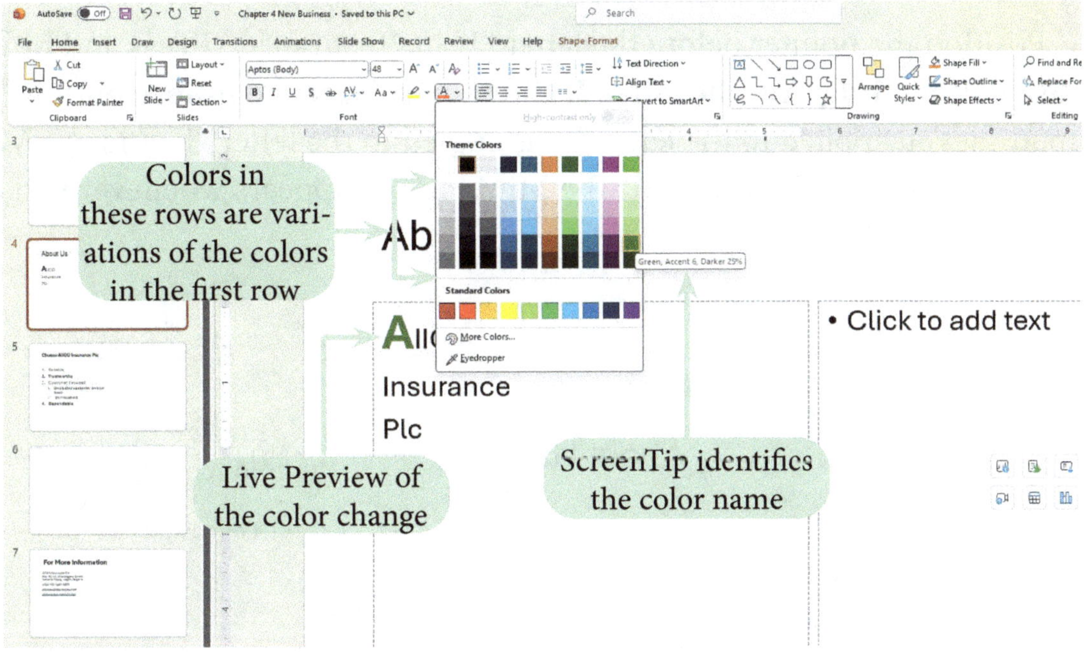

Figure 4-15 Font Color menu

(9). Using the ScreenTips, locate the **Green, Accent 6, Darker 25%** color in the last column, and then **Click** it. The selected text changes to the green color you clicked.

Now you need to format the first letters in the other words in the list to match the letter "A." You can repeat the steps you did when you formatted the letter "A," or you can use the Format Painter to copy all the formatting of the letter "A" to the other letters you need to format.

Also, Oladele wants the text in the unnumbered list to be as large as possible. Because the first letters of each word are larger than the rest of the letters, the easiest way to do this is to select all of the text, and then use the Increase Font Size button. The selected letters will increase in size with each click, and the first letters will still be larger.

To use the Format Painter to copy and apply formatting on Slide 4:

(1). Make sure the letter "**A**" is still selected.

(2). On the **Home Tab**, in the Clipboard group, **click** the **Format Painter** button , and then move the pointer on top of the slide. The button is selected, and the pointer changes to the Format Painter pointer for text .

(3). Position the pointer before the letter "I" in "Insurance," **Press** and **hold** the **mouse** button, **drag** over the **letter I**, and then **release** the **mouse** button. The formatting you applied to the letter "A" is copied to the letter "I," and the Mini toolbar appears. See **Figure 4-16**. The Mini toolbar appears whenever you drag over text to select it.

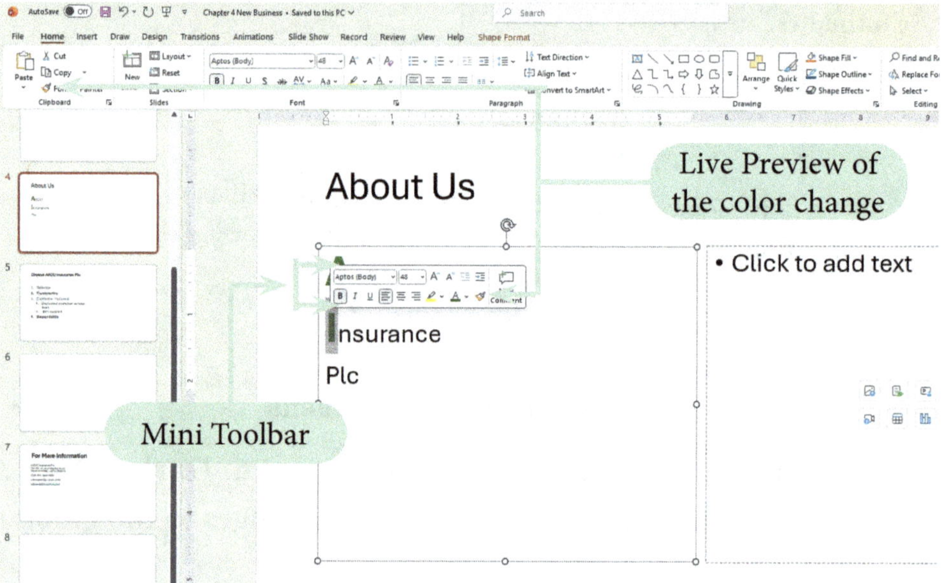

Figure 4-16 Mini toolbar

If you need to copy formatting to more than one location, double-click the Format Painter button to keep it selected until you deselect it by clicking it again.

(4). On the **Mini Toolbar**, **Click** the **Format Painter** button , and then drag across the letter "P" in "Plc."

(5). **Click** the **border** of the text box to select the entire text box, and then in the Font group, **Click** the **Increase Font Size** button five times. In the Font group, the Font Size button indicates that the text is 48+ points. This means that in the selected text box, the text that is the smallest is 48 points and there is some text that is larger.

(6). On the **Quick Access Toolbar**, **Click** the **Save** button to save the changes.

MOVING AND COPYING

You can move and copy text and objects in a presentation using the Clipboard that is part of Windows. The **Clipboard** is a temporary Windows storage area that holds the selections you copy or cut so you can use them later. When you **Cut** something, you remove the text or object from a file and place it on the Clipboard. You can also **Copy** text or an object, which means you select it and place a duplicate of it on the Clipboard, leaving the text or object in its original location. You can then paste the text or object stored on the Clipboard anywhere in the presentation or in any file in any Windows program. To **Paste** something means to place text or an object stored on the Clipboard in a location in a file.

The Clipboard holds only the most recently cut or copied item. As soon as you cut or copy another item, it replaces the previously cut or copied item on the Clipboard. You can paste an item on the Clipboard as many times and in as many locations as you like.

Note that cutting text or an object differs from deleting it. When you press DELETE or BACKSPACE to delete text or objects, they are not placed on the Clipboard and cannot be pasted.

Oladele wants a few changes made to Slides 5 and 2. You'll use the Clipboard as you make these edits.

To cut, copy, and paste text using the Clipboard:

(1). On **Slide 4** ("**About Us**"), **Double-click** the word **Plc** in the list. The word "**Plc**" is selected.

(2). On the **Home Tab**, in the **Clipboard group**, **Click** the **Copy** button . The selected word is copied to the Clipboard.

(3). In the pane containing thumbnails, **Click** the **Slide 5** thumbnail to display that slide, **Click** after the word "**Insurance**" in the title, and then **Press SPACEBAR.**

(4). In the **Clipboard group**, **Click** the **Paste** button. The text appears at the location of the insertion point. The letter "**P**" is still **green** and is larger than the rest of the text. The rest of the text picks up the formatting of its destination, so it is 44 points instead of 48 points as in the list on Slide 4. The Paste Options button (Ctrl) appears below the pasted text.

(5). **Click** the **Paste Options** button (Ctrl). A menu opens with four buttons on it. See **Figure 4-17.**

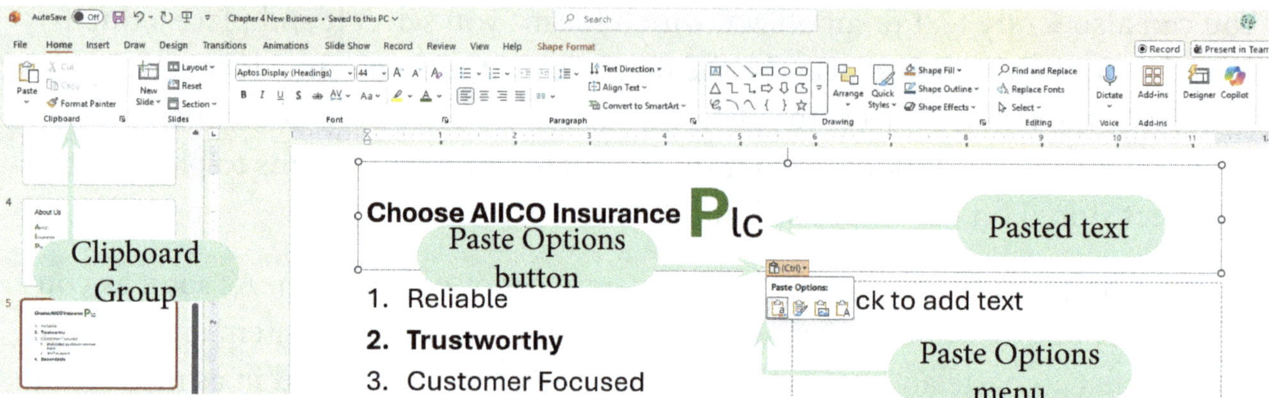

Figure 4-17 Buttons on the Paste Options menu when text is on the Clipboard

(6). **Point** to each **button** on the **menu**, reading the ScreenTips and watching to see how the pasted text changes in appearance. The first button is the Use Destination Theme button, which is the default choice when you paste text.

(7). On the **Paste Options menu, Click** the **Keep Text Only** button. The pasted text changes so that its formatting matches the rest of the title text.

(8). Display **Slide 2** ("Types of Plans"). The last bulleted item ("**Stable**") belongs on **Slide 5.**

(9). In the last bulleted item, **position** the **pointer** on top of the **bullet symbol** so that the **pointer changes** to the **four-headed arrow pointer**, and then **Click**. The entire bulleted item is selected.

(10). In the **Clipboard group, Click** the **Cut** button. The last bulleted item is removed from the slide and is placed on the Clipboard.

(11). Display **Slide 5** ("**Choose AIICO Insurance Plc**"), **Click** after the last item ("**Dependable**"), and then **press ENTER** to create a fifth first-level item.

(12). In the **Clipboard group, Click** the **Paste** button. The bulleted item you cut becomes the fifth first-level item on **Slide 5** using the default paste option of Use Destination Theme. The insertion point appears next to a sixth first-level item.

(13). **Press BACKSPACE twice** to delete the extra line, and then on the **Quick Access Toolbar**, **Click** the **Save** button to save the changes.

SMARTSKILLS Using the Office Clipboard

The **Office Clipboard** is a temporary storage area in the computer's memory that lets you collect text and objects from any Office document and then paste them into other Office documents. Once you activate the Office Clipboard, you can store up to 24 items on it and then select the item or items you want to paste. To activate the Office Clipboard, click the Home tab. In the Clipboard group, click the Dialog Box Launcher (the small square in the lower-right corner of the Clipboard group) to open the Clipboard pane to the left of the displayed slide.

As you play with the Clipboard features you will discover other ways to interact with contents that you're manipulating in Microsoft Office.

MANIPULATING SLIDES

You can manipulate the slides in a presentation to suit your needs. For example, if you need to create a slide that is similar to another slide, you can duplicate the existing slide and then modify the copy. If you no longer want to include a slide in your presentation, you can delete it. You can also reorder slides as necessary.

To duplicate, rearrange, or delete slides, you select the slides in the pane containing the thumbnails in Normal view or switch to **Slide Sorter View**. In Slide Sorter view all the slides in the presentation are displayed as thumbnails in the window.

Oladele wants to display a slide that shows the name of the company at the end of the presentation. To create this slide, you will duplicate Slide 4 ("About Us").

To duplicate Slide 4:

(1). Display **Slide 4** ("**About Us**").

(2). On the **Home Tab**, in the **Slides Group**, **Click** the **New Slide Arrow**, and then **Click Duplicate Selected Slides**. A duplicate of Slide 4 appears as a new Slide 5 and is the current slide. If you had selected more than one slide, they would all be duplicated. The duplicate slide doesn't need the title; Oladele just wants to reinforce the company's name.

(3). On **Slide 5**, **Click** anywhere on the title **About Us**, **Click** the text box border to select the text box, and then press **DELETE**. The title and the title text box are deleted, and the title text placeholder reappears.

You could delete the title text placeholder, but you do not need to. When you display a presentation to an audience as a slide show, any unused placeholders do not appear.

Next you need to rearrange the slides. You need to move the duplicate of the "About Us" slide so it becomes the last slide in the presentation because Oladele wants it to remain displayed after the presentation is over. He hopes this visual will reinforce the company's name for the audience. Oladele also wants Slide 6 ("Choose AIICO Insurance Plc") moved before the "Types of Plans" slide (Slide 2), and he wants the original "About Us" slide (Slide 4) to be the second slide in the presentation.

To rearrange the slides in the presentation:

(1). In the pane containing the thumbnails, **scroll**, if necessary, so that you can see **Slides 2** and **6**, and then **Click** the **Slide 6** ("**Choose AIICO Insurance Plc**") thumbnail. **Slide 6** ("**Choose AIICO Insurance Plc**") is the current slide.

(2). **Point** to the **Slide 6** thumbnail, **press** and **hold** the **mouse** button, **drag** the **Slide 6** thumbnail up above the **Slide 2** ("**Types of Plans**") thumbnail, and then **release** the **mouse** button. As you drag, the Slide 6 thumbnail follows the pointer and the other slides move down to make room for the slide you are dragging. The "Choose AIICO Insurance Plc." slide becomes Slide 2 and "Types of Plans" becomes Slide 3. You'll move the other two slides in Slide Sorter view.

(3). On the status bar, **Click** the **Slide Sorter** button ⊞. The view switches to Slide Sorter view. Slide 2 appears with an orange border, indicating that it is selected.

(4). On the **Status Bar, Click** the **Zoom Out** button [-] as many times as necessary until you can see all nine slides in the presentation. See **Figure 4-18**.

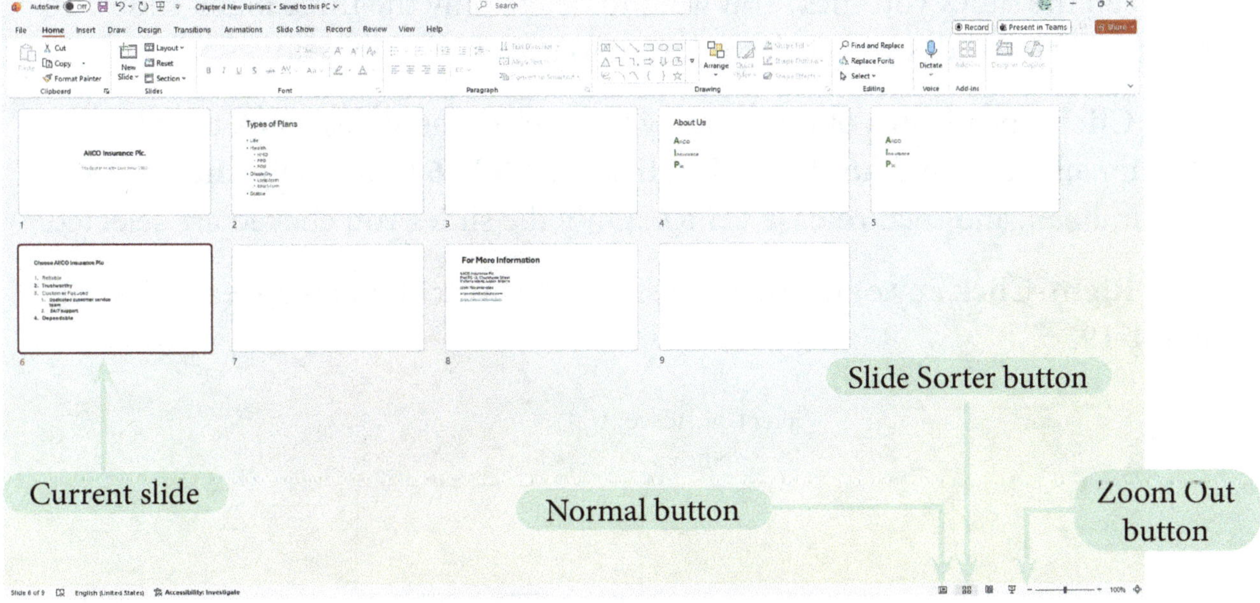

Figure 4-18 Slide Sorter view

(5). **Drag** the **Slide 5** ("About Us") thumbnail to between **Slides 1** and **2**. As you drag, the other slides move out of the way. The slides are renumbered so that the "About Us" slide is now **Slide 2**.

(6). **Drag** the **Slide 6** thumbnail (the slide containing just the name of the company) after the last slide in the presentation (Slide 9).

Now you need to delete the blank slides. To delete a slide, you right-click its thumbnail to display a shortcut menu, and then click Delete Slide on that menu.

You already know that to select a single slide you click its thumbnail. You can also select more than one slide at a time. To select sequential slides, click the first slide, press and hold SHIFT, and then click the last slide you want to select. To select nonsequential slides, click the first slide, press and hold CTRL, and then click any other slides you want to select. When more than one slide is selected, you can delete or duplicate all of the selected slides with one command.

To delete the blank slides:

(1). **Click** the **Slide 5** thumbnail (the first blank slide), **press** and hold **SHIFT**, **click** the **Slide 8** thumbnail (the last blank slide), and then **release SHIFT**. The two slides you clicked are selected, as well as the slides between them. Holding SHIFT when you click items selects the slides you click as well as the slides between the slides you click. You want to delete only the three blank slides. To select only the slides you click, you need to hold CTRL instead.

(2). **Click** a blank area of the window to deselect the slides, **click** the **Slide 5** thumbnail, **press** and **hold CTRL**, **Click** the **Slide 6** thumbnail, **click** the **Slide 8** thumbnail, and then **release CTRL**. Only the slides you clicked are selected.

(3). **Right-Click** any of the selected slides. A shortcut menu appears. See **Figure 4-19**.

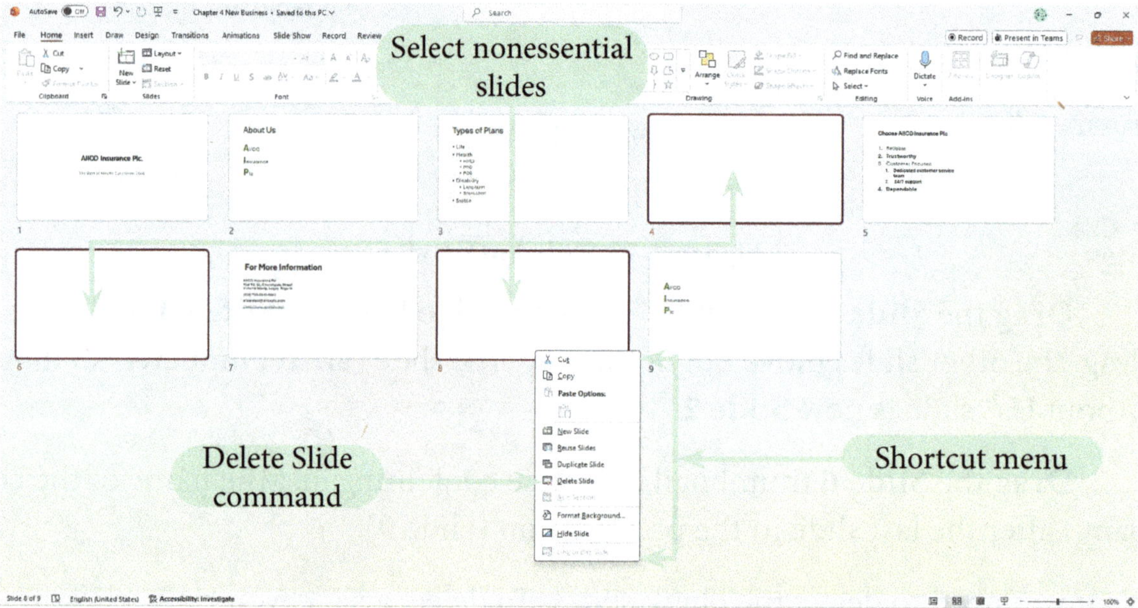

Figure 4-19 Shortcut menu for selected slides

(4). On the shortcut menu, **Click Delete Slide**. The shortcut menu closes, and the three selected slides are deleted. The presentation now contains six slides.

(5). On the **Status Bar**, **Click** the **Normal** button . The presentation appears in Normal view.

(6). On the **Quick Access Toolbar**, **Click** the **Save** button to save the changes to the presentation.

CHANGING THE THEME

A **Theme** is a predefined, coordinated set of colors, fonts, graphical effects, and other formats that can be applied to a presentation. In PowerPoint, most themes have variants that have different coordinating colors and sometimes slightly different backgrounds. All presentations have a theme. If you don't choose one, the default Office theme is applied; that is the theme currently applied to the presentation you created.

Every theme has a palette of 12 coordinated colors. You saw the Office theme colors when you changed the color of the text on the "About Us" slide. If you don't like the color palette of the theme you chose, you can change to a different one.

Themes also have a font set. One font, called the **Headings Font**, which is used for slide titles. The other font, called the Body font, is used for the rest of the text on a slide. In the Office theme, the Headings font is Aptos, and the Body font is Aptos. In some themes, the Headings and Body font are the same font.

Oladele wants you to try changing the theme colors and fonts.

To examine the current theme and then change the theme color and theme fonts:

(1). Display **Slide 5** ("For More Information"). Notice that the link is blue.

(2). Display **Slide 6**, and then, in the unnumbered list, select the green letter P.

(3). On the **Home Tab**, in the Font group, **Click** the Font **Color arrow** A. Look at the colors under Theme Colors. The last column contains shades of green. In that column, the second to last color is selected. The colors in the Theme Colors section change depending on the selected theme and the selected theme colors. The colors in the row of Standard Colors do not change when you choose a different theme or theme color palette.

(4). In the **Font Group**, **Click** the **Font Arrow**. A menu of fonts installed on the computer opens. At the top under Theme Fonts, Aptos (Body) is selected because the letter "P" that you selected is in a content text box. See **Figure 4-20** on the next page. If the selected text was in the title text box, the first font in the list, Aptos Light (Headings) would be selected.

In Figure 4-20 on the next page, Slide 6 is displayed in Normal view. The left content placeholder for Slide 6 is selected. The Font arrow in the Font group on the Home tab was clicked, displaying a menu of fonts. The two fonts at the top of the list are identified as Theme Fonts. The first reads "Aptos Body left parenthesis Headings right parenthesis", indicating the font used for titles. The second reads "Aptos left parenthesis Body right parenthesis", indicating the font used for slide content.

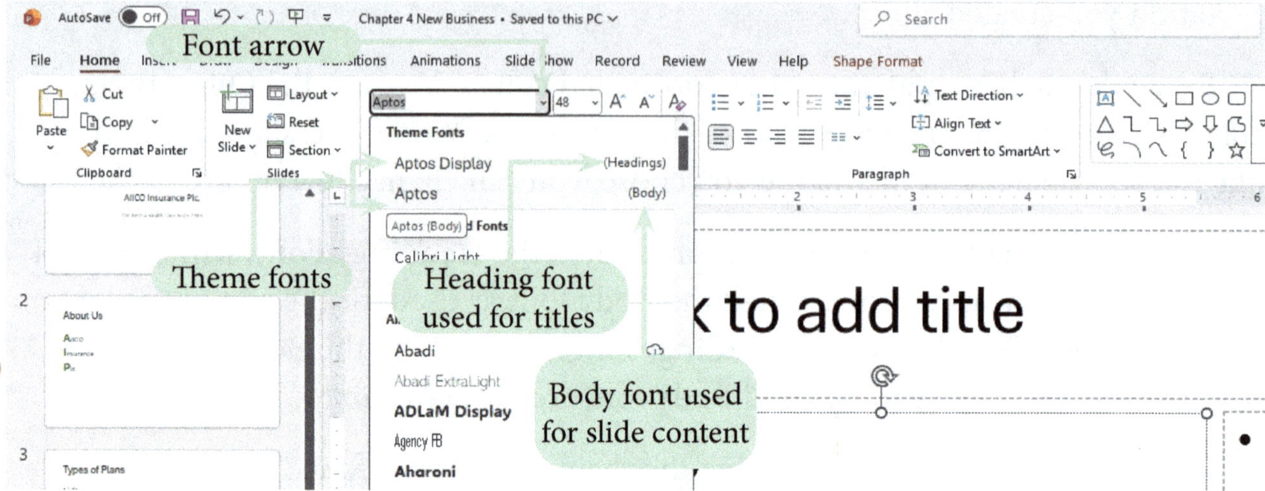

Figure 4-20 Theme fonts on the Font menu

(5). On the **Ribbon**, **Click** the **Design Tab**. The Design tab is active. See **Figure 4-21**. In the Themes group, the first theme is the theme applied to the presentation. In this case, it is the Office theme. The second theme is also the Office theme and it is shaded to indicate that it is selected. In the Variants group, the first variant is shaded to indicate that it is selected.

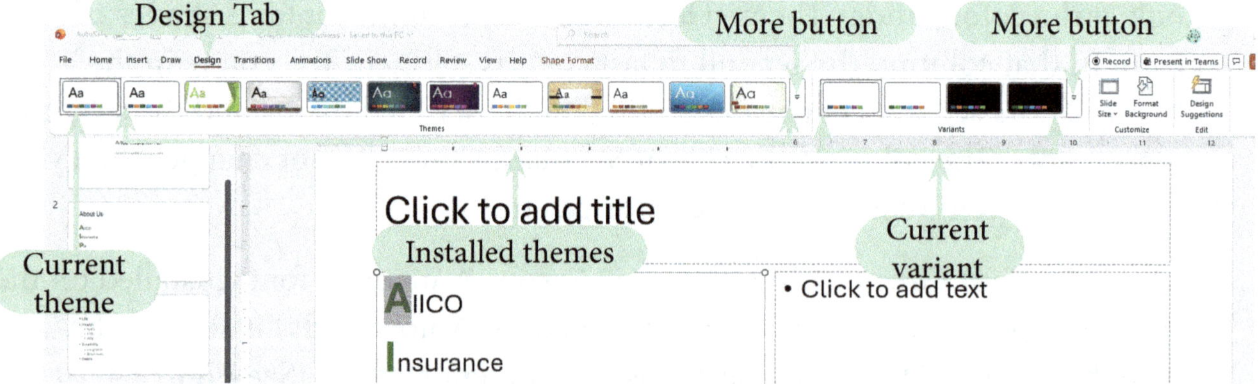

Figure 4-21 Themes and variants on the Design tab

(6). In the **Variants Group**, **Click** the **More** button ⊡. A menu opens containing commands for changing the theme colors and the theme fonts. See **Figure 4-22** on the next page.

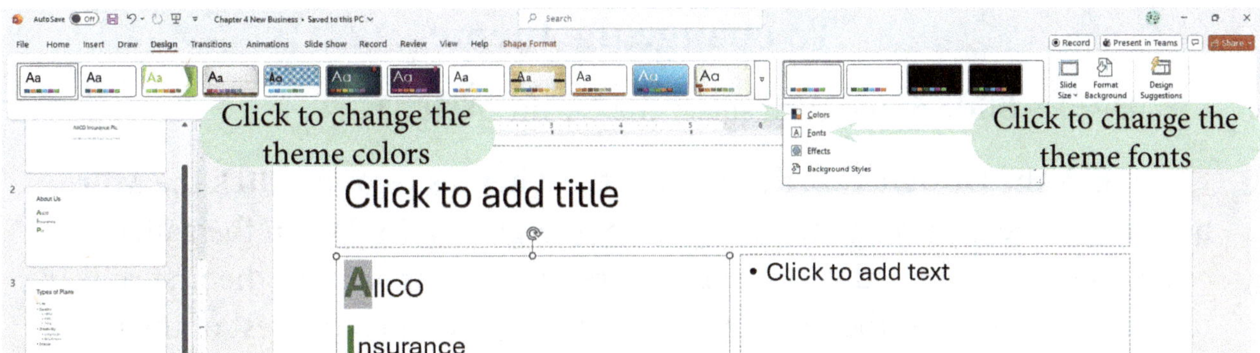

Figure 4-22 More button menu in the Variants group

(7). On the **Menu**, point to **Colors** to open a submenu of color palettes, and then **Click** the **Blue Warm** palette. The colored letters on **Slide 6** change to a shade of grayish brown.

(8). In the **Variant Group**, **Click** the **More** button, point to **Fonts**, **scroll down**, and then **Click Tw Cen MT-Rockwell**. The font of the list and the title text placeholder on **Slide 6** changes.

(9). **Click** the **Home Tab**, and then in the **Font group**, **Click** the **Font Color Arrow**. The second to last color in the last column is still selected, but now that column contains shades of **grayish brown**. The row of Standard Colors is the same as it was when the Office theme colors were applied.

(10). In the **Font Group**, **Click** the **Font Arrow**. The Headings and Body font have changed to the **Tw Cen MT** and the **Rockwell fonts**.

(11). Display **Slide 5**. The link that was blue before you changed the theme colors is now **purple**.

PowerPoint comes with several installed themes, and many more themes are available online at Office.com. In addition, you can use a custom theme stored on your computer or network.

You can select a different installed theme when you create a new presentation by clicking one of the themes on the New or Home screen in Backstage view. If you want to change the theme of an open presentation, you can choose an installed theme on the Design tab.

Oladele still thinks the presentation could be more interesting, so he asks you to apply a different

To change the theme

(1). Display **Slide 6**, and then select the "A" in "**AIICO**."

(2). **Click** the **Design Tab**, and then in the **Themes Group**, **Click** the **More** button. The gallery of themes opens. See **Figure 4-23**. When the gallery is open, the theme applied to the current presentation appears in the first row. In the next row, the first theme is the Office theme, and then the rest of the installed themes appear. Some of these themes also appear on the Home and New screens in Backstage view.

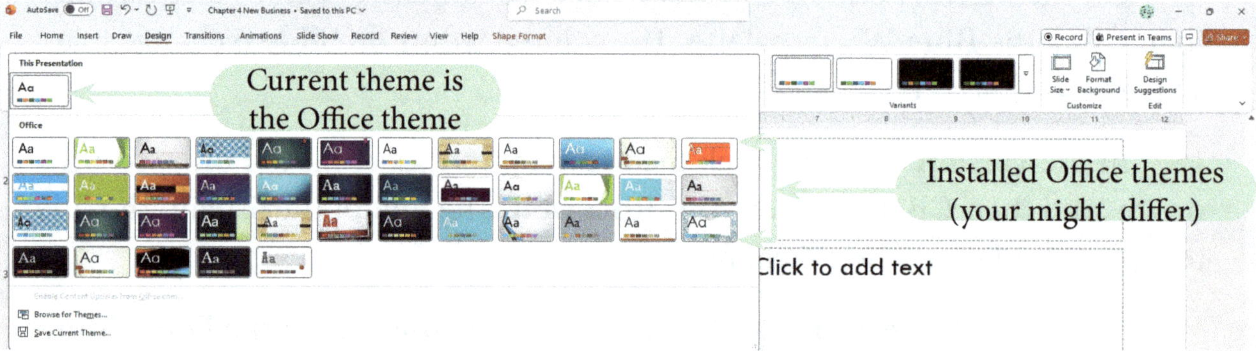

Figure 4-23 Themes gallery expanded

(3). Point to several of the themes in the gallery to display their ScreenTips and to see a Live Preview of the theme applied to the current slide.

(4). **Click** the **Wisp Theme**. The gallery closes, and all the slides have the Wisp theme with the default variant (the first variant in the Variants group) applied. The background of the slides changes from white to tan, and the letters that you had colored green on Slide 6 change to a shade of grayish green. In the empty content placeholder on Slide 6, the bullet symbol changed from a circle to an arrow.

(5). In the Variants group, point to the other three variants to see a Live Preview of each of them, and then click the third variant (the blue one). The letters on Slide 6 change to purple. You prefer the original colors. On the Quick Access Toolbar, click the **Undo** button.

(6). **Click** the **Home Tab**, and then in the Font group, **Click** the **Font Color Arrow**. The selected color is still the second to last color in the last column, but now the last column contains shades of purple. Again, the row of Standard Colors is the same as it was before you made changes.

(7). In the **Font Group, Click** the **Font Arrow**. You can see that the Theme Fonts are now Century Gothic for both Headings and the Body.

(8). **Press ESC**. The Font menu closes.

After you apply a new theme, you should examine your slides to make sure that they look the way you expect them to. Slide 6 looks fine.

To examine the slides with the new theme and adjust font sizes:

(1). **Display Slides 5, 4, 3**, and **Slide 2**. These slides look fine.

(2). **Display Slide 1** (the title slide). The title text is too large with the Wisp theme applied.

(3). **Click** anywhere on the title text, and then click the text box border. The entire text box is selected.

(4). In the **Font group, Click** the **Decrease Font Size** button as many times as necessary until the font size of the title text decreases to **44 points**.

(5). On the **Quick Access Toolbar, Click Save**. The changes to the presentation are saved.

SMARTSKILLS Understanding the Difference between Themes and Templates

As explained earlier, a theme is a coordinated set of colors, fonts, backgrounds, and effects. A **Template** is a file that has a theme applied and contains text, graphics, and placeholders that direct you in creating content for a presentation. You can create and save your own custom templates or find everything from calendars to marketing templates among the thousands of templates available on Office.com. To find a template on Office.com, display the Home or New screen in Backstage view, type keywords in the "Search for online templates and themes" box, and then click the Search button in the box to display templates related to the search terms. To create a new presentation based on the template you find, click the template and then click Create.

If a template is stored on your computer, you can apply the theme used in the template to an existing presentation. If you want to apply the theme used in a template on Office.com to an existing presentation, you need to download the template to your computer first.

CLOSING A PRESENTATION

When you are finished working with a presentation, you can close it and leave PowerPoint open. To do this, you click the File tab to open Backstage view, and then click the Close command. If you have only one presentation open, if you click the Close button in the upper-right corner of the PowerPoint window, you will not only close the presentation, you will exit PowerPoint as well.

You're finished working with the presentation for now, so you will close it. First you will add your name to the title slide.

To add your name to Slide 1 and close the presentation:

(1). On **Slide 1** (the title slide), **Click** the subtitle, position the insertion point after "**1963**" **Press ENTER**, and then **Type** your **Full Name**.

(2). **Click** the **File tab**. Backstage view appears with the Home screen displayed. See **Figure 4-24**.

Figure 4-24 Home screen in Backstage view

(3). In the navigation pane, **Click Close**. Backstage view closes, and a dialog box opens, asking if you want to save your changes.

(4). In the dialog box, **Click Save**. The dialog box and the presentation close, and the empty presentation window appears.

Are You Having Trouble? If you want to take a break, you can exit PowerPoint by clicking the Close button in the upper-right corner of the PowerPoint window.

You've created a presentation that includes slides to which you added bulleted, numbered, and unnumbered lists. You also formatted text, manipulated slides, and applied a theme. You are ready to give the presentation draft to Oladele to review.

Quick Review

1. Define "presentation."
2. How do you display Backstage view?
3. What is a slide layout?
4. In addition to a title text placeholder, what other type of placeholder do many layouts contain?
5. What is the term for an object that contains text?
6. What is the difference between the Clipboard and the Office Clipboard?
7. Explain what a theme is and what changes with each variant.

Opening a Presentation and Saving It with a New Name

If you have closed a presentation, you can always reopen it to modify it. To do this, you can double-click the file in a File Explorer window, or you can open Backstage view in PowerPoint and use the Open command.

Oladele Kanu reviewed the presentation you created and made a few changes. You will continue modifying the presentation using his version.

To open the revised presentation:

(1). **Click** the **File Tab** on the ribbon to display the Home screen in Backstage view.

Are You Having Trouble? If PowerPoint is not running, start PowerPoint, and then in the left pane, **Click Open**.

(2). In the navigation pane, **Click Open** to display the Open screen. Recent is selected, and you might see a list of recently opened presentations on the right.

(3). **Click Browse**. The **Open dialog** box appears. It is similar to the Save As dialog box.

Are You Having Trouble? If you store your files on your OneDrive, click OneDrive, and then log in if necessary.

(4). Navigate to the drive that contains your Data Files, navigate to the **PowerPoint folder** > **Chapter 4 folder**, **Click Chapter 4 New Business.pptx** to select it, and then **Click Open**. The Open dialog box closes, and the presentation opens in the PowerPoint window, with Slide 1 displayed.

If you want to edit a presentation without changing the original, you need to create a copy of it. To do this, you use the Save As command to open the Save As dialog box, which is the same dialog box you saw when you saved your presentation for the first time. When you save a presentation with a new name, you create a copy of the original presentation, the original presentation closes, and the newly named copy appears in the PowerPoint window.

To save the revised presentation with a new name:

(1). **Click** the **File tab**, and then in the navigation pane, **Click Save As**. The Save As screen in Backstage view appears.

(2). **Click Browse** to open the **Save As dialog** box.

(3). If necessary, navigate to the drive and folder where you are storing your Lab's Data Files.

(4). In the **File Name** box, change the **filename** to **Chapter 4 New Business Revised**, and then **Click Save**. The Save As dialog box closes, and a copy of the file is saved with the new name **Chapter 4 New Business Revised** and appears in the PowerPoint window.

Inserting Pictures and Adding Alt Text

In many cases, graphics are more effective than words for communicating an important point or invoking an emotional reaction. For example, if a sales force has reached its sales goals for the year, including a photo in your presentation of a person reaching the top of a mountain can convey a sense of accomplishment to your audience. To add a graphic to a slide, you can use the buttons in a content placeholder or buttons on the Insert tab.

When you insert a graphic and when specific built-in layouts are applied to the slide, the Design Ideas pane opens containing suggestions for interesting layouts for the slide. You can click one of these layouts to apply it or close the Design Ideas pane without accepting any of the suggestions.

Oladele has a photo that he wants you to insert on Slide 2.

To insert a photo on Slide 2 and view the Design Ideas:

(1). **Display Slide 2** ("About Us"), and then in the content placeholder on the right, **Click** the **Pictures** button. The Insert Picture dialog box opens. This dialog box is similar to the Open dialog box.

(2). Navigate to the **PowerPoint folder** > **Chapter 4 folder** included with your Data Files, **Click Family.jpg**, and then **Click Insert**. The dialog box closes, and a picture of a family, a father and his three daughters in a kitchen appears in the placeholder and is selected. Text that describes the picture might appear briefly at the bottom of the picture. Also, the Design Suggestions pane might open listing suggestions for interesting layouts for this slide. On the ribbon, the contextual Picture Format tab appears and is the active tab. See **Figure 4-25**.

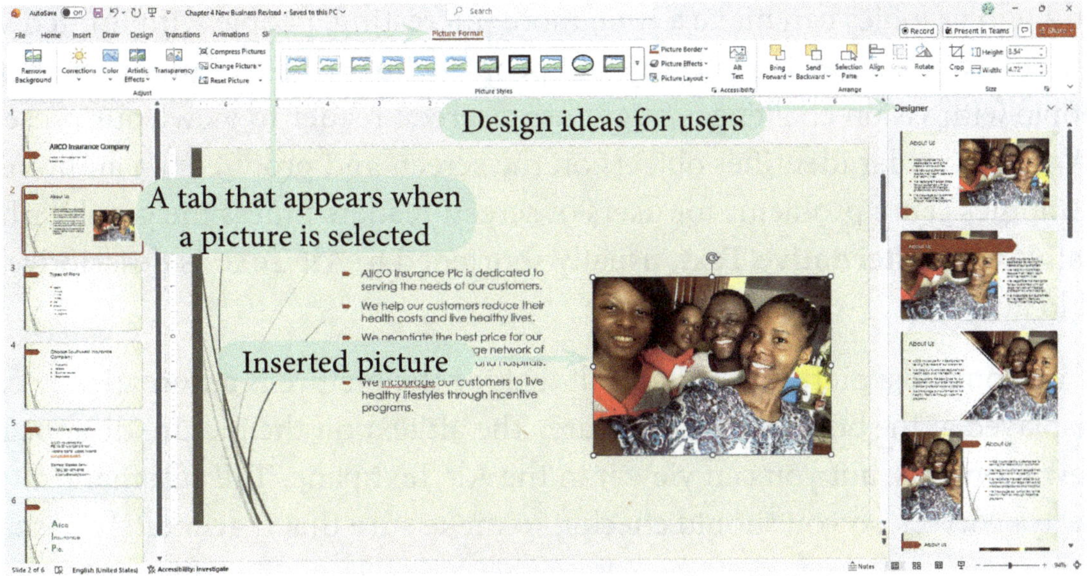

Figure 4-25 Picture inserted on Slide 2

Are You Having Trouble? If the Design Suggestions pane does not appear, click the Design tab, and then in the Designer group, click the Design Suggestions button.

Are You Having Trouble? If the descriptive text does not appear below the picture, do not be concerned. You will display it later.

(3). In the **Design Ideas pane**, **Click** each of the thumbnails to see the effect on the slide. Although Oladele likes some of the layouts suggested in the Design Ideas pane, he wants you to apply the Two Content layout again. First, you need to undo the change you made.

(4). On the **Quick Access Toolbar**, **Click** the **Undo Arrow**. No matter how many thumbnails you clicked in the Design Ideas pane, only one "Apply Design Idea" action is listed in the Undo menu.

(5). On the menu, **Click Apply Design Idea**. The slide is reset to its original layout.

(6). In the **Design Ideas pane**, in the top-right corner, **Click** the **Close** button. The pane closes.

The layout suggestions in the Design Suggestions pane can help you create interesting slides. If you open the Design Suggestions pane and it does not contain any suggestions, make sure you are using one of the themes that is included with PowerPoint, and change the slide layout to Title Slide or Title and Content.

Although graphics can make a slide more interesting, people with limited vision might not be able to see them clearly and people who are blind cannot see them at all. People with vision challenges might use a screen reader to view your presentation. A screen reader identifies objects on the screen and produces an audio of the text. Graphics cause problems for users of screen readers unless the graphics have alternative text. **Alternative Text**, usually shortened to **Alt Text**, is descriptive text added to an object.

When you add a picture to a slide, alt text for the picture is automatically created and displayed at the bottom of the picture. The alt text on the picture disappears after a few moments, but you can view it in the Alt Text pane. The automatic alt text is not always correct, so you should check it to make sure that it accurately describes the image.

You will examine and edit the alt text of the photo you added to Slide 2.

To modify the alt text of the photo on Slide 2:

(1). On **Slide 2** ("About Us"), **Click** the **Picture** to **Select** it if necessary, **Click** the **Picture Format tab**, and then in the **Accessibility group**, **Click** the **Alt Text** button. The Alt Text pane appears. See **Figure 4-26.**

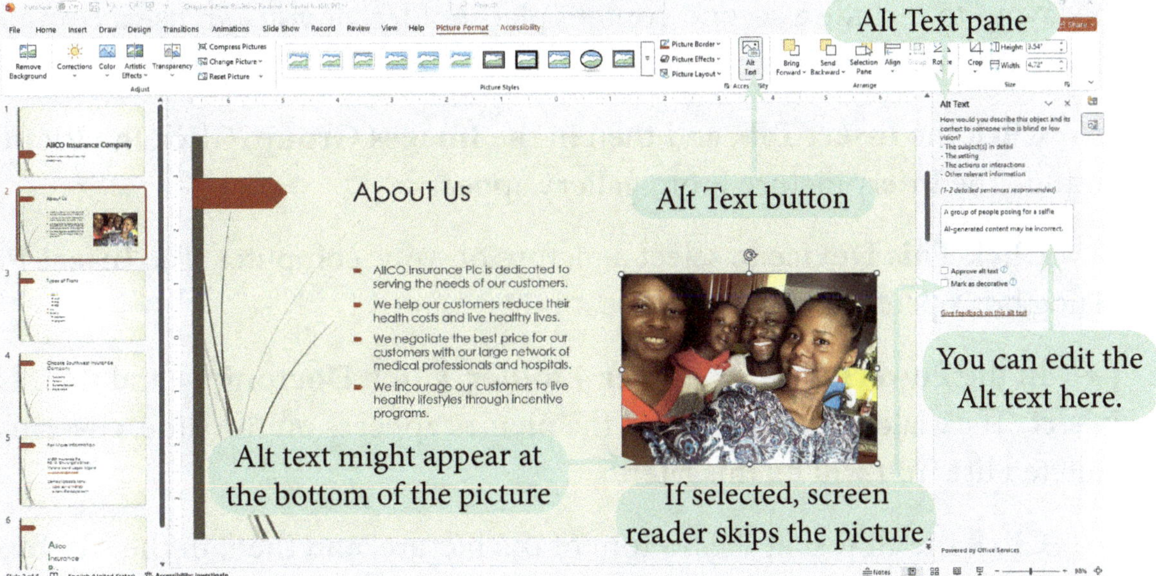

Figure 4-26 Alt Text pane open showing automatically generated alt text

Are You Having Trouble? If alt text is not automatically generated, this feature might be turned off on your computer or your computer might not have been able to connect to the Microsoft server. Click in the white box in the Alt Text pane, and then skip **Step 2.**

(2). In the **Alt Text Pane**, in the white box, **select all of the text**, including the phrase "**Description automatically generated.**"

(3). **Type A happy father with his three young daughters taking a selfie in their kitchen** in the white box. The text you type replaces the selected text. The next time you select the picture, a new command—Generate a description for me—will appear. You could click that command to have the alt text generated again, replacing the text you typed.

(4). In the **Alt Text Pane**, in the top-right corner, **Click** the **Close** button ☒.

Oladele has two more photos that he wants you to add to the presentation. He asks you to add the photos to **Slides 3** and **6**.

To insert photos on Slides 3 and 6:

(1). **Display Slide 3** ("Types of Plans"). This slide has the Title and Content layout applied, so it does not have a second content placeholder. You can change the layout to include a second content placeholder, or you can use a command on the ribbon to insert a photo.

(2). **Click** the **Insert Tab**, and then in the **Images Group**, **Click** the **Pictures** button. The Insert Picture From gallery appears.

(3). **Click This Device** to select a picture on your computer. The **Insert Picture** dialog box opens.

(4). In the **PowerPoint** > **Chapter 4 folder**, **Click Doctor.jpg**, and then **Click Insert**. The dialog box closes, and the picture appears on the slide, covering the bulleted list. You will fix this later.

(5). **Click** the alt text at the bottom of the picture, and then, in the **Alt Text Pane**, **Select** all of the text in the white box.

Are You Having Trouble? If the alt text disappeared before you could click it or doesn't appear at all, click the Alt Text button in the Accessibility group on the Picture Tools tab. If there is no text in the white box, click in the white box.

(6). **Type A smiling female doctor in a white coat with a stethoscope around her neck showing** in the white box.

(7). **Display Slide 6** (the last slide). Slide 6 has the Two Content layout applied, but you can still use the Pictures command on the Insert tab.

(8). On the **Ribbon**, **Click** the **Insert Tab**.

(9). In the **Images Group**, **Click** the **Pictures** button, **Click This Device**, **Click Handshake.jpg** in the **PowerPoint** > **Chapter 4 folder**, and then **Click Insert**. The picture replaces the empty content placeholder on the slide.

(10). In the **Alt Text Pane**, **Select** all of the text in the white box, and then

Type Close-up of two hands with white coat sleeves clasping another hand in the white box.

(11). **Close** the **Alt Text Pane**, and then **Close** the **Design Ideas Pane** ☒, if necessary.

SMARTSKILLS Decision Making: Deciding Whether to Allow Alt Text to Be Generated

People are becoming much more aware of privacy concerns when posting information to the cloud or to social media. When you insert a picture on a slide, the picture is sent to Microsoft's servers in order to generate alt text. This means that you are sharing the picture in the cloud. If you are concerned about sharing your private pictures, you can turn this feature off. To do this, click the File tab, and then click Options to open the PowerPoint Options dialog box. On the left, click Ease of Access to display the options for making PowerPoint more accessible. In the Automatic Alt Text section, click the Automatically generate alt text for me check box to deselect it. If you change your mind and you want alt text generated for a specific picture, you can still click the command to generate new alt text in the Alt Text pane.

CROPPING PICTURES

Sometimes you want to display only part of a photo. For example, if you insert a photo of a party scene that includes a bouquet of colorful balloons, you might want to show only the balloons. To do this, you can crop the photo. To **Crop** means to trim away part of a picture. In PowerPoint, you can crop a picture to any size you want, crop it to a preset ratio, or crop it to a shape.

It can be helpful to display rulers and gridlines in the window to help you crop photos to specific sizes. There are two rulers. One is horizontal and appears above the slide. The other is vertical and appears to the left of the slide. **Gridlines** are evenly spaced horizontal and vertical lines on the slide that help you align objects.

Oladele wants you to crop the photo on Slide 3 ("Types of Plans") to make the dimensions of the final photo smaller without making the images in the photo smaller.

To crop the photo on Slide 3:

(1). **Click** the **View Tab**, and then in the **Show Group**, **Click** the **Ruler** and the **Gridlines** check boxes. Rulers appear above and to the left of the displayed slide, and the gridlines appear on the slide.

(2). **Display Slide 3** ("Types of Plans"), **Click** the photo to **Select** it, and then **Click** the **Picture Format Tab**, if necessary.

(3). In the **Size Group**, **Click** the **Crop** button. The Crop button is selected, and crop handles appear around the edges of the photo just inside the sizing handles. See **Figure 4-27**.

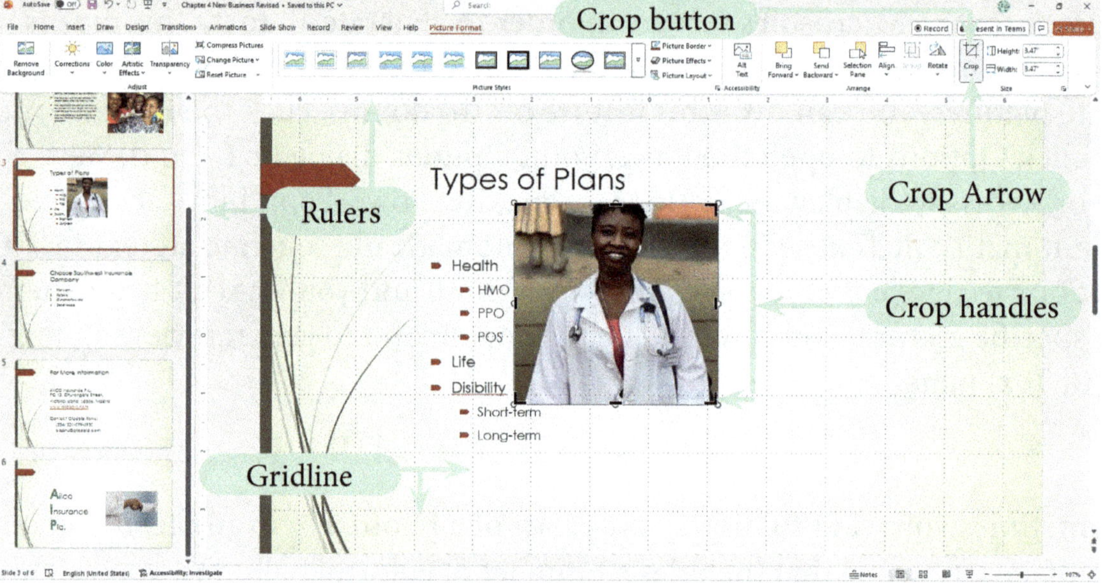

Figure 4-27 Photo with crop handles

Figure 4-27 above displays a photo with crop handles. The Picture Format tab at the top of the screen on the ribbon is active, and Slide 3 is selected. The Crop button in the Size group has been selected, and crop handles appear at the corners and sides of the inserted photo of a female doctor speaking to a young girl with a ponytail. Rulers are displayed above and to the left of the slide pane, and dotted gridlines display on the slide. The left edge of the inserted photo is aligned with the gridline at the negative 4-inch mark on the horizontal ruler. The right edge of the inserted photo is aligned with the gridline at the 4-inch mark on the horizontal ruler. The Crop arrow at the top right of the screen on the ribbon is called out.

(4). **Position** the **pointer** directly on top of the middle **Crop Handle** on the left side of the picture so that it changes to the left-middle crop pointer ⊣ . On the rulers, a red dotted line shows the position of the pointer.

(5). **Press** and hold the **Mouse** button, **drag** the **crop handle** to the **right** until the left cropped edge is on the gridline that aligns with the negative 3-inch mark on the horizontal ruler, and then release the mouse button. The part of the photo that will be cropped off is shaded dark gray. See **Figure 4-28**.

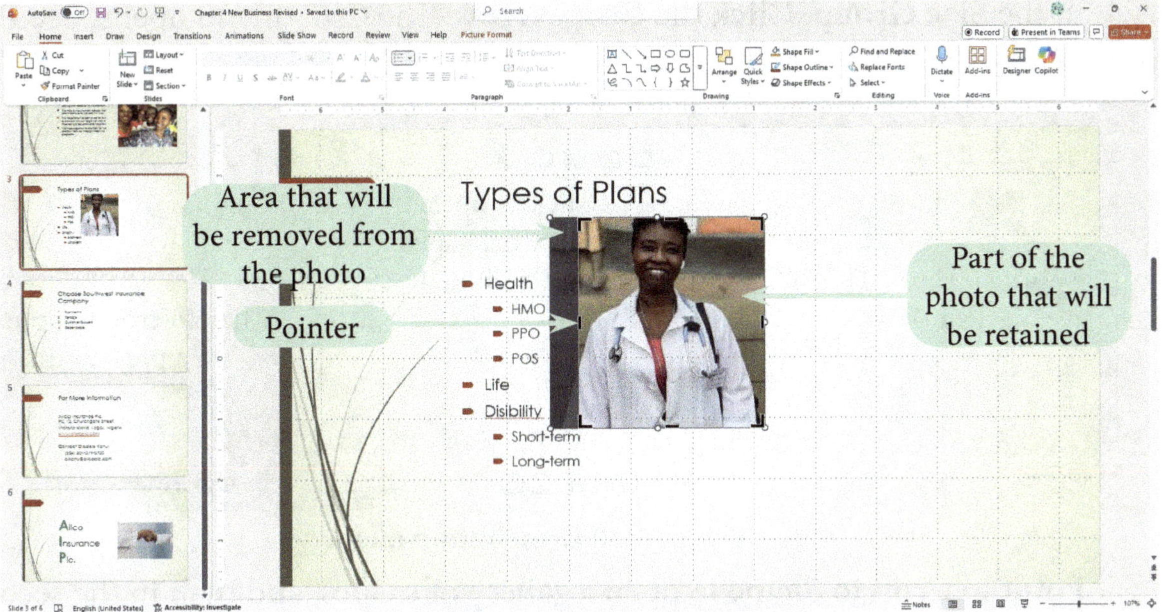

Figure 4-28 Cropped photo

(6). **Move** the **Pointer** on top of the photo so that the pointer changes to the move pointer ✥, **Press** and **Hold** the **Mouse** button, and then **Drag** the photo to the **Right** until the right side of the doctor's arm is next to the right edge of the visible part of the photo. See **Figure 4-29**.

(7). **Click** the **Crop** button again. The Crop feature turns off, but the photo is still selected, and the Picture Format tab is still the active tab.

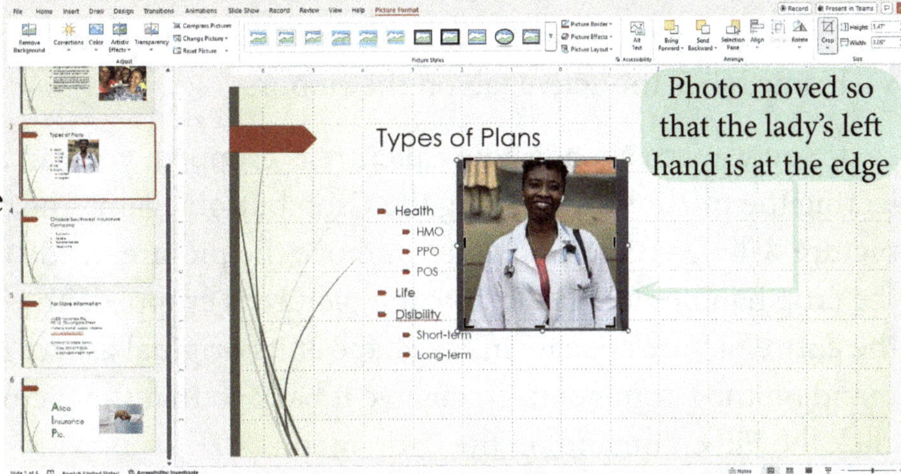

Figure 4-29 Photo moved inside cropped area

When you crop a picture to a shape, the picture fills the shape. Oladele wants you to crop the photo on Slide 6 (the last slide) so that it fills a cross shape.

To crop the photo on Slide 6 to a shape:

(1). **Display Slide 6** (the last slide), **Click** the photo to select it, and then **Click** the **Picture Format Tab**, if necessary.

(2). In the **Size Group**, **Click** the **Crop Arrow**. The Crop menu opens. See **Figure 4-30**.

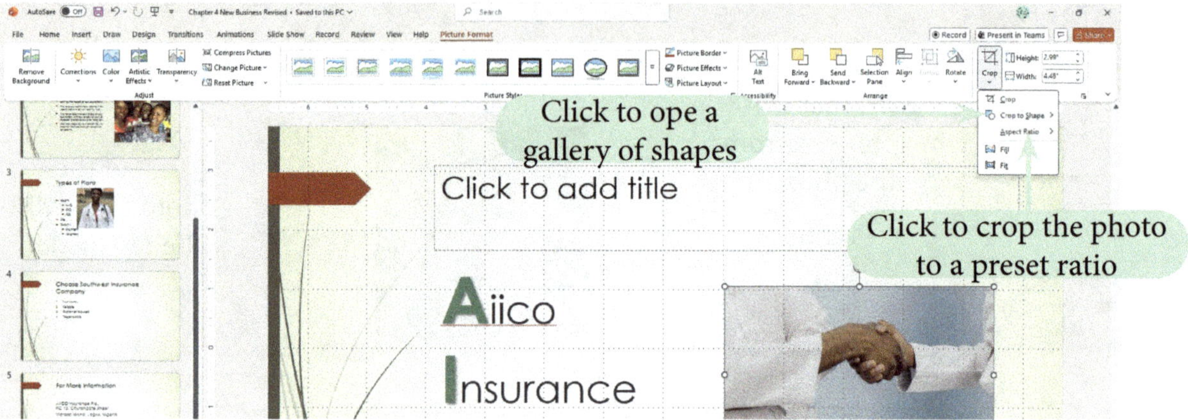

Figure 4-30 Crop button menu

(3). **Point** to **Crop to Shape** to open a gallery of shapes, and then in the second row under Basic Shapes, **Click** the **Cross shape**. The photo is cropped to a cross shape. Notice that the rectangular selection border of the original photo is still showing.

(4). In the **Size Group**, **Click** the **Crop** button. You can now see the cropped portions of the original, rectangle photo that are shaded gray.

(5). **Click** a **blank area** of the slide. The picture is no longer selected, and the Home tab is the active tab on the ribbon.

If you want to change how a picture is cropped, you can adjust the crop by selecting the picture and clicking the Crop button again. The cropped portion of the picture will be visible, and you can move the picture inside the crop marks or drag the crop handles to change how the picture is cropped. To change the crop so that the entire picture appears in the shape at its original aspect ratio, click the Fit command on the Crop menu. To change it back so that the picture fills the shape again, click the Fill command on the Crop menu.

RESIZING AND MOVING OBJECTS

You can resize and move any object to best fit the space available on a slide. One way to resize an object is to drag a sizing handle. **Sizing Handles** are small circles at the corners, and often the edges, of a selected object. When you use this method, you can adjust the size of the object so that it best fits the space visually. If you need to size an object to exact dimensions, you can modify the measurements in the Size group on the Format tab that appears when you select the object.

The **Aspect Ratio** of an object is the proportional relationship between an object's height and width. Pictures and other objects that cause the Picture Format tab to appear when you select them have their aspect ratios locked. This means that if you resize the object by dragging a corner sizing handle or by changing the measurement in either the Height or the Width box in the Size group on the Picture Format tab, both the height and the width of the object will change by the same proportions. However, if you drag one of the sizing handles on the side of the object, you will override the locked aspect ratio setting and resize the object only in the direction you drag. Generally, you do not want to do this with photos because the images will become distorted.

If you want to reposition an object on a slide, you drag it. If you need to move a selected object just a very small distance on the slide, you can press one of the ARROW keys to nudge it in the direction of the arrow. To move it in even smaller increments, press and hold CTRL while you press an ARROW key. When you drag an object on a slide, smart guides, dashed red lines, appear as you drag to indicate the center and the edges of the object, other objects, and the slide itself. Smart guides can help you position objects so that they are aligned and spaced evenly.

You need to resize and move the photos you inserted so the slides are more attractive.

To move and resize the photos on Slides 2, 3, and 6:

(1). **Display Slide 2** ("About Us"), **Click** the photo, and then **position** the **pointer** on the **top-middle sizing handle** so that the pointer changes to the **Double-Headed Vertical Pointer** ⇕ .

(2). **Press** and **Hold** the **Mouse** button so that the pointer changes to the thin cross pointer +, drag the top-middle sizing handle up approximately two inches, and then release the mouse button. The photo is two inches taller, but the image is distorted. You can undo the change you made.

(3). On the **Quick Access Toolbar**, **Click** the **Undo** button. The photo returns to its original size. You need to resize the photo by dragging a corner sizing handle to maintain the aspect ratio.

(4). **Click** the **Picture Format tab** if necessary, and then note the measurements in the Height and Width boxes in the Size group. The photo is 2.02 inches high and 4.72 inches wide (yours might differ).

(5). **Position** the **Pointer** on the T**op-Left corner sizing handle** so that it changes to the double-headed **Diagonal Pointer**, **Press** and **Hold** the **Mouse** button so that the pointer changes to the thin **Cross I+nter** , and then **drag** the **top-left sizing handle up**. Even though you are dragging in only one direction, because you are dragging a corner sizing handle, both the width and height change proportionately to maintain the aspect ratio of the photo.

(6). When the left edge of the photo is aligned with the gridline dots that are below the **negative 0.5-inch mark** on the **Horizontal Ruler**, **Release** the **Mouse** button. In the Height and Width boxes, the measurements changed to reflect the picture's new size.

(7). If the measurement in the **Shape Height** box in the Size group is not **2.5**, **Click** in the **Shape Height** box to **Select** the **current measurement**, **Type 2.5**, and then **Press ENTER**.

(8). **Drag** the photo so that the **right edge** of the photo aligns with the **right edge** of the **slide**, and the **top** of the photo is aligned with the **top** of the **bulleted list**, as shown in **Figure 4-31** on the next page.

In Figure 4-31 on the next page, we see the Picture Format tab at the top of screen on the ribbon is active and Slide 2 is displayed. Gridlines are visible on the slide. In the Size group, the measurements of the photo are set to a height of 2.5 inches and a width of 5.83 inches. Dashed lines called smart guides indicate the following: the alignment of the middle of the photo with the middle of the slide, the alignment of the top of the photo with the top of the bulleted list, and the alignment of the right edge of the photo with the right edge of the slide.

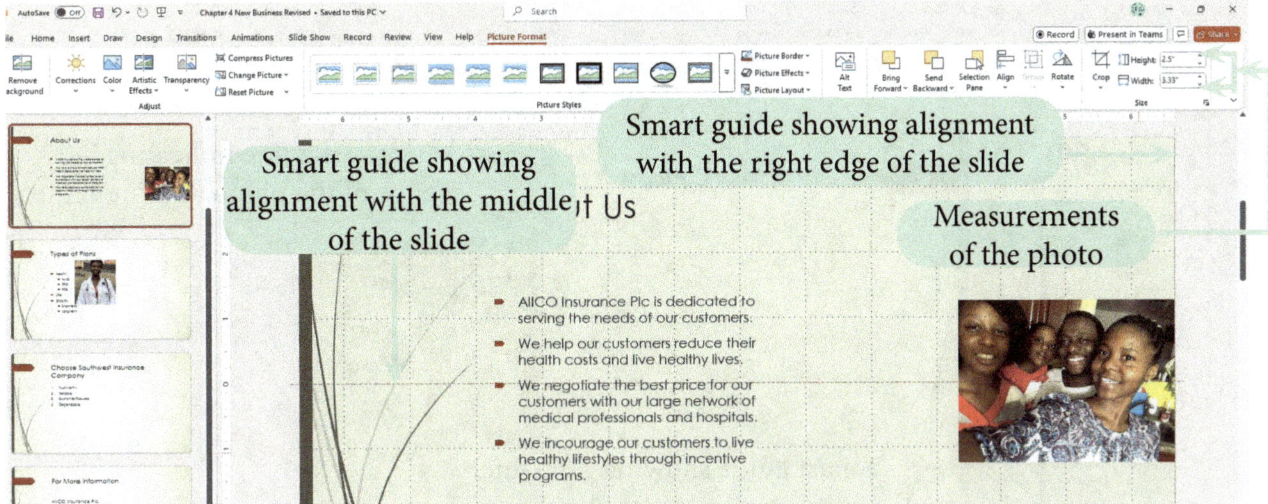

Figure 4-31 Repositioning the photo on Slide 2 using smart guides and gridlines

Are You Having Trouble? If the smart guides do not appear, click the View tab, and then in the Show group, click the Dialog Box Launcher to open the Grid and Guides dialog box. Click the “Display smart guides when shapes are aligned” check box to select it.

(9). **Release** the **Mouse** button. The photo is in its new location.

(10). **Display Slide 3** (“Types of Plans”), **Click** the photo to select it, and then **Click** the **Picture Format Tab** if necessary.

(11). In the **Size group**, **Click** in the **Shape Height** box to select the current measurement, **Type 4**, and then **Press ENTER**. The measurement in the Shape Width box in the Size group changes proportionately to maintain the aspect ratio, and the new measurements are applied to the photo.

(12). **Drag** the **photo down** and to the **right** until a **horizontal smart guide** appears indicating that the top of the photo is about **one-eighth** of an inch above the top of the bulleted list and aligned with the gridline dots at the 1⅜-inch mark on the vertical ruler, and a vertical smart guide appears showing that the left edge of the photo aligns with the center of the slide and with the gridline at the 0-inch mark on the horizontal ruler, as shown in **Figure 4-32** on the next page.

Are You Having Trouble? If a menu appears after you release the mouse button, you clicked the right mouse button when you dragged. Click Move Here on the menu, and then check to make sure the photo is positioned as described in **Step 12**.

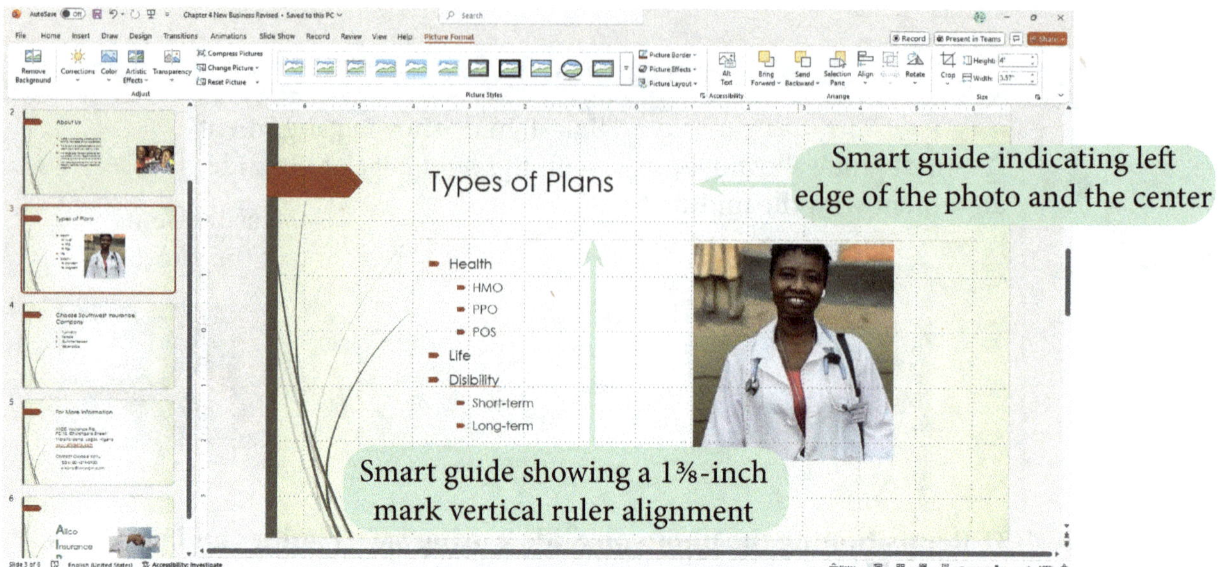

Figure 4-32 Moving the resized photo on Slide 3

(13). When the photo is aligned as shown in Figure 4-32 on the previous page, release the mouse button.

(14). **Display Slide 6** (the last slide), and then **position** the **photo** so that the top of the photo aligns with the **smart guide** that indicates the middle of the title text box and the right edge aligns with the smart guide that indicates the right edge of the title text box.

Text boxes, like other objects that cause the Shape Format tab to appear when selected, do not have their aspect ratios locked. This means that when you resize a text box by dragging a corner sizing handle or changing one measurement in the Shape Height box or the Shape Width box in the Size group, the other dimension does not change.

Like any other object on a slide, you can reposition text boxes. To do this, you must position the pointer on the text box border, anywhere except on a sizing handle, to drag it to its new location.

To improve the appearance of Slide 6, you will resize the text box containing the unnumbered list so that it vertically fills the slide.

To resize the text box on Slide 6.

(1). On **Slide 6** (the last slide), **Click** the unnumbered list to display the text box border.

(2). **Position** the **pointer** on the **top-middle sizing handle** so that it changes to the **Double-Headed Vertical pointer** ↕ , and then **drag** the **sizing handle up** until the **top edge** of the text box aligns with the top edge of the title text placeholder.

(3). **Drag** the **right-middle sizing handle** to the **left** so that the right border of the text box is aligned with the positive **1.5-inch mark** on the ruler. Next, you will shift the text box a little to the **right**.

(4). **Position** the **pointer** on top of the **border** of the text box so that it changes to the move pointer ✥ , and then **drag** the **text box** to the **right** so that the **right border** of the text box aligns with the **smart guide** that indicates the center of the slide. Even though the title text placeholder will not appear during a slide show, you will delete it to see how the final slide will look.

(5). **Click** the **border** of the title text placeholder, and then **Press DELETE**. See **Figure 4-33**.

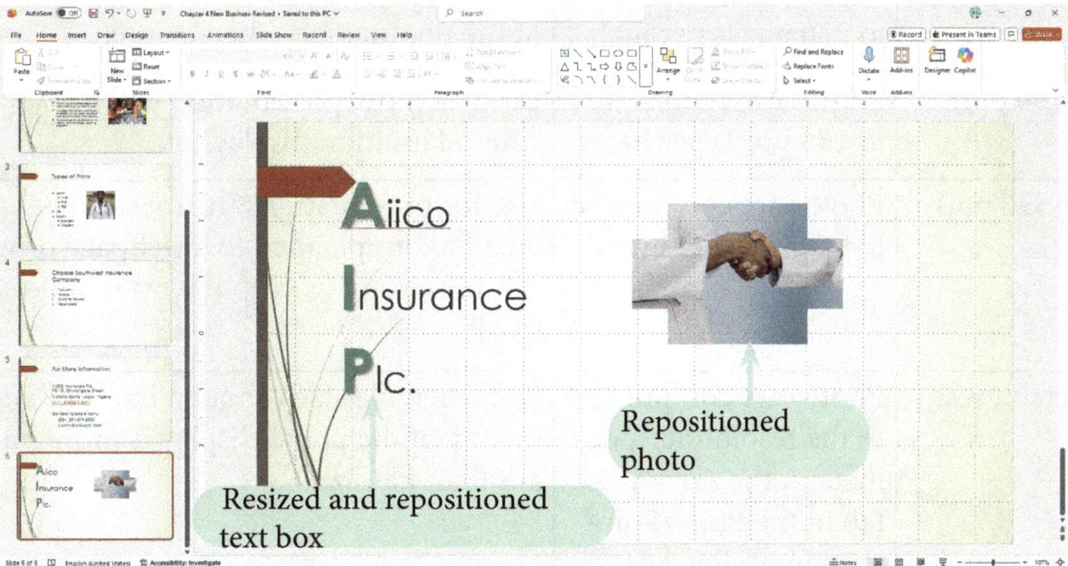

Figure 4-33 Slide 6 with resized text box

(6). **Click the View Tab**, and then **Click** the **Ruler and Gridlines** check boxes to deselect them. The ruler and the gridlines disappear.

(7). **Save** the changes to the presentation.

COMPRESSING PICTURES

When you save a presentation that contains pictures, you can choose to compress the pictures to make the size of the PowerPoint file smaller. See **Figure 4-34** for a description of the compression options available.

Compression Setting	Compression Value	When to Use
High fidelity	Photos are compressed very minimally, and only if they are larger than the slide.	Use when a presentation will be viewed on a high-definition (HD) display, when photograph quality is of the highest concern, and file size is not an issue. This is the default setting.
HD (330 ppi)	Photos are compressed to 330 pixels per inch.	Use when slides need to maintain the quality of the photograph when displayed on HD displays, but file size is of some concern.
Print (220 ppi)	Photos are compressed to 220 pixels per inch.	Use when slides need to maintain the quality of the photograph when printed.
Web (150 ppi)	Photos are compressed to 150 pixels per inch.	Use when the presentation will be viewed on a low-definition display.
E-mail (96 ppi)	Photos are compressed to 96 pixels per inch.	Use for presentations that need to be emailed or uploaded to a webpage or when it is important to keep the overall file size small.
Use default resolution	Photos are compressed to the resolution specified on the Advanced tab in the PowerPoint Options dialog box. (The default setting is High fidelity.)	Use when file size is not an issue, or when quality of the photo display is more important than file size.
No compression	Photos are not compressed at all.	Use when it is critical that photos remain at their original resolution.

Figure 4-34 Photo compression settings

Compressing photos reduces the size of the presentation file, but it also reduces the quality of the photos. Often this trade-off is acceptable because most monitors cannot display high-resolution photos at high-fidelity resolution.

You can change the compression setting for each photo that you insert, or you can change the settings for all the photos in the presentation. If you cropped photos, you also can discard the cropped areas of the photo to make the presentation file size smaller. (Note that when you crop to a shape, the cropped portions are not discarded.) If you insert additional photos or crop a photo after you apply the new compression settings to all the slides, you will need to apply the new settings to the new photos.

Smart Tips

Modifying Photo Compression Settings and Removing Cropped Areas

- After you have added all photos to the presentation file, click any photo in the presentation to select it.
- Click the Picture Format tab. In the Adjust group, click the Compress Pictures button.
- In the Compress Pictures dialog box, click the option button next to the resolution you want to use.
- To apply the new compression settings to all the photos in the presentation, click the Apply only to this picture check box to deselect it.
- To keep cropped areas of photos, click the Delete cropped areas of pictures check box to deselect it.
- Click OK.

You will adjust the compression settings to make the file size of the presentation as small as possible so that Oladele can easily send it or post it for others without worrying about file size limitations on the receiving server.

To modify photo compression settings and remove cropped areas from photos:

(1). On **Slide 6** (the last slide), **Click** the **photo**, and then **Click** the **Picture Format Tab**, if necessary.

(2). In the **Adjust group**, **Click** the **Compress Pictures** button. The Compress Pictures dialog box opens. See **Figure 4-35** on the next page. Under Resolution, the Use default resolution option button is selected. (If an option in this dialog box is gray and is not available for you to select, the photo is a lower resolution than that option.)

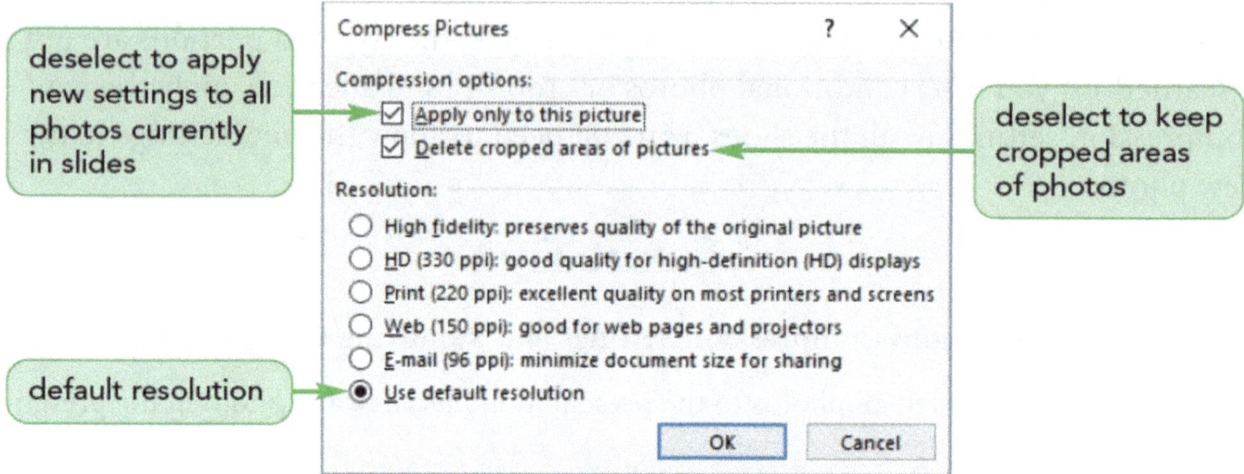

Figure 4-35 Compress Pictures dialog box

(3). **Click the E-mail (96 ppi)** option button. This setting compresses the photos to the smallest size available. At the top of the dialog box under Compression options, the Apply only to this picture check box is selected. You want the settings applied to all the photos in the file.

(4). **Click** the **Apply only to this picture** check box to deselect it. The Delete cropped areas of pictures check box is also selected. You want the presentation file size to be as small as possible, so you'll leave this option selected.

Are You Having Trouble? If the Delete cropped areas of pictures check box is not selected, click it.

(5). **Click OK.**

The dialog box closes and the compression settings are applied to all the photos in the presentation. You can confirm that the cropped areas of photos were removed by examining the photo on Slide 3. (The photo on Slide 6 was cropped to a shape, so the cropped areas on it were not removed, in case you later change to a different shape cropping.)

(6). **Display Slide 3** ("Types of Plans"), **Click** the **photo**, and then **Click** the **Picture Format Tab**, if necessary.

(7). In the **Size group**, **Click** the **Crop** button. The Crop handles appear around the photo, but the portions of the photo that you cropped out no longer appear.

(8). **Click** the **Crop** button again to deselect it, and then save the changes to the presentation.

SMARTSKILLS Changing the Default Compression Settings for Pictures

In PowerPoint, the default compression setting for pictures is High fidelity. This means that High fidelity compression is automatically applied to pictures when the file is saved. You can change this setting if you want. To change the settings, click the File tab to open Backstage view, click Options in the navigation pane to open the PowerPoint Options dialog box, click Advanced in the navigation pane, and then locate the Image Size and Quality section. To choose a different compression setting, click the Default resolution arrow, and then select a setting in the list. To prevent pictures from being compressed at all, click the Do not compress images in file check box. Note that these changes affect only the current presentation.

CONVERTING A LIST TO A SMARTART GRAPHIC

A **SmartArt Graphic** is a diagram that shows information or ideas visually using a combination of shapes and text. Some SmartArt shapes also contain pictures. SmartArt is organized into the following categories:

- **List**—Shows a list of items
- **Process**—Shows a sequence of steps in a process or a timeline
- **Cycle**—Shows a process that is a continuous cycle
- **Hierarchy**—Shows the relationship between individuals or units, such as an organization chart for a company or information organized into categories and subcategories
- Relationship (including Venn diagrams, radial diagrams, and target **diagrams**)—Shows the relationship between two or more elements
- **Matrix**—Shows information placed around two axes
- **Pyramid**—Shows foundation-based relationships

- **Picture**—Provides a location for a picture or pictures that you insert

When you create a SmartArt graphic, you need to choose a SmartArt layout. In SmartArt, a layout is the shapes and arrangement of the shapes in the SmartArt graphic. Once you create a SmartArt graphic, you can easily change the layout to another one if you want.

A quick way to create a SmartArt graphic is to convert an existing list. There are two ways to do this. First, you can try displaying the Design Ideas pane. When the Design Ideas pane shows options for a slide that contains a list, some of the layouts include the list transformed into a SmartArt graphic. The other way you can create a SmartArt graphic from a list is to click the Convert to SmartArt Graphic button in the Paragraph group on the Home tab.

When you change a list to SmartArt, each first-level item in the list is converted to a shape in the SmartArt. If the list contains subitems, you might need to experiment with different layouts to find one that best suits the information in your list.

Smart Tips

Converting a List into SmartArt

- Display the slide containing the list you want to convert to SmartArt.
- If the Design Ideas pane does not open, click the Design tab on the ribbon, and then in the Designer group, click the Design Ideas button.
- In the Design Ideas pane, select the SmartArt and slide layout that you want to use.

or

- Click anywhere in the list that you want to convert.
- In the Paragraph group on the Home tab, click the Convert to SmartArt button, and then click More SmartArt Graphics.
- In the Choose a SmartArt Graphic dialog box, select the desired SmartArt category in the list on the left.
- In the center pane, click the SmartArt you want to use.
- Click OK.

Oladele wants you to change the numbered list on Slide 4 into a SmartArt diagram.

To convert the list on Slide 4 into SmartArt:

(1). **Display Slide 4** ("Choose AIICO Insurance Plc.").

(2). If the **Design Suggestions Pane** does not open, **Click** the **Design tab** on the ribbon, and then, in the **Designer group**, **Click** the **Design Suggestions** button. The Design Suggestions pane opens on the right. Your screen will look similar to the one shown in **Figure 4-36**.

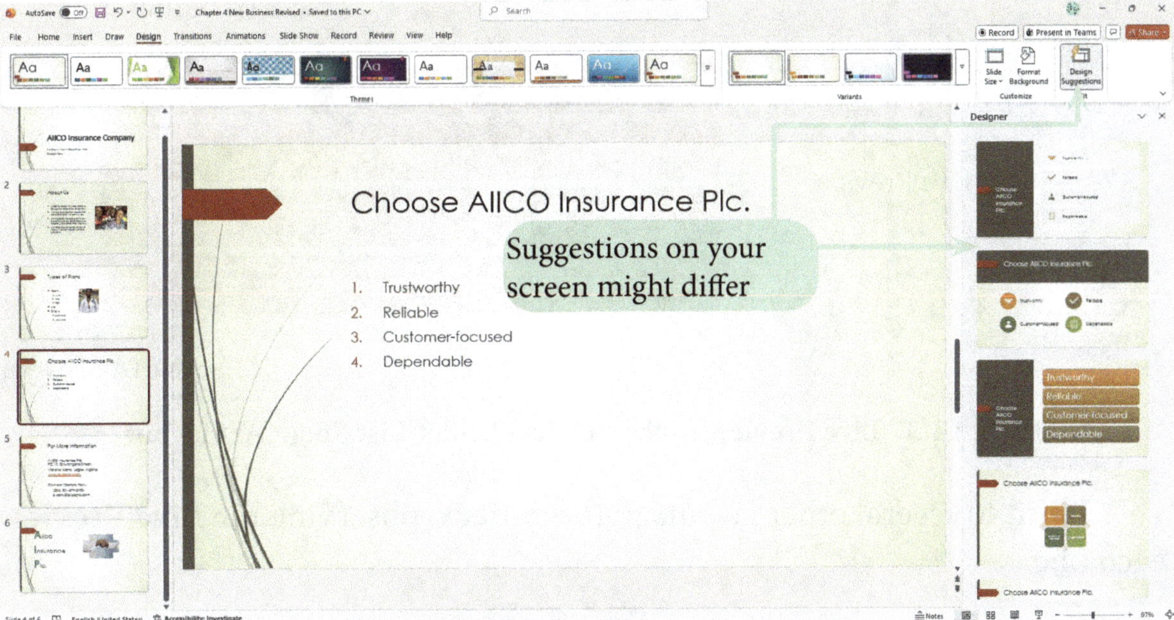

Figure 4-36 Compress Pictures dialog box

(3). In the **Design Suggestions Pane, Click** several of the thumbnails to see the effect on the slide.

(4). After you are finished exploring the layouts in the **Design Suggestions Pane, Click** the **Undo** button on the **Quick Access Toolbar**. The slide resets to its original layout.

(5). **Close** the **Design Suggestions Pane**, and then on the slide, **Click** anywhere in the list.

(6). On the **Ribbon, Click** the **Home Tab**, and then in the **Paragraph Group**,

Click the **Convert to SmartArt Graphic** button. A gallery opens listing SmartArt layouts.

(7). **Point** to the **first layout**. The ScreenTip identifies this layout as the **Vertical Bullet List layout**, and Live Preview shows you what the list will look like with that layout applied. See **Figure 4-37**.

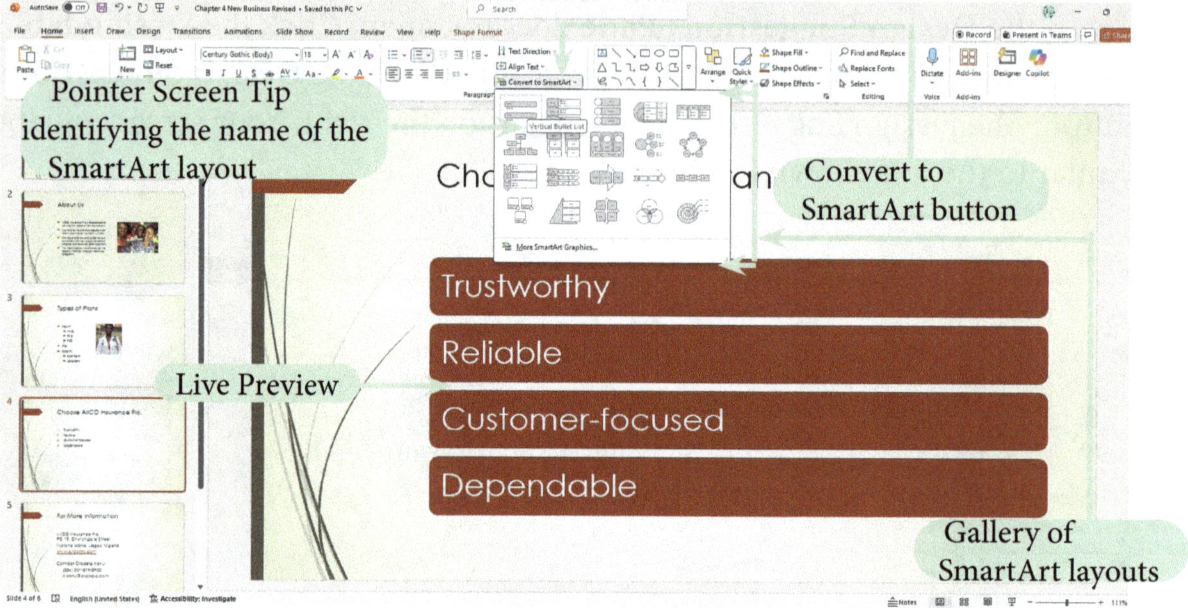

Figure 4-37 Live Preview of the Vertical Bullet List SmartArt layout

(8). **Point** to several other layouts in the **gallery**, observing the **Live Preview** of each one.

(9). In the **Gallery**, **Click** the **Horizontal Bullet List** layout (the last layout in the first row). The list is changed to a **SmartArt graphic** with each first-level item in the top part of a box. On the ribbon, the SmartArt contextual tabs appear, and the SmartArt Design tab is selected. In the Create Graphic group, the Text Pane button is not selected.

Are You Having Trouble? If the Text Pane button is selected, skip **Step 10**.

(10). On the **SmartArt Design Tab**, in the **Create Graphic Group**, **Click** the **Text Pane** button. The button is selected, and the Text pane appears to the left of the SmartArt graphic. Oladele doesn't like the layout you chose and wants you to use a different layout.

(11). On the **SmartArt Design Tab**, in the **Layouts Group**, **Click** the **More** button ▾. A gallery of SmartArt layouts appears, which is very different from the one seen earlier in Figure 4-37 on the previous page.

(12). At the bottom of the **Gallery**, **Click More Layouts**. The **Choose** a **SmartArt Graphic** dialog box opens. See **Figure 4-38**. You can click a category in the left pane to filter the middle pane to show only the layouts in that category.

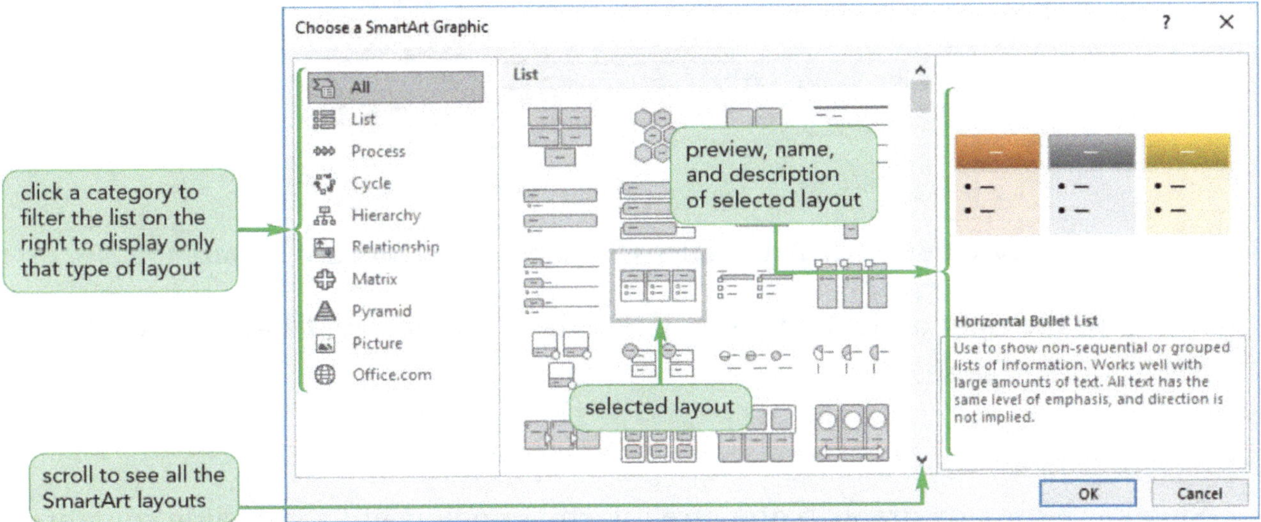

Figure 4-38 Choose a SmartArt Graphic dialog box

(13). In the left pane, **Click List**, and then in the **middle pane**, **Click** the **Vertical Box List Layout**, using the ScreenTips to identify it. The right pane changes to show a description of that layout. Your screen might differ, though.

(14). **Click OK**. The dialog box closes, and each of the first-level items in the list appears in the colored shapes in the diagram. The items also appear as a bulleted list in the Text pane. See **Figure 4-39** on next page.

In Figure 4-39 on the next page, Slide 4 shows that the SmartArt diagram is selected. The SmartArt Tools contextual tabs (Design and Format) appear on the ribbon. The SmartArt Took Design tab is active. In the Create Graphic group, the Text Pane button was clicked, displaying the Text Pane to the left of the SmartArt diagram. Clicking the button again would close the pane. The text from the SmartArt graphic appears in the Text Pane, and the name of the SmartArt layout (Vertical Box List) appears at the bottom of the Text Pane. The Collapse text pane button, which closes the text pane which clicked is also visible.

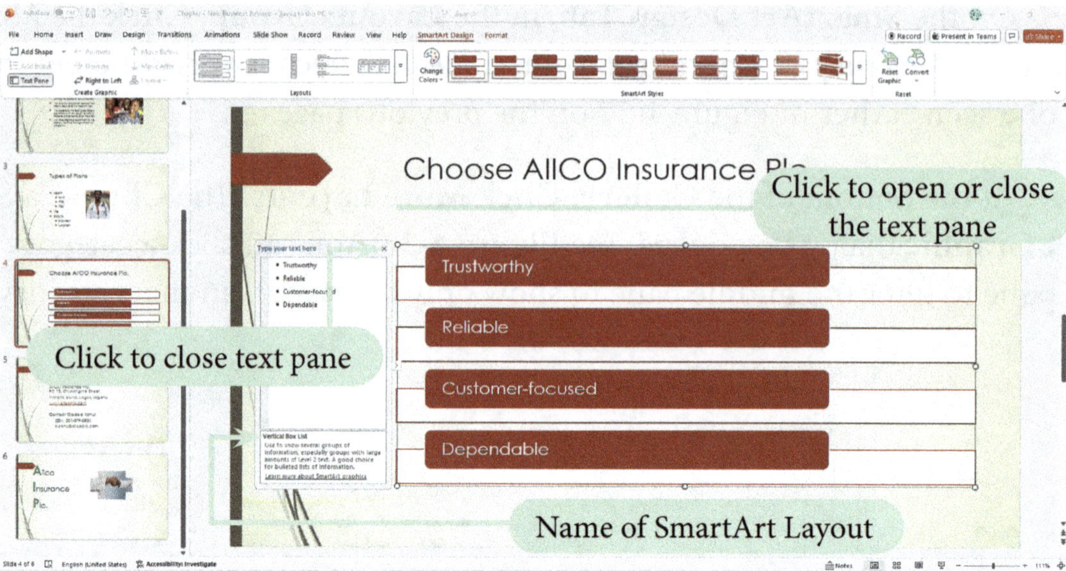

Figure 4-39 SmartArt graphic with the Vertical Box List layout

(15). To the right of the **text pane**, **Click** the **Collapse Text Pane** button [>]. The text pane closes.

ADDING SPEAKER NOTES

Speaker Notes, or simply **Notes**, are information you add about slide content to help you remember to bring up specific points during the presentation. Speaker notes should not contain all the information you plan to say during your presentation, but they can be a useful tool for reminding you about facts and details related to the content on specific slides.

You add notes in the **Notes Pane**, which is an area at the bottom of the window that you can use to type speaker notes. The notes are not visible when you present a slide show.

You also can switch to **Notes Page View**, in which a reduced image of the slide appears in the top half of the window and the notes for that slide appear in the bottom half. Notes are not visible to the audience during a slide show.

To add notes to Slides 3 and 5:

(1). **Display Slide 5** ("For More Information"), and then, on the status bar, **Click** the **Notes** button. The Notes pane appears below **Slide 5** with "**Click to add notes**" as placeholder text. See **Figure 4-40** on the next page.

Figure 4-40 Notes pane below Slide 5

(2). **Click** in the **Notes Pane**. The placeholder text disappears, and the insertion point is in the **Notes pane**.

(3). **Type Hand out contact information to audience. Use the link to demonstrate how to use the website.** in the Notes pane.

(4). **Display Slide 3** (**"Types of Plans"**), and then **Click** in the **Notes pane**.

(5). **Type Briefly describe the differences among HMO, PPO, and POS plans.** in the Notes pane.

(6). **Click** the **View Tab** on the **ribbon**, and then in the **Presentation Views Group**, **Click** the **Notes Page** button. Slide 3 appears in Notes Page view. **See Figure 4-41**.

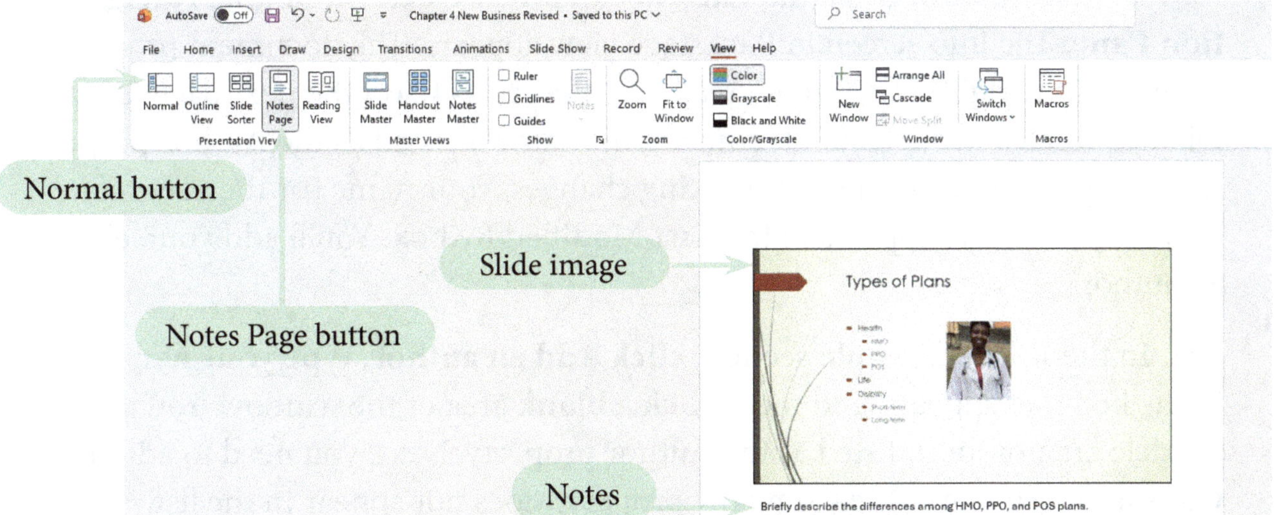

Figure 4-41 Notes pane below Slide 5

(7). In the **Note**, **Click** after the period at the end of the sentence, **press SPACEBAR**, and then **type Be sure to describe the cost differences in premi-**

ums. (including the period).

(8). On the **View Tab**, in the **Presentation Views Group**, **Click** the **Normal button** to return to **Normal View**. The Notes pane stays open until you close it again.

(9). On the **status bar**, **Click** the **Notes** button to close the Notes pane, and then **Save** the changes to the presentation.

EDITING COMMON FILE PROPERTIES

File properties are identifying information—characteristics—about a file that is saved along with the file that help others understand, identify, and locate the file. Common properties are the title, the author's name, and the date the file was created. You can use file properties to organize presentations or to search for files that have specific properties. To view or modify properties, you need to display the Info screen in Backstage view.

Oladele wants you to modify the Author property by adding yourself as an author and he wants you to add the Company property.

To add common file properties:

(1). On the **Ribbon**, **Click** the **File Tab**, and then **Click Info** in the **Navigation Pane**. The Info screen in Backstage view appears. The document properties appear on the right side of the screen. See **Figure 4-42** on the next page. Because Oladele created the original document, his name is listed as the Author property. Because you saved the file after making changes, your name (or the user name on your computer) appears in the Last Modified By box. You'll add yourself as an author.

(2). In the **Related People** section, **click Add an author**, **type your name** in the box that appears, and then **Click** a **blank** area of the window. You and Oladele are now both listed as the Author property. Next, you need to add the **Company property**. The Company property does not appear in the list.

(3). **Scroll down**, and then at the bottom of the **Properties list**, **Click Show All Properties**. The Properties list expands to include all of the common document properties, including the Company property.

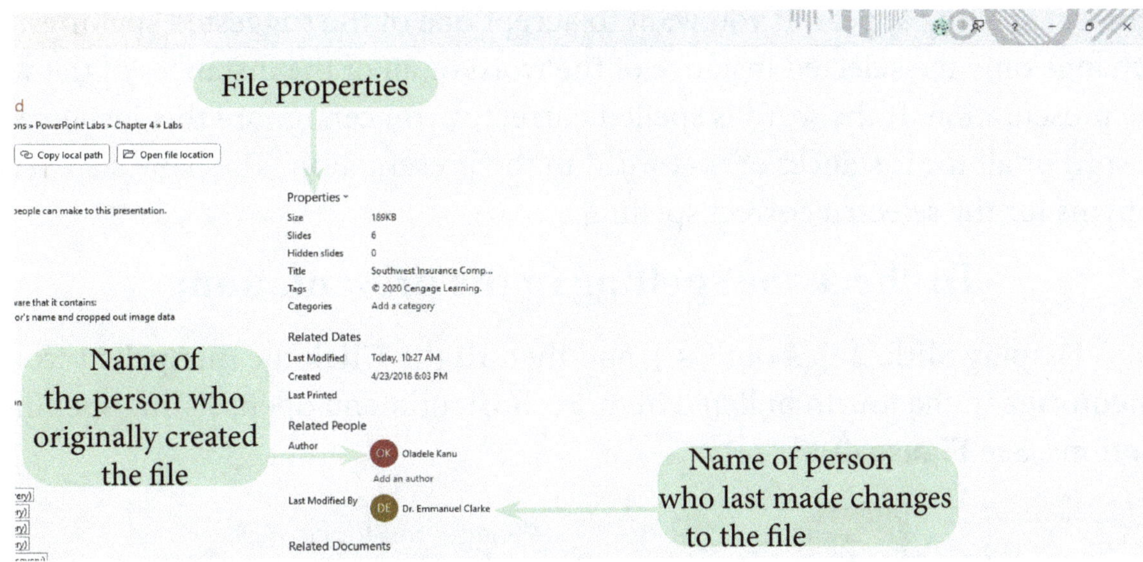

Figure 4-42 File properties on the Info screen in Backstage view

(4). Next to Company, **Click Specify the company**, **Type AIICO Insurance Plc.**, and then **Click** a **blank area** of the screen. You are finished adding properties to the file.

(5). **Scroll up** if necessary, and at the top of the **navigation pane**, **Click** the **Back** button to return to **Slide 3** in **Normal view**.

CHECKING SPELLING

You should always check the spelling and grammar in your presentation before you finalise it. To make this task easier, you can use PowerPoint's spelling checker. You can quickly tell if there are words on slides that are not in the built-in dictionary by looking at the Spelling button at the left end of the status bar. If there are no words flagged as possibly misspelled, the button is ; if words are flagged, the button changes ton . To indicate that a word might be misspelled, a wavy red line appears under it.

To correct misspelled words, you can right-click a flagged word to see a list of suggested spellings on the shortcut menu, or you can check the entire presentation to locate possible misspellings. To check the spelling of all the words in the presentation, you click the Spelling button in the Proofing group on the Review tab. This opens the Spelling pane to the right of the displayed slide and starts the spell check from the current slide. When a possible misspelled word is found, suggestions for the correct spelling appear. If you want to accept one of the suggested spellings, you can change only the selected instance of the word or all of the instances of the word in the presentation. If the word is spelled correctly, you can ignore this instance of that word or all the instances of that word in the presentation. The pane also lists synonyms for the selected correct spelling.

To check the spelling in the presentation:

(1). **Display Slide 2** ("About Us"), and then **Right-Click** the misspelled word **incourage** in the fourth bulleted item. A shortcut menu opens listing spelling options. See **Figure 4-43**.

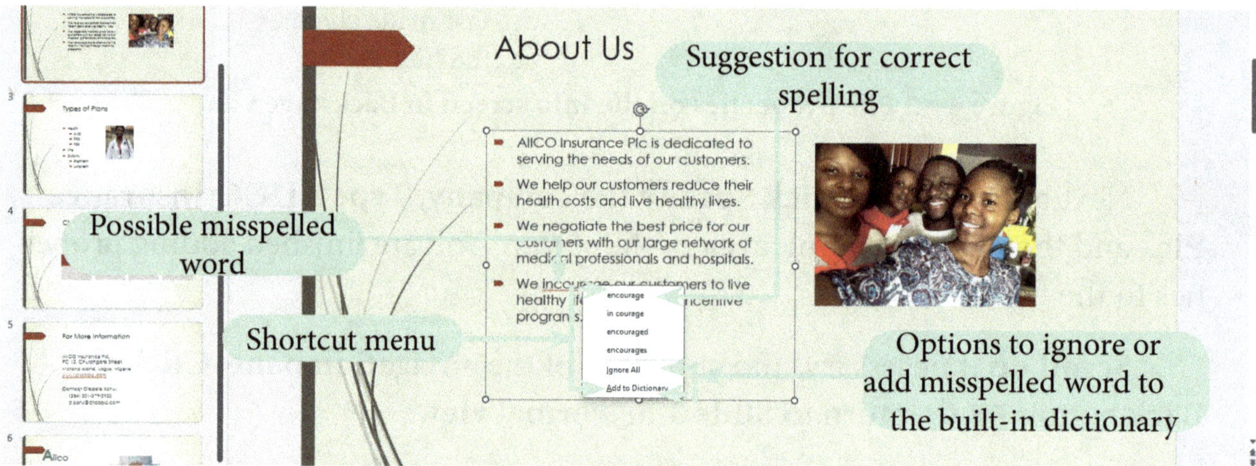

Figure 4-43 Shortcut menu for a misspelled word

Are You Having Trouble? If the word "incourage" does not have a wavy red line under it, click the Review tab, and then in the Proofing group, click the Spelling button. The wavy red line should now appear. Right-click "incourage," continue with **Step 2**, and then do not do **Step 3**.

(2). On the **Shortcut Menu, Click encourage**. The menu closes, and the spelling is corrected.

(3). **Click** the **Review tab**, and then in the **Proofing Group**, **Click** the **Spelling** button. The **Spelling pane** opens to the right of the displayed slide, and the next possible misspelled word, "**Disibility**" on **Slide 3** ("Types of Plans"), is selected on the slide and in the Spelling pane. See **Figure 4-44**. In the Spelling pane, the first suggested correct spelling is selected. The selected correct spelling also appears at the bottom of the pane, with synonyms for the word listed below it and a speaker icon next to it.

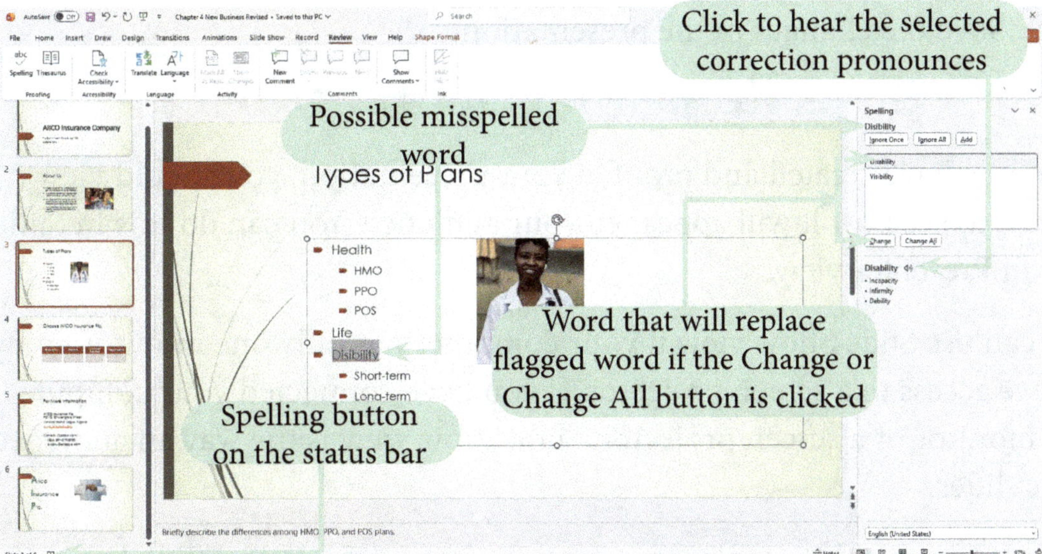

Figure 4-44 Shortcut menu for a misspelled word

(4). In the **Spelling pane**, **Click** the speaker icon . A voice says the word "**Disability**."

(5). In the list of suggested corrections, **click Visibility**. The word at the bottom of the pane changes to "**Visibility**," and the synonyms change also.

(6). In the list of suggested corrections, **Click Disability**, and then **click Change**. The word is corrected, and the next slide containing a possible misspelled word, **Slide 5** ("**For More Information**"), appears with the flagged word, "**Kanu**," selected and listed in the Spelling pane. This is Oladele's last name, so you want the spell checker to ignore every instance of this word, not just this instance.

(7). In the pane, **Click Ignore All**. Because that was the last flagged word in the presentation, the Spelling pane closes, and a dialog box opens telling you that the spell check is complete.

Are You Having Trouble? If the spell checker finds any other misspelled words, correct them.

(8). **Click OK**. The dialog box closes.

(9). **Display Slide 1** (the title slide). Oladele's last name no longer has a wavy red line under it because you clicked Ignore All when it was flagged as a possible misspelled word on Slide 5.

(10). **Save** the changes to the presentation.

RUNNING A SLIDE SHOW

After you have created and proofed your presentation, you should view it as a slide show to see how it will appear to your audience. You can do this in Slide Show view or in Presenter view.

You can use Slide Show view if your computer has only one monitor and you don't have access to a screen projector. If you have connected your computer to a second monitor or a screen projector, Slide Show view is the way an audience will see your slides.

In Slide Show and Presenter views, you can move from one slide to another in several ways. **Figure 4-45** describes the methods you can use to move from one slide to another during a slide show.

Desired Result	Method
To display the next slide	• Press SPACEBAR. • Press ENTER. • Press RIGHT ARROW. • Press DOWN ARROW. • Press PGDN. • Press N. • Click the slide. • In Slide Show view, move the pointer to display the buttons in the lower-left corner of the slide, and then click the Advance to the next animation or slide button. • In Presenter view, click the Advance to the next animation or slide button. • Right-click the slide, and then on the shortcut menu, click Next.

To display the previous slide	• Press BACKSPACE. • Press LEFT ARROW. • Press UP ARROW. • Press PGUP. • Press P. • In Slide Show view, move the pointer to display the buttons in the lower-left corner of the slide, and then click the Return to the previous animation or slide button . • In Presenter view, click the Return to the previous animation or slide button • Right-click the slide, and then on the shortcut menu, click Previous.
To display a specific slide	In Slide Show view, move the pointer to display the buttons in the lower-left corner of the slide, click the See all slides button and then click the thumbnail of the slide you want to display. In Presenter view, click the See all slides button , and then click the thumbnail of the slide you want to display. Type the number of the slide you want to display, and then press ENTER. Right-click the slide, and then on the shortcut menu, click See all slides.
To display the first slide	Press HOME.
To display the last slide	Press END.
To end the slide show	Press ESC. Right-click the slide, and then on the shortcut menu, click End Show.

Figure 4-45 Methods of moving from one slide to another during a slide show

Oladele asks you to review the slide show in Slide Show view to make sure the slides look professional.

To use Slide Show view to view the final presentation:

(1). On the **Quick Access Toolbar**, **Click** the **Start From Beginning** button. **Slide 1** appears on the screen in Slide Show view. Now you need to advance the slide show.

(2). **Press SPACEBAR**. **Slide 2** ("About Us") appears on the screen.

(3). **Click** the **Mouse** button. The next slide, **Slide 3** ("Types of Plans"), appears on the screen.

(4). **Press BACKSPACE**. The previous slide, **Slide 2**, appears again.

(5). **Move** the **mouse** to display the buttons in the **lower-left corner** of the slide, and then **Click** the **See all slides** button. All of the slides in the file are displayed as thumbnails on the screen, similar to Slide Sorter view.

(6). **Click** the **Slide 5** thumbnail. **Slide 5** ("For More Information") appears on the screen.

(7). **Move** the **mouse** to display the pointer, and then **position** the **pointer** on the **website address** https://www.aiicoplc.com. The pointer changes to the **pointing finger pointer** to indicate that this is a link, and the ScreenTip that appears shows the full website address including "http://". **Click** the link to open your web browser and display the **AIICO's** website to your audience. Because you moved the pointer, the faint row of buttons appears in the lower-left corner. See **Figure 4-46** on the next page.

(8). **Move** the **Pointer** again, if necessary, to display the buttons that appear in the lower-left corner of the screen, and then **Click** the **Return** to the **previous animation** or **slide** button twice to redisplay **Slide 3** ("Types of Plans").

Are You Having Trouble? If you can't see the buttons at the bottom of the screen, move the pointer to the lower-left corner so it is on top of the first button to darken that button, and then move the pointer to the right to see the rest of the buttons.

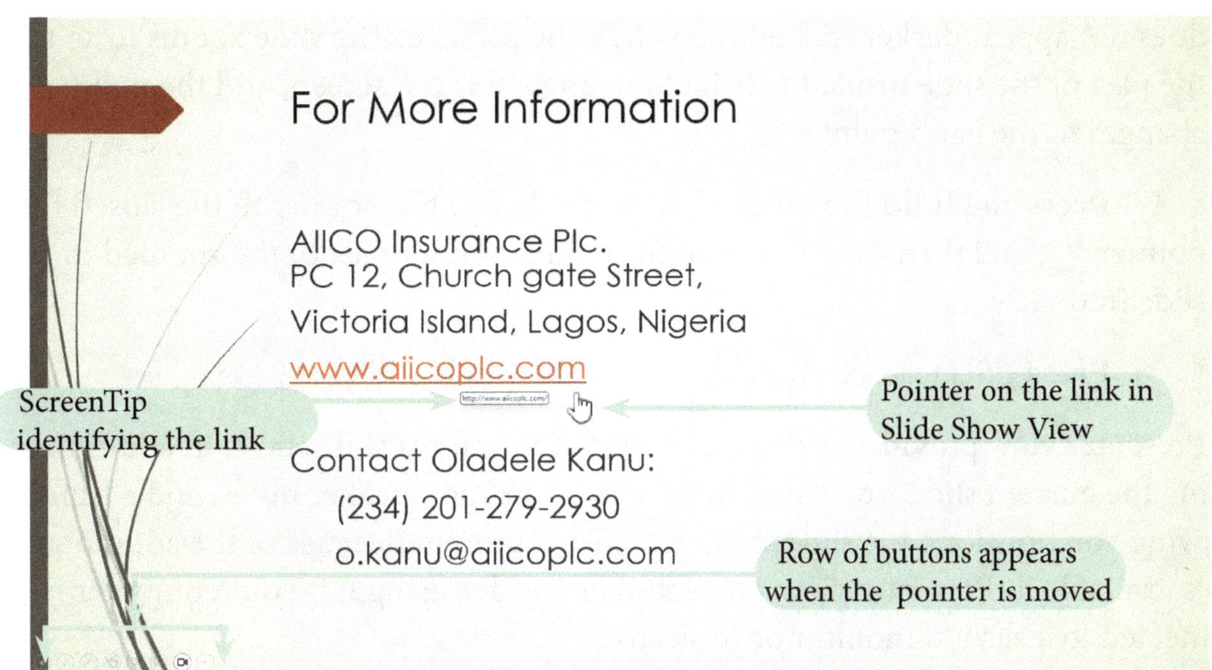

Figure 4-46 Link and row of buttons in Slide Show view

(9). **Display** the buttons at the **bottom** of the screen again, and then **Click** the **Zoom** into the slide button ⊕. The **pointer changes** to the **zoom** in pointer ⊕, and three-quarters of the slide is darkened. See **Figure 4-47**.

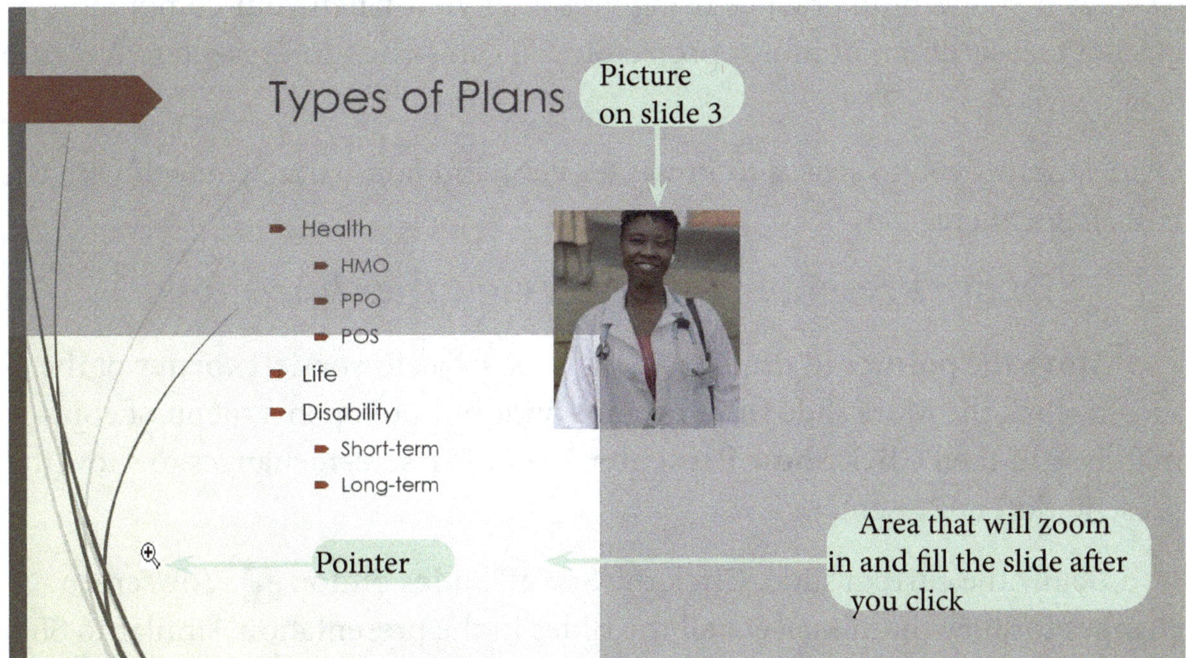

Figure 4-47 Zoom feature activated in Slide Show view

(10). **Move** the **pointer** on top of the **picture** so that the top part of the picture does not appear darkened, and then **click** the **picture**. The slide zooms in so that the part of the slide inside the bright rectangle fills the screen, and the pointer changes to the hand pointer .

(11). **Press** and **hold** the **mouse** button to change the pointer to the closed fist pointer , and then **drag** to the **right** to pull another part of the zoomed in slide into view.

(12). **Press ESC** to zoom back out to see the whole slide.

Presenter view provides additional tools for running a slide show. In addition to seeing the current slide, you can also see the next slide, speaker notes, and a timer showing you how long the slide show has been running. Because of the additional tools available in Presenter view, you should consider using it if your computer is connected to a second monitor or projector.

If your computer is connected to a projector or second monitor, and you start a slide show in Slide Show view, Presenter view starts on the computer and Slide Show view appears on the second monitor or projection screen. If, for some reason, you don't want to use Presenter view in that circumstance, you can switch to Slide Show view. If you want to practice using Presenter view when your computer is not connected to a second monitor or projector, you can switch to Presenter view from Slide Show view.

Oladele wants you to switch to Presenter view and familiarize yourself with the tools available there.

To use Presenter view to review the slide show:

(1). **Move** the **pointer** to display the buttons in the **lower-left corner** of the screen, click the More slide show options button to open a menu of commands, and then **Click Show Presenter View**. The screen changes to show the presentation in Presenter view.

(2). Below the current slide, **Click** the **See all slides** button . The screen changes to show thumbnails of all the slides in the presentation, similar to Slide Sorter view.

(3). **Click** the **Slide 4** thumbnail. Presenter view reappears, displaying **Slide 4**

("**Choose AIICO Insurance Plc.**") as the current slide.

(4). **Click** anywhere on **Slide 4**. The slide show advances to display **Slide 5** ("For More Information").

(5). At the bottom of the screen, **click** the **Advance to the next animation or slide** button. **Slide 6** (the last slide) appears.

(6). **Click** the **More slide show options** button, and then **click Hide Presenter View**. **Slide 6** appears in Slide Show view.

(7). **Press SPACEBAR**. A black slide appears displaying the text "**End of slide show, click to exit.**"

(8). **Press SPACEBAR** again. **Presenter view closes**, and you return to Normal view.

SmartSkills **Decision Making: Displaying a Blank Slide During a Presentation**

Sometimes during a presentation, the audience has questions about the material and you want to pause the slide show to respond. Or you might want the audience to focus its attention on you instead of on the visuals on the screen. In these cases, you can display a black or white blank slide. Some presenters plan to use blank slides and insert them at specific points during their slide shows. Planning to use a blank slide can help you keep your presentation focused. It can also remind you that the purpose of the PowerPoint slides is to provide visual aids to enhance your presentation; the slides themselves are not the presentation.

If you did not create blank slides in your presentation file, but during your presentation you feel you need to display a blank slide, you can easily do this in Slide Show or Presenter view. To display a blank black slide, press B. To display a blank white slide, press W. You can also click the More slide show options button in Slide Show view, click Screen, and then Black Screen or White Screen. In Presenter view, you can click the More slide show options button, point to Screen, and then click Black Screen or White Screen. Or you can right-click the screen, point to Screen on the menu, and then click Black Screen or White Screen. To remove the black or white slide and redisplay the slide that had been on the screen before you displayed the blank slide, press any key on the keyboard or click anywhere on the screen. In Presenter view, you can also click the Black or unblack slide show button to toggle a blank slide on or off.

PRINTING A PRESENTATION

Before you deliver your presentation, you might want to print it. PowerPoint provides several printing options. For example, you can print the slides in color, grayscale (white and shades of gray), or pure black and white, and you can print one, some, or all of the slides in several formats.

You use the Print screen in Backstage view to set print options such as specifying a printer and color options. First, you will add your name to the title slide.

To add your name to the title slide and choose a printer and color options:

(1). **Display Slide 1**, **Click** after "**Kanu**" in the subtitle, **Press ENTER**, and then **Type your name**.

(2). **Click** the **File tab** to display Backstage view, and then **Click Print** in the navigation pane. Backstage view changes to display the Print screen. The Print screen contains options for printing your presentation, and a preview of the first slide as it will print with the current options. See **Figure 4-48**.

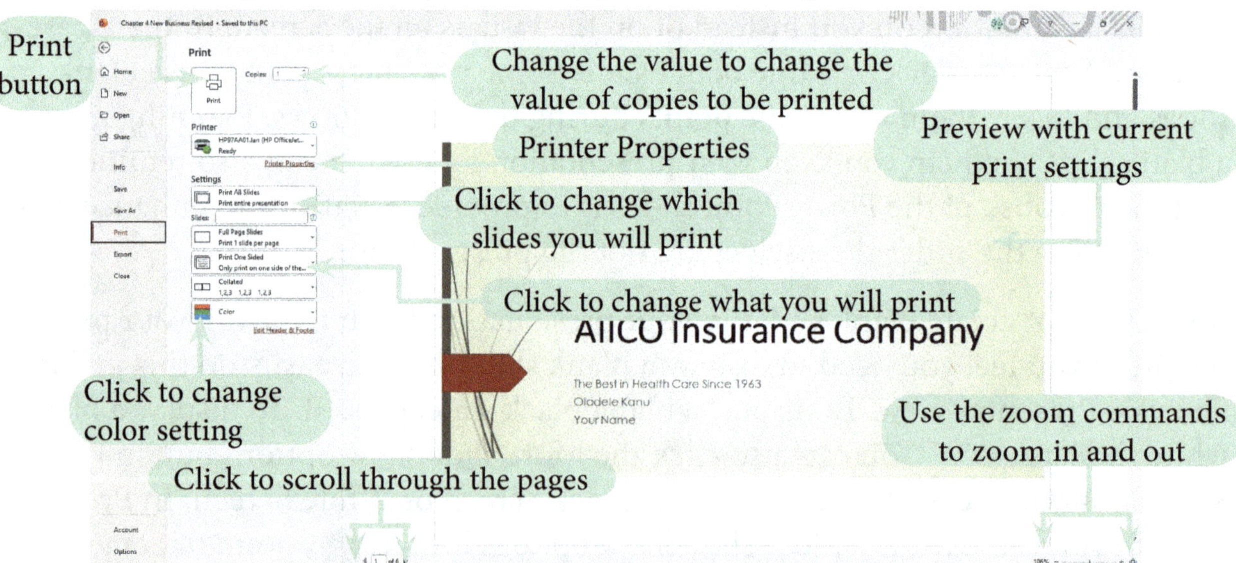

Figure 4-48 Print screen in Backstage view

Are You Having Trouble? If your screen does not match Figure 4-48, click the first button below Settings, click Print All Slides, click the second button below Settings, and then click Full Page Slides.

(3). If you are connected to a network or to more than one printer, make sure the printer listed in the Printer box is the one you want to use; if it is not, **Click** the **Printer** button, and then **click** the correct printer in the list.

(4). **Click** the **Printer Properties** link to open the Properties dialog box for your printer. Usually, the default options are correct, but you can change any printer settings, such as print quality or the paper source, in this dialog box.

(5). **Click Cancel** to close the Properties dialog box. Now you can choose whether to print the presentation in color, black and white, or grayscale. If you plan to print in black and white or grayscale, you should change this setting so that you can see what your slides will look like without color and to make sure they are legible.

(6). **Click** the **Color** button, and then **click Grayscale**. The preview changes to grayscale.

(7). At the bottom of the **preview pane**, **click** the **Next Page** button ▸ twice to display **Slide 3** ("Types of Plans"). The slides are legible in grayscale.

(8). If you will be printing in color, **click** the **Grayscale** button, and then **click Color**.

In the Settings section on the Print screen, you can click the Full Page Slides button to choose from among several choices for printing the presentation, as described below:

- **Full Page Slides**—Prints each slide full size on a separate piece of paper.
- **Notes Pages**—Prints each slide as a notes page.
- **Outline**—Prints the text of the presentation as an outline.
- **Handouts**—Prints the presentation with one or more slides on each piece of paper. When printing four, six, or nine slides, you can choose whether to order the slides from left to right in rows (horizontally) or from top to bottom in columns (vertically).

Oladele wants you to print the slides as a one-page handout, with all eight slides on a single sheet of paper. In the rest of the steps in this section, you can follow the instructions in each set of steps up to the step that tells you to click Print. You should click Print only if your instructor wants you to actually print the presentation in the various formats.

To print the slides as a handout:

(1). In the **Settings** section, **Click** the **Full Page Slides** button. A menu opens listing the various ways you can print the slides. See **Figure 4-49**.

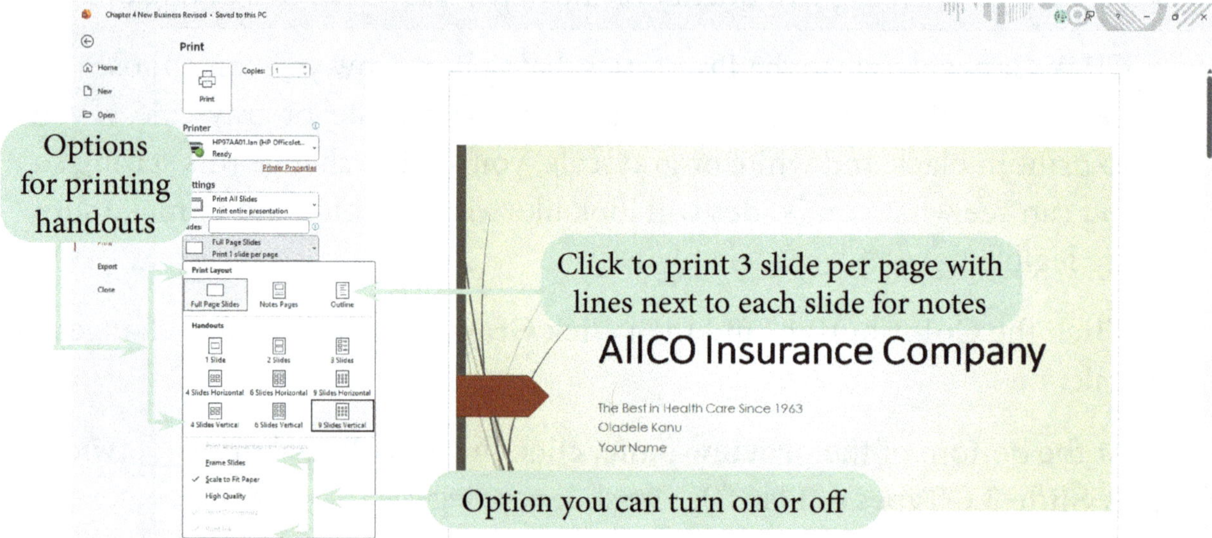

Figure 4-49 Print screen in Backstage view with print options menu open

(2). In the **Handouts section, Click 6 Slides Horizontal**. The preview changes to show all six slides in the preview pane, arranged in order horizontally in three rows from left to right. The current date appears in the top-right corner, and a page number appears in the bottom-right corner.

(3). At the top of the **Print section, click Print**. Backstage view closes and the handout prints.

Next, Oladele wants you to print the title slide as a full-page slide so that he can use it as a cover page for his handouts.

To print the title slide as a full-page slide:

(1). **Click** the **File Tab**, and then **Click Print** in the navigation pane. The Print screen appears in Backstage view. The preview still shows all six slides on one page. "**6 Slides Horizontal**" appears on the second button in the Settings section because that was the last printing option you chose.

(2). In the **Settings section, Click 6 Slides Horizontal**, and then **Click Full Page Slides. Slide 1** (the title slide) appears as the preview. Below the preview of Slide 1, it indicates that you are viewing Slide 1 of six slides to print.

(3). In the **Settings section, click** the **Print All Slides** button. Note on the menu that opens that you can print all the slides, selected slides, the current slide, or a custom range. You want to print just the title slide as a full-page slide.

(4). **Click Print Current Slide. Slide 1** appears in the **Preview Pane**, and at the bottom, it now indicates that you will print only one slide.

(5). **Click** the **Print** button. Backstage view closes and **Slide 1 prints**.

Recall that you created speaker notes on Slides 3 and 5. Oladele would like you to print these slides as notes pages.

To print the nonsequential slides containing speaker notes:

(1). **Open** the **Print Screen** in Backstage view again, and then **Click** the **Full Page Slides** button. The menu opens.

(2). In the **Print Layout section** of the menu, **Click Notes Pages**. The menu closes, and the **Preview** displays **Slide 1** as a **Notes Page**.

(3). In the **Settings section, Click** in the **Slides box, Type 3,5** and then **Click** a blank area of the Print screen.

(4). **Scroll** through the **Preview** to confirm that **Slides 3** ("Types of Plans") and 5 ("For More Information") will print, and then **Click Print**. Backstage view closes, and **Slides 3** and **5 print** as notes pages.

Finally, Oladele would like you to print the outline of the presentation. Remember, Slide 6 is designed to be a visual that Anthony can leave displayed at the end of the presentation, so you don't need to include it in the outline.

To print Slides 1 through 5 as an outline:

(1). **Open** the **Print screen** in Backstage view, **Click** the **Notes Pages** button, and then in the **Print Layout section, Click Outline**. The text on **Slides 3** and **5** appears as an **outline** in the **Preview Pane**.

(2). **Click** in the **Slides box, type 1-5**, and then **click** a blank area of the Print screen. See **Figure 4-50**.

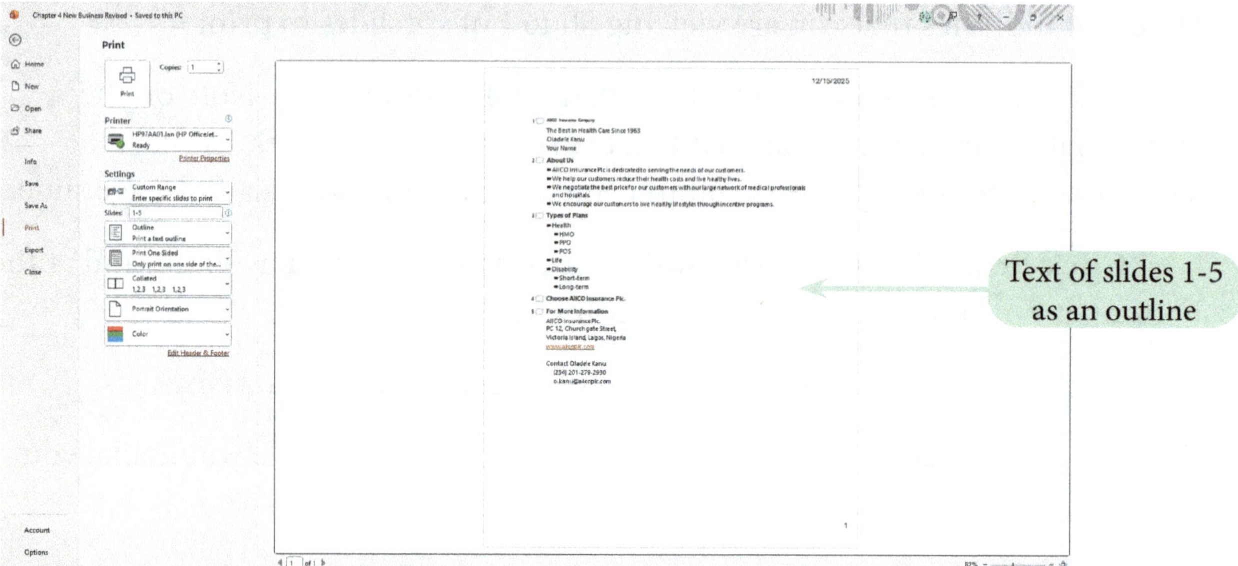

Figure 4-50 Print screen in Backstage view with Slides 1–5 previewed as an outline

(3). At the top of the **Print section, click** the **Print** button. Backstage view closes, and the text of **Slides 1–5 prints**.

CLOSING POWERPOINT

When you are finished working with your presentation, you can close PowerPoint. If you only have one presentation open, you click the Close button ☒ in the upper-right corner of the program window. If you have more than one presentation open, clicking this button will close only the current presentation; to close PowerPoint, you need to click the Close button in each of the open presentation's windows.

To close PowerPoint:

(1). In the **Upper-Right corner** of the program window, **Click** the **Close** button ☒. A **dialog box opens**, asking **if you want to save your changes**. This is because you did not save the file after you added your name to the title slide.

(2). In the **dialog box, Click Save**. The dialog box closes, the changes are saved, and PowerPoint closes.

Are You Having Trouble? If any other PowerPoint presentations are still open, click the Close button ☒ on each open presentation's program window until no more presentations are open to exit PowerPoint.

In this session, you opened an existing presentation and saved it with a new name, changed the theme, added and cropped photos and adjusted the photo compression, and resized and moved objects. You have also added speaker notes and checked the spelling. Finally, you printed the presentation in several forms and exited PowerPoint. Your work will help Anthony give an effective presentation to potential customers of AIICO Insurance Plc.

Quick Review

1. What is alt text?
2. Explain what happens when you crop photos.
3. Describe sizing handles.
4. How do you use smart guides?
5. Why is it important to maintain the aspect ratio of photos?
6. How do you convert a list to a SmartArt diagram without using the Design Ideas pane?
7. What is the difference between Slide Show view and Presenter view?
8. List the four formats for printing a presentation.

Note Below

CHAPTER REVIEW

Lab Project 1

CHAPTER REVIEW LABS

These are the Data Files needed for the Project Assignments: Chapter 4.pptx, Clarke and Family.jpg, Meeting.jpg, Standing.jpg, Doctor.jpg.

Oladele Kanu, which is the combination of two executives at AIICO, a sales manager in the Lagoe office of AIICO Insurance Plc, is preparing a presentation for the upcoming summer sales meeting. Because his team has sold so many new policies, he has been asked to give a presentation to the branch managers. He will focus on how they can create new business from their current customers by actively selling the new policy types the company sells. He asks you to begin creating the presentation. Complete the following steps:

(1). **Start PowerPoint** and create a **new, blank presentation**. On the title slide, **type New Sales Leads** as the title, and then **Type your name** as the subtitle. **Save** the presentation as **Project 1 Leads** to the drive and folder where you are storing your files.

(2). Edit the slide title by **typing Developing** before "New" so that the title is now "**Developing New Sales Leads**."

(3). Add a **new Slide 2** with the Title and Content layout, **type Contact Your Existing Clients** as the slide title, and then in the content placeholder **type** the following:

- **Offer new plans**
  - **Agric Insurance Plan**
  - **Travel Insurance Plan**
  - **Marine Insurance Plan**
- **Offer competitive package pricing**
- **Emphasize your personal connection and service**
  - **Contact Oladele Kanu**

CHAPTER REVIEW

Lab Project 1

CHAPTER REVIEW LABS

(4). **Create** a **new Slide 3** with the **Two Content layout**, then create a **new Slide 4** with the **Two Content layout**. On **Slide 4, add Recipe for Success** as the slide title, and then **type** the following as a numbered list in the left content placeholder:

1. **Offer new products to existing clients**
2. **Present to local organizations**
    1. **Chamber of Commerce**
    2. **Service organizations**
    3. **Professional organizations**
3. **Reach out to self-insured with customized packages**

(5). **Create** a **new Slide 5** using the Title and Content layout, and then create a **new Slide 6** with the Title and Content layout. On **Slide 6**, add **For More Information** as the slide title.

(6). Use the **Cut** and **Paste** commands to move the last bulleted item on **Slide 2** ("Contact Oladele Kanu") to Slide 6 as the first bulleted item in the content placeholder.

(7). On **Slide 6**, remove the **bullet symbol** from the text you pasted, and then add the following as the next two items in the unnumbered list:

Email: o.kanu@aiicoplc.com

Cell: (234) 201-279-2930

(8). **Click** after "**Kanu**" in the first item in the list, and then create a new line below it without creating a new item in the list so that there is no extra space above the new line. **Type Sales Manager, Lagos Office** on the new line.

CHAPTER REVIEW

Lab Project 1

CHAPTER REVIEW LABS

(9). **Remove** the link formatting from the email address.

(10). **Duplicate Slide 2** ("Contact Your Existing Clients"). On the new Slide 3, do the following:

- Edit the title so it is **Introduce New Products**
  - Edit the first bulleted item so it is **New Products**
  - Delete the second and third first-level bulleted items

(11). **Delete** the **blank Slides 4** and **6**.

(12). **Move Slide 3** ("Introduce New Products") so it becomes Slide 2.

(13). Change the theme to Banded and choose the second variant.

(14). Save your changes, and then close the presentation.

(15). **Open** the file **Project 1 Leads.pptx**, located in the **PowerPoint folder** > **Chapter 4 folder** included with your Data Files, add your name as the subtitle on the title slide, and then **save** it as **Project 1 Leads Updated** to the drive and folder where you are storing your files.

(16). **Change** the theme **colors** to **Orange Red**. **Change** the **theme fonts** to **Cambria**.

(17). **Change** the **layout** of **Slide 3** ("Contact Your Existing Clients") to **Two Content**.

(18). On **Slide 3**, **Insert** the picture **Woman.jpg**, located in the **PowerPoint1** > **Review folder**. **Add Woman on the phone conducting business**. as the alt text for this picture.

(19). **Open** the **Design Ideas** pane if necessary, and then click several of the

CHAPTER REVIEW

Lab Project 1

CHAPTER REVIEW LABS

suggested layouts. When you are finished, close the pane, and then on the Quick Access Toolbar, click the Undo button to reset the slide, and then close the Design Ideas pane.

(20). **Resize** the picture on **Slide 3** while maintaining the aspect ratio so that the picture height is **4 inches. Reposition** the picture so that its **right edge** aligns with the **right edge** of the **slide** and its **top edge** aligns with the **top edge** of the **text box** containing the list.

(21). **Change** the **layout of Slide 4** ("Create Custom Packages for Self-Employed") to Title and Content.

(22). On **Slide 4, insert** the photo Support_PPT_1_**Standing.jpg**, located in the **PowerPoint1 > Review folder. Add Business people standing and chatting in a group, holding glasses of water, in an office setting**. as the alt text for this picture.

(23). **Resize** the **picture** on **Slide 4** while maintaining the aspect ratio so that the picture height is 4 inches. **Reposition** the **picture** so that its **right edge aligns** with the **right edge** of the slide and its **bottom edge** aligns with the **bottom** of the **slide**.

(24). On **Slide 5, insert** the photo Support_PPT_1_**Meeting.jpg**, located in the **PowerPoint1 > Review folder. Add People in business casual attire at a conference table in an office setting listening to a man speak**. as the alt text for this picture.

(25). **Resize** the picture on **Slide 5** while maintaining the aspect ratio so that the picture height is 5 inches.

(26). **Display** the **rulers** and the **gridlines**, and then **crop 1 inch** off the **bottom** of the **picture** on **Slide 5**. Resize the cropped part of the picture so that the

CHAPTER REVIEW

Lab Project 1

CHAPTER REVIEW LABS

head of the man standing is about one-eighth of an inch from the top of the picture. Then crop one-half inch off the right side of the picture.

(27). **Reposition** the **picture** on **Slide 5** so that its **right edge** aligns with the **right edge** of the **slide**, its **top edge** aligns with the **gridline** at the **1-inch mark** on the **vertical ruler**, and its **bottom edge** aligns with the **gridline** at the **negative 3-inch mark** on the vertical ruler.

(28). On **Slide 5**, **change** the **width** of the **text box** containing the bulleted list by **dragging** the **sizing handle** in the **middle** of the **right border** of the **text box** so that the **right border** aligns with the **gridline** at the negative **1-inch mark** on the horizontal ruler. Then change the **height** of the **text box** by **dragging** the **sizing handle** in the **middle** of the **top border down** so that the **top** of the **text box** aligns with the top of the picture.

(29). On **Slide 6** ("Recipe for Success"), **open** the **Design Ideas pane** and **click** several of the suggested layouts. Then, on the **Quick Access Toolbar**, **click** the **Undo** button and **close** the **Design Ideas pane**.

(30). On **Slide 6**, **convert** the **numbered list** to **SmartArt** using the **Vertical Block List layout** on the **Convert to SmartArt menu**.

(31). On Slide 6, change the SmartArt layout to the Segmented Process layout.

(32). On **Slide 6**, **display** the **Notes pane**, and then **type Some local organizations to consider are the Chamber of Commerce, service organizations, and professional organizations**. as a **speaker note**. When you are finished, **close** the **Notes pane**.

(33). On **Slide 7** ("For More Information"), **increase** the **size** of the **text** in the **unnumbered list** to **24 points**. Then, in the **first bulleted** item, **select** the text "**Oladele Kanu**." and **format** it as **bold** and **28 points**.

CHAPTER REVIEW

Lab Project 1

CHAPTER REVIEW LABS

(34). On **Slide 7**, **insert** the picture **Oladele.jpg**, and then **Portrait of Oladele Kanu** as the alt text for this picture.

(35). **Crop** the **photo** to the **Oval shape**. **Click** the **Crop** button, and then **drag** the **bottom-middle** crop handle **up one inch**. **Reposition** the picture so that the **top** of the picture aligns with the **horizontal gridline** at the **1-inch mark** on the **vertical ruler** if necessary.

(36). **Hide** the **rulers** and **gridlines**.

(37). **Compress** all the **photos** in the **slides** to **E-mail (96 ppi)** and **delete cropped areas** of pictures.

(38). **Add your name** as an **author property**, and add **AIICO Insurance Plc.** as the **Company property**.

(39). **Check** the **spelling** in the presentation. **Correct** the **spelling error** on **Slide 2** by **selecting** "**Liability**" as the correct spelling, and the **error** on **Slide 3** by **selecting** "**Emphasize**" as the correct spelling. **Ignore all instances** of Oladele's first and last names. If you made any additional spelling errors, correct them as well. If your name on **Slide 1** is flagged as misspelled, **ignore** this error. **Save** the changes to the presentation.

(40). **Review** the **slide show** in **Slide Show** and **Presenter views**.

(41). **View** the **slides** in **grayscale**, and then **print** the following in color or in **grayscale** depending on your printer: the **title slide** as a **full-page-sized slide**; **Slides 2** through **7** as a **handout** on a single piece of paper with the slides in **order horizontally**; **Slide 6** as a **notes page**; and **Slides 2** through **5** and **Slide 7** as an **outline**. **Save** and **close** the **presentation** and **PowerPoint** when you are finished.

CHAPTER REVIEW

Lab Project 2

CHAPTER REVIEW LABS

Self Challenge Lab

Data Files needed for this Self Driven Problem: Project 2.pptx, Application.jpg, Building.jpg, Key.jpg, Rose.jpg

EcoBank Liberia Limited has branches all over Liberia. Rose Williams (a combination of two executive's names), the Vice President of Commercial and Residential Lending at the 11th Street Sinkor branch in Monrovia, Liberia and across Africa, hired you as her executive assistant. Rose wants to create a simple presentation that will help her explain some of the details about applying for a mortgage to commercial buyers and to first-time home buyers in Liberia. She asks you to help complete the slides. Complete the following steps:

(1). **Open** the presentation named **Project 2.pptx**, located in the **PowerPoint** > **Chapter 4 folder** included with your Data Files, and then **save** it as **Mortgage** to the drive and folder where you are storing your files.

(2). **Insert** a **new slide** with the Title Slide layout. **Add Mortgage Essentials** as the presentation title on the title slide. In the subtitle text placeholder, **type your name**. **Move** this slide so it is the first slide in the presentation.

(3). **Apply** the **Frame theme**, and then **apply** the **third theme variant**.

(4). **Change** the **theme fonts** to **Garamond-TrebuchetMs**.

(5). **On Slide 1** (the title slide), **change** the **font size** of the title text to **36 points**. Then **resize** the title **text box** so it is **2.25 inches wide**. If necessary, **reposition** the title text box so that the **left edge** of the **text box** is aligned with the **left edge** of the **subtitle text box** and so that there is the same amount of space between the top of the text box and the top slide edge as there is between the bottom of the subtitle text box and the bottom of the slide. **Resize** the **subtitle text box** so it is **3.3 inches wide**, and then align its **left edge** with the **left edge** of the **title text box**.

CHAPTER REVIEW

Lab Project 2

CHAPTER REVIEW LABS

(6). **On Slide 1**, **insert** the picture **Application.jpg**, located in the **PowerPoint > Chapter 4 folder. Add Picture of a mortgage application form with a red "Approved" stamp on it and the wooden stamp next to it.** as the alt text.

(7). **On Slide 1**, **resize** the photo, maintaining the aspect ratio, so that it is **5.84 inches high**. **Position** the **photo** so that its **middle aligns** with the **middle** of the **tan rectangle** and its **right edge aligns** with the **right edge** of the **slide**.

(8). **On Slides 2** through **6**, increase the size of the text in the bulleted list so the **first-level** items are **24 points** and any **second-level** items are **20 points**.

(9). **On Slide 4** ("What Are Closing Costs?"), **cut** the last bulleted item ("$200,000 loan"), and then **paste** it in on **Slide 3** ("What Are Points?") as the **third bulleted** item. If a blank line is added below the pasted text, delete it.

(10). On Slide 3, add the following as second-level items below "$200,000 loan", adjusting the font size to 20 points if necessary:

2 point (2%) = $4,000

3 point (3%) = $6,000

(11). **On Slide 2** ("Steps"), convert the bulleted list to **SmartArt** using the **Step Down Process layout**. (Hint: You need to **click More SmartArt Graphics** to open the **Choose a SmartArt Graphic** dialog box.)

(12). **On Slide 5** ("Documents Needed"), **change** the **layout** to **Two Content**, then **insert** the **picture** Support_PPT_1_Key.jpg, located in the **PowerPoint > Chapter 4 folder**. **Add Drawing of a hand passing an approved mortgage towards another person's hand holding a key on a key chain shaped like a house**. as the alt text.

(13). **On Slide 5**, **resize** the **picture**, maintaining the aspect ratio, so that it is **4.5 inches square**, and then **position** it so that its **middle aligns** with the

CHAPTER REVIEW

Lab Project 2

CHAPTER REVIEW LABS

middle of the **text box** containing the bulleted list and its **right edge aligns** with the **left edge** of the **gray rectangle** on the **right side** of the slide.

(Hint: Position the picture as close as possible to the edge of the gray rectangle. Then with the picture selected, press RIGHT ARROW or LEFT ARROW to nudge it into the correct position.)

(14). **On Slide 5**, in the last bulleted item, format "**and**" with **italics. Enter Make sure applicants understand that they need two forms of ID**. as a speaker note, and then **close** the **Notes pane.**

(15). **On Slide 6** ("Contact Information"), **Remove** the **link formatting** from both the **email address** and the **Internet address** of the Mortgages page for the bank.

(16). **On Slide 6, click** before the word "**Contact**" in the slide title, and then **press ENTER three times. Insert** the photo **Rose.jpg**, located in the **PowerPoint1 > Case1 folder. Add Portrait** of smiling **Rose Williams**. as the alt text.

(17). **On Slide 6, crop 1.5 inches** off the **bottom** of the picture, then **crop** the **photo** to the **Rectangle: Rounded Corners** shape.

(18). **On Slide 6, resize** the photo so it is **2.8 inches high**, maintaining the aspect ratio. **Reposition** the **photo** in the **tan rectangle** above the title so that the **vertical smart guide** that appears shows that the photo aligns with the **center** of the **tan rectangle**, and the **bottom** of the **photo aligns** with the **middle** of the **slide.**

(19). **Add** a **new Slide 7** with the **Content** with **Caption layout**. In the title text placeholder, **type Upper**, and then **create** a **new line** without creating a new paragraph. **Type Coast** on the new line, **create** another **new line**, and then **type**

CHAPTER REVIEW

Lab Project 2

CHAPTER REVIEW LABS

Bank. In the text placeholder below the title, **type The Friendly Bank**.

(20). **On Slide 7, change** the **size** of the **title text to 48 points** and **bold**, and **change** its **color** to **Brown, Accent 1, Darker 50%. Change** the **size** of the text below the title to **24 points** and make it **italic**.

(21). **On Slide 7**, add the picture **Building.jpg**, located in the **PowerPoint** > **Case1 folder. Add Photo** of EcoBank Headquarters in Lagos. as the alt text.

(22). **Compress** all the photos in the presentation to **E-mail (96 ppi)** and **delete cropped portions** of photos.

(23). **Add your name** as an **author property** and add EcoBank as the Company property.

(24). **Check** the **spelling** in the presentation and correct all misspelled words.

(25). **Save** the changes to the presentation, **view** the **slide show in Presenter view**, and then **print Slides 1–6** as a **handout** using the **6 Slides Horizontal arrangement**, and **print Slide 5** as a **notes page**.

(26). **Close** the presentation and PowerPoint.

CHAPTER 5 HIGHLIGHT

Microsoft Access: Creating a Database

After completing this chapter, you will be able to:

1. Plan and Define basic database concepts and terms
2. Start and exit Access
3. Identify the Microsoft Access window and Backstage view
4. Create a blank database
5. Create and save a table in Datasheet view and Design view
6. Add fields to a table in Datasheet view and Design view
7. Set a table's primary key in Design view
8. Open an Access database
9. Open a table using the Navigation Pane
10. Copy and paste records from another Access database
11. Navigate a table datasheet and enter records
12. Create and navigate a simple query
13. Create and navigate a simple form
14. Create, preview, navigate, and print a simple report
15. Use Help in Access
16. Identify how to compact, back up, and restore a database

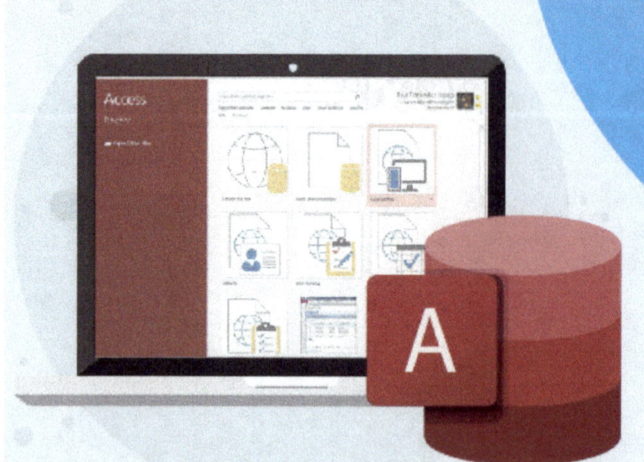

Microsoft Access

CHAPTER 5 | MICROSOFT ACCESS

Abuja Monrovia Community Health Services

Abuja Monrovia Community Health Services, is a nonprofit health clinic located in both the greater Abuja, Nigeria area, and in Central Monrovia on Randall Street, Liberia. Both facilities provide a range of medical services to patients of all ages. The clinics specialize in chronic disease management, cardiac care, and geriatrics. Chioma Taylor, the office manager for Abuja Monrovia Community Health Services, oversees a small staff and is responsible for maintaining records for the clinics' patients.

In order to best manage the clinic, Chioma and her staff rely on electronic medical records for patient information, billing, inventory control, purchasing, and accounts payable. Several months ago, the clinic upgraded to **Microsoft Access 2024** (or simply **Access**), a computer program used to enter, maintain, and retrieve related data in a format known as a database. Chioma and her staff want to use Access to store information about patients, billing, vendors, and products. She asks for your help in creating the necessary Access database.

INTRODUCTION TO DATABASE CONCEPTS

Before you begin using Access to create the database for Chioma, you need to understand a few key terms and concepts associated with databases.

ORGANIZING DATA

Data is a valuable resource to any business. At Abuja Monrovia Community Health Services, for example, important data includes the patients' names and addresses, visit dates, and billing information. Organizing, storing, maintaining, retrieving, and sorting this type of data are critical activities that enable a business to find and use information effectively. Before storing data on a computer, however, you must organize the data.

Your first step in organizing data is to identify the individual fields. A **Field** is a single characteristic or attribute of a person, place, object, event, or idea. For example, some of the many fields that Abuja Monrovia Community Health Services tracks are the patient ID, first name, last name, address, phone number, visit date, reason for visit, and invoice amount.

Next, you group related fields together into tables. A **Table** is a collection of fields that describes a person, place, object, event, or idea. **Figure 5-1** shows an example of a US clinic Patient table that contains the following four fields: PatientID, FirstName, LastName, and Phone. Each field is a column in the table, with the field name displayed as the column heading. The database we will be constructing in the hands on exercise of this chapter will be similar to this database.

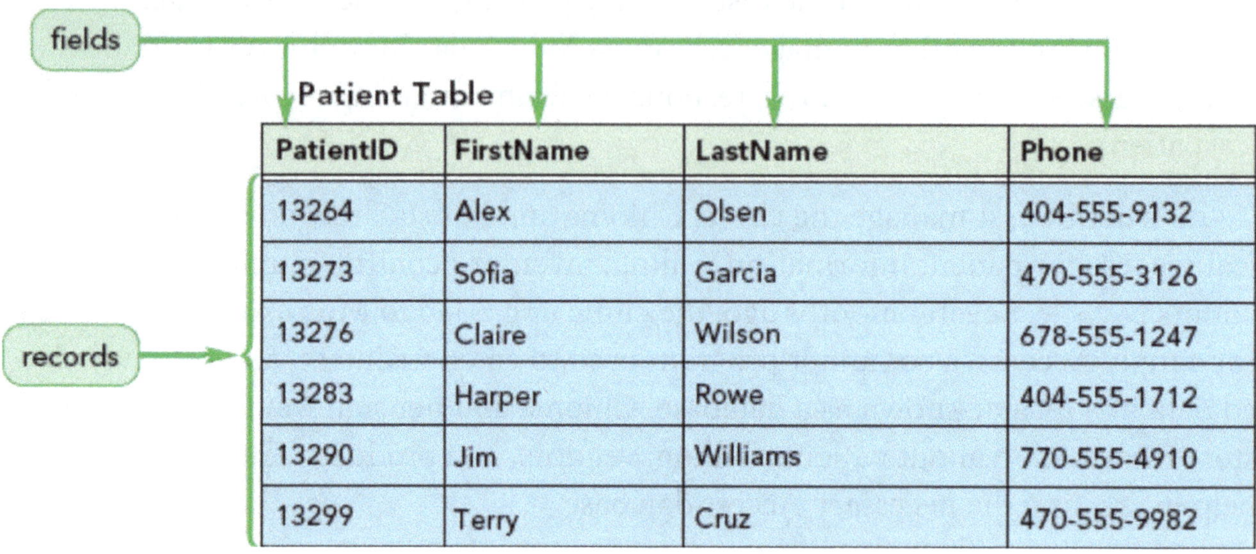

PatientID	FirstName	LastName	Phone
13264	Alex	Olsen	404-555-9132
13273	Sofia	Garcia	470-555-3126
13276	Claire	Wilson	678-555-1247
13283	Harper	Rowe	404-555-1712
13290	Jim	Williams	770-555-4910
13299	Terry	Cruz	470-555-9982

Figure 5-1 Data organization for a table of patients

The specific content of a field is called the **Field Value**. In **Figure 5-1**, the first set of field values for PatientID, FirstName, LastName, and Phone are, respectively: 13264, Alex, Olsen, and 404-555-9132. This set of field values is called a **Record**. In the Patient table, the data for each patient is stored as a separate record. **Figure 5-1** above shows six records; each row of field values in the table is a record.

DATABASES AND RELATIONSHIPS

A collection of related tables is called a **Database**, or a **Relational Database**. In this module, you will create the database for Abuja Monrovia Community Health Services, and within that database, you'll create a table named Visit to store data about patient visits. Later on in this hands on lab, you'll create two more tables, named Patient and Billing, to store related information about patients and their invoices.

As Chioma and her staff use the database that you will create, they will need to access information about patients and their visits. To obtain this information, you must have a way to connect records in the Patient table to records in the Visit table.

You connect the records in the separate tables through a **Common Field** that appears in both tables.

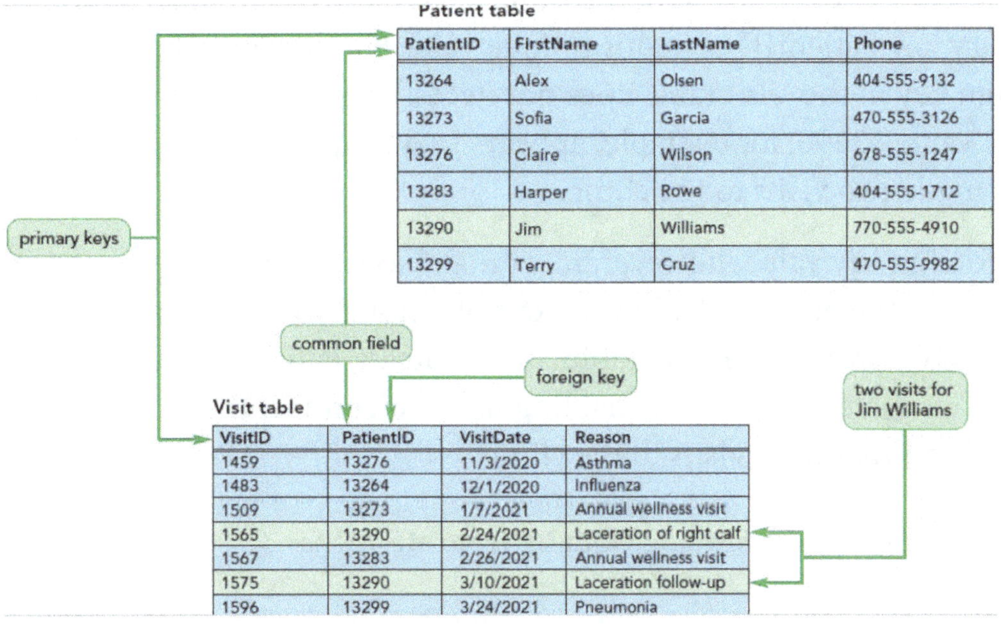

PatientID	FirstName	LastName	Phone
13264	Alex	Olsen	404-555-9132
13273	Sofia	Garcia	470-555-3126
13276	Claire	Wilson	678-555-1247
13283	Harper	Rowe	404-555-1712
13290	Jim	Williams	770-555-4910
13299	Terry	Cruz	470-555-9982

VisitID	PatientID	VisitDate	Reason
1459	13276	11/3/2020	Asthma
1483	13264	12/1/2020	Influenza
1509	13273	1/7/2021	Annual wellness visit
1565	13290	2/24/2021	Laceration of right calf
1567	13283	2/26/2021	Annual wellness visit
1575	13290	3/10/2021	Laceration follow-up
1596	13299	3/24/2021	Pneumonia

Figure 5-2 Database relationship between tables for patients and visits

In this example of the same U.S. clinic's Patient Database shown in **Figure 5-2** above, you see that each record in the Patient table has a field named PatientID, which is also a field in the Visit table. For example, Jim Williams is the fifth patient in the Patient table and has a PatientID field value of 13290. This same PatientID field value, 13290, appears in two records in the Visit table. Therefore, Jim Williams is the patient that was seen at these two visits.

Each ID value in the Patient table must be unique so that you can distinguish one patient from another. These unique PatientID values also identify each patient's specific visits in the Visit table. The PatientID field is referred to as the primary key of the Patient table. A **Primary Key** is a field, or a collection of fields, whose values uniquely identify each record in a table. No two records can contain the same value for the primary key field. In the Visit table, the VisitID field is the primary key because Abuja Monrovia Community Health Services assigns each visit a unique identification number.

When you include the primary key from one table as a field in a second table to form a relationship between the two tables, it is called a **Foreign Key** in the second table, as shown in **Figure 5-2** above. For example, PatientID is the primary key in the Patient table and a foreign key in the Visit table.

The PatientID field must have the same characteristics in both tables. Although the primary key PatientID contains unique values in the Patient table, the same field as a foreign key in the Visit table does not necessarily contain unique values. The PatientID value 13290, for example, appears two times in the Visit table because Jim Williams made two visits to the clinic.

Each foreign key value, however, must match one of the field values for the primary key in the other table. In the example shown in **Figure 5-2** on th previous, each PatientID value in the Visit table must match a PatientID value in the Patient table. The two tables are related, enabling users to connect the facts about patients with the facts about their visits to the clinic.

SMARTSKILLS Storing Data in Separate Tables

When you create a database, you must create separate tables that contain only fields that are directly related to each other. For example, in the Abuja Monrovia database, the patient and visit data should not be stored in the same table because doing so would make the data difficult to update and prone to errors. Consider Jim Williams and his visits to the clinic, and assume that he has many more than just two visits. If all the patient and visit data was stored in the same table, so that each record (row) contained all the information about each visit and the patient, the patient data would appear multiple times in the table. This causes problems when the data changes. For example, if the phone number for Jim Williams changed, you would have to update the multiple occurrences of the phone number throughout the table. Not only would this be time-consuming, it would increase the likelihood of errors or inconsistent data.

RELATIONAL DATABASE MANAGEMENT SYSTEMS

To manage its databases, a company uses a database management system. A **Database Management System (DBMS)** is a software program that lets you create databases, and then manipulate the data they contain. Most of today's database management systems, including Access, are called relational database management systems (RDBMS). In a **Relational Database Management System (RDBMS)**, data is organized as a collection of tables. As stated earlier, a relationship between two tables in a relational DBMS is formed through a common field.

A relational DBMS controls the storage of databases and facilitates the creation, manipulation, and reporting of data, as illustrated in **Figure 5-3**.

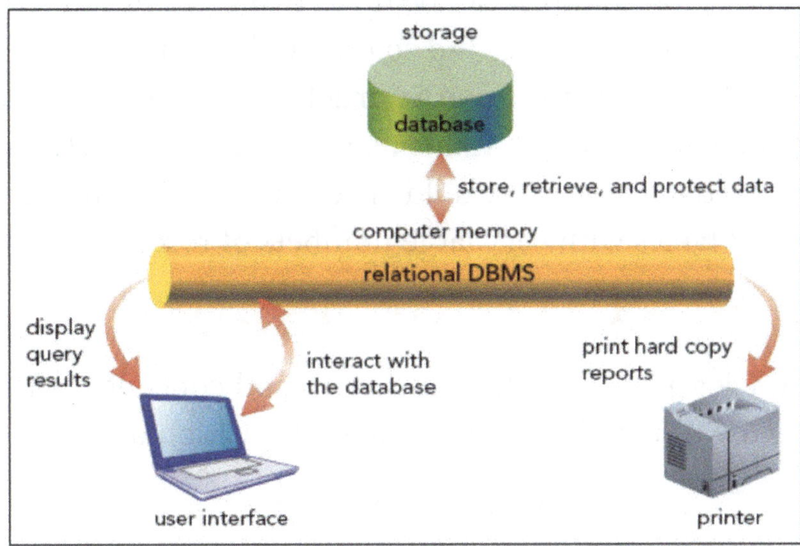

Figure 5-3 Relational database management system

Specifically, a relational DBMS provides the following functions:

- It allows you to create database structures containing fields, tables, and table relationships.
- It lets you easily add new records, change field values in existing records, and delete records.
- It contains a built-in query language, which lets you obtain immediate answers to the questions (or queries) you ask about your data.
- It contains a built-in report generator, which lets you produce professional-looking, formatted reports from your data.
- It protects databases through security, control, and recovery facilities.

An organization such as Abuja Monrovia Community Health Services benefits from a relational DBMS because it allows users working in different groups to share the same data. More than one user can enter data into a database, and more than one user can retrieve and analyse data that other users have entered. For example, the database for Abuja Monrovia Community Health Services will contain only one copy of the Visit table, and all employees will use it to access visit information.

Finally, unlike other software programs, such as spreadsheet programs, a DBMS can handle massive amounts of data and allows relationships among multiple tables. Each Access database, for example, can be up to two gigabytes in size, can contain up to 32,768 objects (tables, reports, and so on), and can have up to 255 people using the database at the same time. For instructional purposes, the databases you will create and work with throughout this text contain a relatively small number of records compared to databases you would encounter outside the classroom, which would likely contain tables with very large numbers of records.

STARTING ACCESS AND CREATING A DATABASE

Now that you've learned some database terms and concepts, you're ready to start Access and create the Abuja Monrovia Database for Chioma.

To start Access:

(1). On the **Windows Taskbar, Click** the **Start** button. The Start menu opens.

(2). On the **Start Menu, scroll down** the list of apps, and then **Click Access**. Access starts and displays the Recent screen in Backstage view. See **Figure 5-4**.

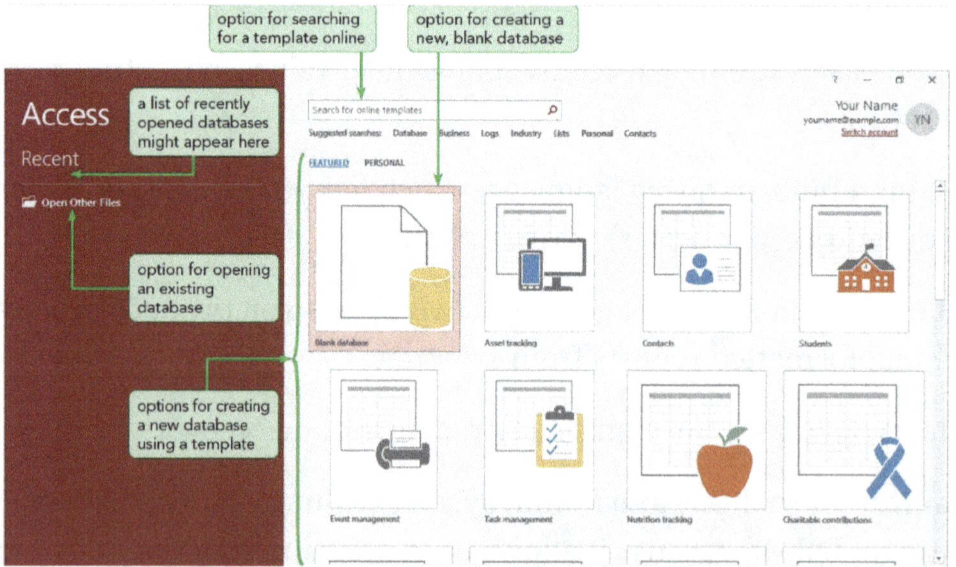

Figure 5-4 Recent screen in Backstage view

When you start Access, the first screen that appears is Backstage view, which is the starting place for your work in Access. **Backstage View** contains commands that allow you to manage Access files and options. The Recent screen in Backstage view provides options for you to create a new database or open an existing database. To create a new database that does not contain any data or objects, you use the Blank database option. If the database you need to create contains objects that match those found in common databases, such as databases that store data about contacts or tasks, you can use one of the templates provided with Access. A **Template** is a predesigned database that includes professionally designed tables, reports, and other database objects that can make it quick and easy for you to create a database. You can also search for a template online using the Search for online templates box.

In this case, the templates provided do not match Chioma's needs for the clinic's database, so you need to create a new, blank database from scratch.

To create the new Abuja Monrovia Database:

(1). Make sure you have the Access starting Data Files on your computer.

Are You Having Trouble? If you don't have the starting Data Files, you need to get them before you can proceed. Your instructor will either give you the Data Files or ask you to obtain them from a specified location (such as a network drive). If you have any questions about the Data Files, see your instructor or technical support person for assistance.

(2). On the **Recent screen, click Blank database** (see **Figure 5-4**). The Blank database screen opens.

(3). In the **File Name box, Type Abuja Monrovia** to replace the selected database name provided by Access, Database1. **Next** you need to **specify the location** for the file.

(4). **Click** the **Browse** button to the right of the File Name box. The File New Database dialog box opens.

(5). Navigate to the drive and folder where you are storing your files, as specified by your instructor, or to your desired drive.

(6). **Make sure** the **Save as type box** displays "**Microsoft Access 2007–2016 Databases.**"

Are You Having Trouble? If your computer is set up to show file name extensions, you will see the Access file name extension ".accdb" in the File name box.

(7). **Click OK**. You return to the **Blank database screen**, and the File Name box now shows the name **Abuja Monrovia.accdb**. The filename extension "**.accdb**" identifies the file as an Access 2007–2016 database.

(8). **Click Create. Access creates** the new database, saves it to the specified location, and then opens an empty table named **Table1**.

Are You Having Trouble? If you see only ribbon tab names and no buttons, click the Home tab to expand the ribbon, and then in the lower-right corner of the ribbon, click the Pin this pane button to pin the ribbon.

SMARTSKILLS Understanding the Database File Type

Access 2019 uses the .accdb file extension, which is the same file extension used for databases created with Microsoft Access 2007, 2010, 2013, and 2016. To ensure compatibility between these earlier versions and the Access 2019 software, new databases created using Access 2019 have the same file extension and file format as Access 2007, Access 2010, Access 2013, and Access 2016 databases.

This file extension for Microsoft Access databases which is .accdb that stands for (Access database), was introduced almost 20 years ago with the launch of Access 2007 and has remained the default for all subsequent versions of Access, including the latest Microsoft Access 2024 and Microsoft 365 versions.

WORKING IN TOUCH MODE

If you are working on a touch device, such as a tablet, you can switch to Touch Mode in Access to make it easier for you to tap buttons on the ribbon and perform other touch actions. Your screens will not match those shown in the book exactly, but this will not cause any problems.

Note: The following steps assume that you are using a mouse. If you are instead using a touch device, please read these steps but don't complete them, so that you remain working in Touch Mode.

To switch to Touch Mode:

(1). On the **Quick Access Toolbar**, **Click** the **Customize Quick Access Toolbar** button . A menu opens listing buttons you can add to the Quick Access Toolbar as well as other options for customizing the toolbar.

Are You Having Trouble? If the **Touch/Mouse Mode** command on the menu has a checkmark next to it, **Press ESC** to close the menu, and then skip to **Step 3**.

(2). **Click Touch/Mouse Mode**. The Quick Access Toolbar now contains the Touch/Mouse Mode button , which you can use to switch between Mouse Mode, the default display, and Touch Mode.

(3). On the **Quick Access Toolbar**, **Click** the **Touch/Mouse Mode** button . A menu opens with two commands: Mouse, which shows the ribbon in the standard display and is optimized for use with the mouse; and Touch, which provides more space between the buttons and commands on the ribbon and is optimized for use with touch devices. The icon next to Mouse is shaded to indicate that it is selected.

Are You Having Trouble? If the icon next to Touch is shaded red, **Press ESC** to close the menu and skip to **Step 5**.

(4). **Click Touch**. The display switches to **Touch Mode** with more space between the **commands** and **buttons** on the ribbon. See **Figure 5-5**.

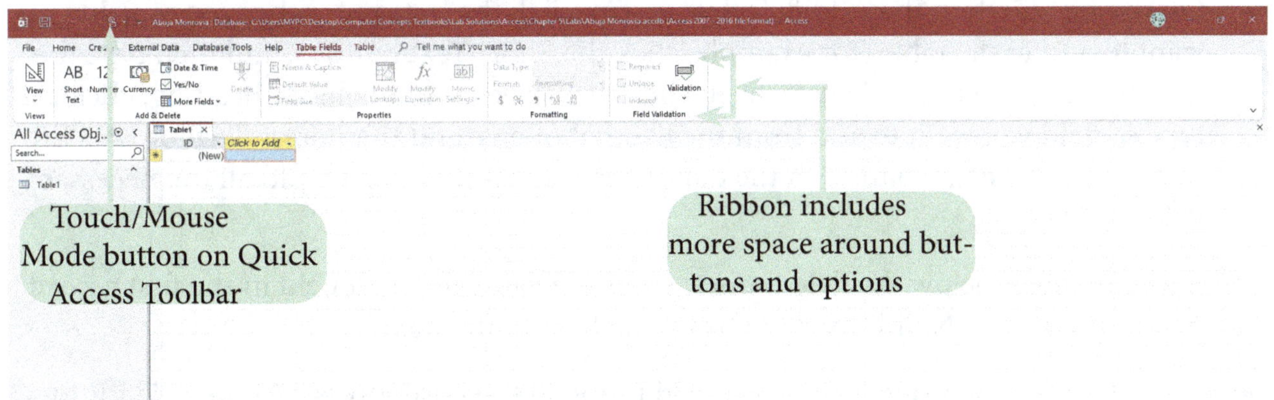

Figure 5-5 Ribbon displayed in Touch Mode

The figures in this text show the standard Mouse Mode display, and the instructions assume you are using a mouse to click and select options, so you'll switch back to Mouse Mode.

Are You Having Trouble? If you are using a touch device and want to remain in Touch Mode, **skip Steps 5** and **6**.

(5). On the **Quick Access Toolbar**, **Click** the **Touch/Mouse Mode** button, and then **Click Mouse**. The ribbon returns to the standard display, as shown in the Session 1.1 Visual Overview.

(6). On the **Quick Access Toolbar**, **Click** the **Customize Quick Access Toolbar** button, and then **Click Touch/Mouse Mode** to **deselect** it. The Touch/Mouse Mode button is removed from the Quick Access Toolbar.

CREATING A TABLE IN DATASHEET VIEW

Tables contain all the data in a database and are the fundamental objects for your work in Access. You can create a table in Access in different ways, including entering the fields and records for the table directly in Datasheet view.

Smart Tips

Creating a Table in Datasheet View

- On the ribbon, click the Create tab.
- In the Tables group, click the Table button.
- Rename the default ID primary key field and change its data type, if necessary; or accept the default ID field with the AutoNumber data type.
- On the Fields tab in the Add & Delete group, click the button for the type of field you want to add to the table (for example, click the Short Text button), and then type the field name; or, in the table datasheet, click the Click to Add column heading, click the type of field you want to add from the list that opens, and then press TAB or ENTER to move to the next column in the datasheet. Repeat this step to add all the necessary fields to the table.
- In the first row below the field names, enter the value for each field in the first record, pressing TAB or ENTER to move from one field to the next.
- After entering the value for the last field in the first record, press TAB or ENTER to move to the next row, and then enter the values for the next record. Continue this process until you have entered all the records for the table.
- On the Quick Access Toolbar, click the Save button, enter a name for the table, and then click OK.

For Abuja Monrovia Community Health Services, Chioma needs to track information about each patient visit at the clinic. She asks you to create the Visit table according to the plan shown in **Figure 5-6**.

Field	Purpose
VisitID	Unique number assigned to each visit; will serve as the table's primary key
PatientID	Unique number assigned to each patient; common field that will be a foreign key to connect to the Patient table
VisitDate	Date on which the patient visited the clinic
Reason	Reason/diagnosis for the patient visit
WalkIn	Whether the patient visit was a walk-in or scheduled appointment

Figure 5-6 Plan for the Visit table

As shown in Chioma's plan, she wants to store data about visits in five fields, including fields to contain the date of each visit, the reason for the visit, and if the visit was a walk-in or scheduled appointment. These are the most important aspects of a visit and, therefore, must be tracked. Also, notice that the VisitID field will be the primary key for the table; each visit at Abuja Monrovia Community Health Services is assigned a unique number, so this field is the logical choice for the primary key. Finally, the PatientID field is needed in the Visit table as a foreign key to connect the information about visits to patients. The data about patients and their invoices will be stored in separate tables, which you will create later.

Notice the name of each field in **Figure 5-6** above. You need to name each field, table, and object in an Access database.

SMARTSKILLS Decision Making: Naming Fields in Access Tables

One of the most important tasks in creating a table is deciding what names to specify for the table's fields. Keep the following guidelines in mind when you assign field names:

- A field name can consist of up to 64 characters, including letters, numbers, spaces, and special characters, except for the period (.), exclamation mark (!), grave accent (`), and square brackets ([ ]).

- A field name cannot begin with a space.
- Capitalize the first letter of each word in a field name that combines multiple words, for example VisitDate.
- Use concise field names that are easy to remember and reference and that won't take up a lot of space in the table datasheet.
- Use standard abbreviations, such as Num for Number, Amt for Amount, and Qty for Quantity, and use them consistently throughout the database. For example, if you use Num for Number in one field name, do not use the number sign (#) for Number in another.
- Give fields descriptive names so that you can easily identify them when you view or edit records.
- Although Access supports the use of spaces in field names (and in other object names), experienced database developers avoid using spaces because they can cause errors when the objects are involved in programming tasks.

By spending time obtaining and analyzing information about the fields in a table, and understanding the rules for naming fields, you can create a well-designed table that will be easy for others to use.

RENAMING THE DEFAULT PRIMARY KEY FIELD

As noted earlier, Access provides the ID field as the default primary key for a new table you create in Datasheet view. Recall that a primary key is a field, or a collection of fields, whose values uniquely identify each record in a table. However, according to Chioma's plan, the VisitID field should be the primary key for the Visit table. You'll begin by renaming the default ID field to create the VisitID field.

To rename the ID field to the VisitID field:

(1). **Right-Click** the **ID** column heading to open the shortcut menu, and then **Click Rename Field**. The column heading ID is selected, so that whatever text you type next will replace it.

(2). **Type VisitID** and then **Click** the row below the heading. The column heading changes to **VisitID**, and the insertion point moves to the row below the heading. The **Insertion Point** is a flashing cursor that shows where text you type will be inserted. In this case, it is hidden within the selected field value (New). See **Figure 5-7**.

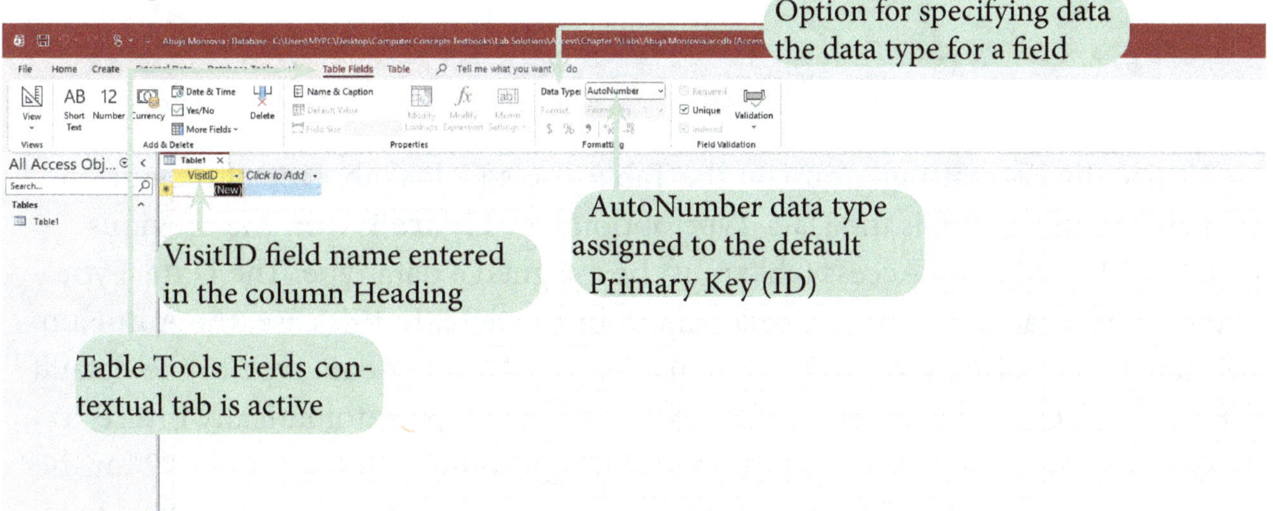

Figure 5-7 ID field renamed to VisitID

Are You Having Trouble? If you make a mistake while typing the field name, use BACKSPACE to delete characters to the left of the insertion point or use DELETE to delete characters to the right of the insertion point. Then type the correct text. To correct a field name by replacing it entirely, press ESC, and then type the correct text.

Notice that the Table Tools Fields tab is active on the ribbon. This is an example of a contextual tab, which is a tab that appears and provides options for working with a specific object that is selected—in this case, the table you are creating. As you work with other objects in the database, other contextual tabs will appear with commands and options related to each selected object.

SMARTSKILLS Buttons and Labels on the Ribbon

Depending on the size of the monitor you are using and your screen resolution settings, you might see more or fewer buttons on the ribbon, and you might not see labels next to certain buttons. The screenshots in these modules were created using a screen resolution setting of 1366 x 768 with the program window maximized. If you are using a smaller monitor or a lower screen resolution, some buttons will appear only as icons, with no labels next to them, because there is not enough room on the ribbon to display the labels.

You have renamed the default primary key field, ID, to VisitID. However, the VisitID field still retains the characteristics of the ID field, including its data type. Your next task is to change the data type of this field.

CHANGING THE DATA TYPE OF THE DEFAULT PRIMARY KEY FIELD

Notice the Formatting group on the Table Tools Fields tab. One of the options available in this group is the Data Type option (see **Figure 1-7** on the previous page). Each field in an Access table must be assigned a data type. The **Data Type** determines what field values you can enter for the field. In this case, the AutoNumber data type is displayed. Access assigns the AutoNumber data type to the default ID primary key field because the **AutoNumber** data type automatically inserts a unique number in this field for every record, beginning with the number 1 for the first record, the number 2 for the second record, and so on. Therefore, a field using the AutoNumber data type can serve as the primary key for any table you create.

Visit numbers at Abuja Monrovia Community Health Services are specific, four-digit numbers, so the AutoNumber data type is not appropriate for the VisitID field, which is the primary key field in the table you are creating. A better choice is the **Short Text** data type, which allows field values containing letters, digits, and other characters, and which is appropriate for identifying numbers, such as visit numbers, that are never used in calculations. So, Chioma asks you to change the data type for the VisitID field from AutoNumber to Short Text.

To change the data type for the VisitID field:

(1). Make sure that the **VisitID column** is **selected**. A column is selected when you click a field value, in which case the background color of the column heading changes to orange (the default color) and the insertion point appears in the field value. You can also click the column heading to select a column, in which case the background color of both the column heading and the field value changes (the default colors are gray and blue, respectively).

(2). On the **Table Tools Fields Tab**, in the **Formatting Group**, **Click** the **Data Type Arrow**, and then **Click Short Text**. The VisitID field is now a Short Text field. See **Figure 5-8** on the next page.

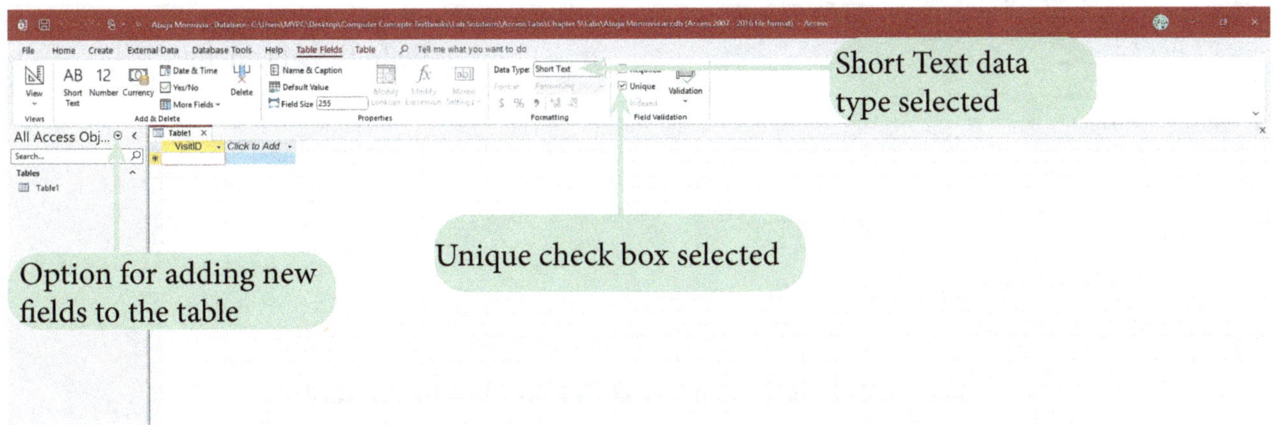

Figure 5-8 Short Text data type assigned to the VisitID field

Note the Unique check box in the Field Validation group. This check box is selected because the VisitID field assumed the characteristics of the default primary key field, ID, including the fact that each value in the field must be unique. Because this check box is selected, no two records in the Visit table will be allowed to have the same value in the VisitID field.

With the VisitID field created and established as the primary key, you can now enter the rest of the fields in the Visit table.

Adding New Fields

When you create a table in Datasheet view, you can use the options in the Add & Delete group on the Table Tools Fields tab to add fields to your table. You can also use the Click to Add column in the table datasheet to add new fields. (See Figure 1-8.) You'll use both methods to add the four remaining fields to the Visit table. The next field you need to add is the PatientID field. Similar to the VisitID field, the PatientID field will contain numbers that will not be used in calculations, so it should be a Short Text field.

To add the rest of the fields to the Visit table:

(1). On the **Table Tools Fields Tab**, in the **Add & Delete group**, **Click** the **Short Text** button. Access adds a new field named "**Field1**" to the right of the VisitID field. See **Figure 5-9** on the next page.

The text "Field1" is selected, so you can simply type the new field name to replace it.

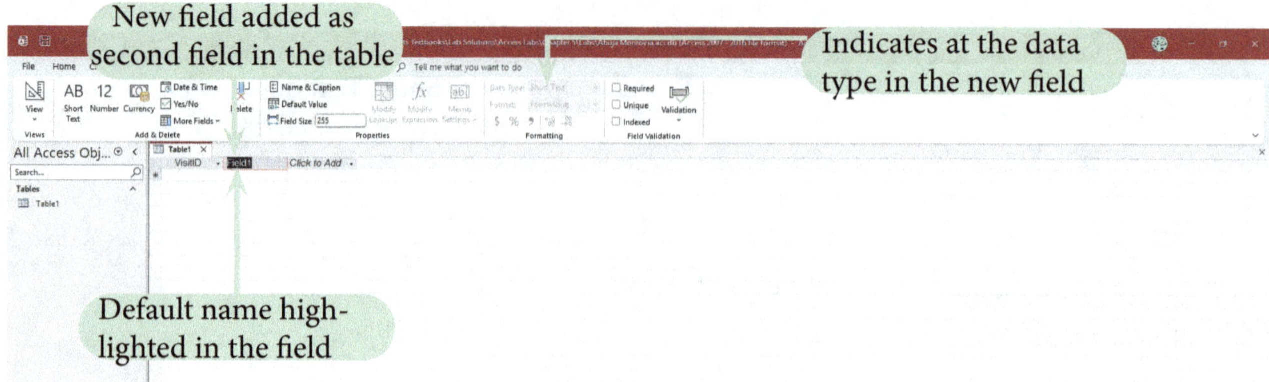

Figure 5-9 New Short Text field added to the table

(2). **Type PatientID**. Access adds the second field to the table. Next, you'll add the **VisitDate** field. Because this field will contain date values, you'll add a field with the Date/Time data type, which allows field values in a variety of date and time formats.

(3). In the **Add & Delete group**, **Click** the **Date & Time** button. Access adds a third field to the table, this time with the Date/Time data type.

(4). **Type VisitDate** to replace the selected name "**Field1**." The fourth field in the Visit table is the **Reason** field, which will contain brief descriptions of the reason for the visit to the clinic. You'll add another Short Text field—this time using the Click to Add column.

(5). **Click** the **Click to Add column** heading. Access displays a list of available data types for the new field.

(6). **Click Short Text** in the list. Access adds a fourth field to the table.

(7). **Type Reason** to replace the highlighted name "**Field1**," and then **Press ENTER**. The Click to Add column becomes active and displays the list of field data types.

The fifth and final field in the Visit table is the WalkIn field, which will indicate whether the patient had a scheduled appointment. The Yes/No data type is suitable for this field because it defines fields that store values representing one of two options—true/false, yes/no, or on/off.

(8). **Click Yes/No** in the list, and then **Type WalkIn** to replace the highlighted name "**Field1**."

Are You Having Trouble? If you pressed TAB or ENTER after typing the WalkIn field name, **Press ESC** to close the **Click** to **Add** list.

(9). **Click** in the row below the **VisitID** column heading. You have entered all five fields for the Visit table. See **Figure 5-10**.

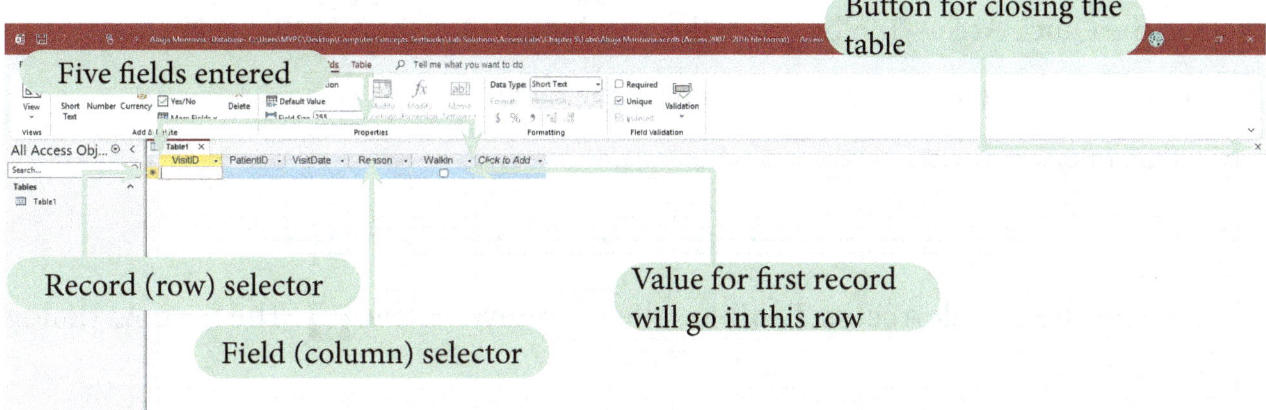

Figure 5-10 Table with all fields entered

The table contains three Short Text fields (VisitID, PatientID, and Reason), one Date/Time field (VisitDate), and one Yes/No field (WalkIn). You'll learn more about field data types in the next module.

As we noted earlier in this module of the chapter, Datasheet view shows a table's contents in rows (records) and columns (fields). Each column is headed by a field name inside a field selector, and each row has a record selector to its left (see **Figure 5-10** above). Clicking a **Field Selector** or a **Record Selector** selects that entire column or row (respectively), which you then can manipulate. A field selector is also called a **Column Selector**, and a record selector is also called a **Row Selector**.

SAVING THE VISIT TABLE STRUCTURE

As you find out later, the records you enter are immediately stored in the database as soon as you enter them; however, the table's design—the field names and characteristics of the fields themselves, plus any layout changes to the datasheet—are not saved until you save the table. When you save a new table for the first time, you should give it a name that best identifies the information it contains. Like a field name, a table name can contain up to 64 characters, including spaces.

According to Chioma's plan, you need to save the table with the name "Visit." By saving the table you will be able to work in it later

Smart Tips

Saving a Table

- Make sure the table you want to save is open.
- On the Quick Access Toolbar, click the Save button. The Save As dialog box opens.
- In the Table Name box, type the name for the table.
- Click OK.

To save, name, and close the Visit table:

(1). On the **Quick Access Toolbar**, click the Save button. The Save As dialog box opens.

(2). With the default name **Table1** selected in the **Table Name** box, **Type Visit** and then **Click OK**. The tab for the table now displays the name "Visit," and the Visit table design is saved in the Abuja Monrovia Database.

(3). **Click** the **Close 'Visit'** button on the object tab (see **Figure 5-10** on the previous page for the location of this button). The Visit table closes, and the main portion of the Access window is now blank because no database object is currently open. The Abuja Monrovia database file is still open, as indicated by the filename in the Access window title bar.

CREATING A TABLE IN DESIGN VIEW

The Abuja Monrovia Database also needs a table that will hold all of the invoices generated by each office visit at both locations. Chioma has decided to call this new table the Billing table. You created the structure for the Visit table in Datasheet view. An alternate method of creating the structure of a table is by using Design view. You will create the new Billing table using Design view.

Creating a table in Design view involves entering the field names and defining the properties for the fields, specifying a primary key for the table, and then saving the table structure. Chioma began documenting the design for the new Billing table by listing each field's name, data type, and purpose, and will continue to refine the design. See **Figure 5-11** on the next page.

Field Name	Data Type	Purpose
InvoiceNum	Short Text	Unique number assigned to each invoice; will serve as the table's primary key
VisitID	Short Text	Unique number assigned to each visit; common field that will be a foreign key to connect to the Visit table
InvoiceAmount	Currency	Dollar amount of each invoice
InvoiceDate	Date/Time	Date the invoice was generated
InvoicePaid	Yes/No	Whether the invoice has been paid or not

Figure 5-11 Initial design for the Billing table

You'll use Chioma's design as a guide for creating the Billing table in the Abuja Monrovia Database.

To begin creating the Billing table:

(1). If the **Navigation Pane** is open, **Click** the **Shutter Bar Open/Close Button** « to close it.

(2). On the ribbon, **Click** the **Create tab**.

(3). In the **Tables Group**, **Click** the **Table Design** button. A new table named Table1 opens in Design view.

DEFINING FIELDS

When you first create a table in Design view, the insertion point is located in the first row's Field Name box, ready for you to begin defining the first field in the table. You enter values for the Field Name, Data Type, and Description field properties (optional), and then select values for all other field properties in the Field Properties pane. These other properties will appear when you move to the first row's Data Type box.

The first field you need to define is the InvoiceNum field. This field will be the primary key for the Billing table. Each invoice at Abuja Monrovia Community Health Services is assigned a specific five-digit number. Although the InvoiceNum field will contain these number values, the numbers will never be used in calculations; therefore, you'll assign the Short Text data type to this field.

Smart Tips

Defining a Field in Design View

- In the Field Name box, type the name for the field, and then press TAB.
- Accept the default Short Text data type, or click the arrow and select a different data type for the field. Press TAB.
- Enter an optional description for the field, if necessary.
- Use the Field Properties pane to type or select other field properties, as appropriate.

Any time a field contains number values that will not be used in calculations—such as phone numbers, postal codes, and so on—you should use the Short Text data type instead of the Number data type.

To define the InvoiceNum field:

(1). **Type InvoiceNum** in the first row's Field Name box, and then **press TAB** to advance to the **Data Type** box. The default data type, **Short Text**, appears highlighted in the Data Type box, which now also contains an arrow, and the field properties for a Short Text field appear in the Field Properties pane. See **Figure 5-12**.

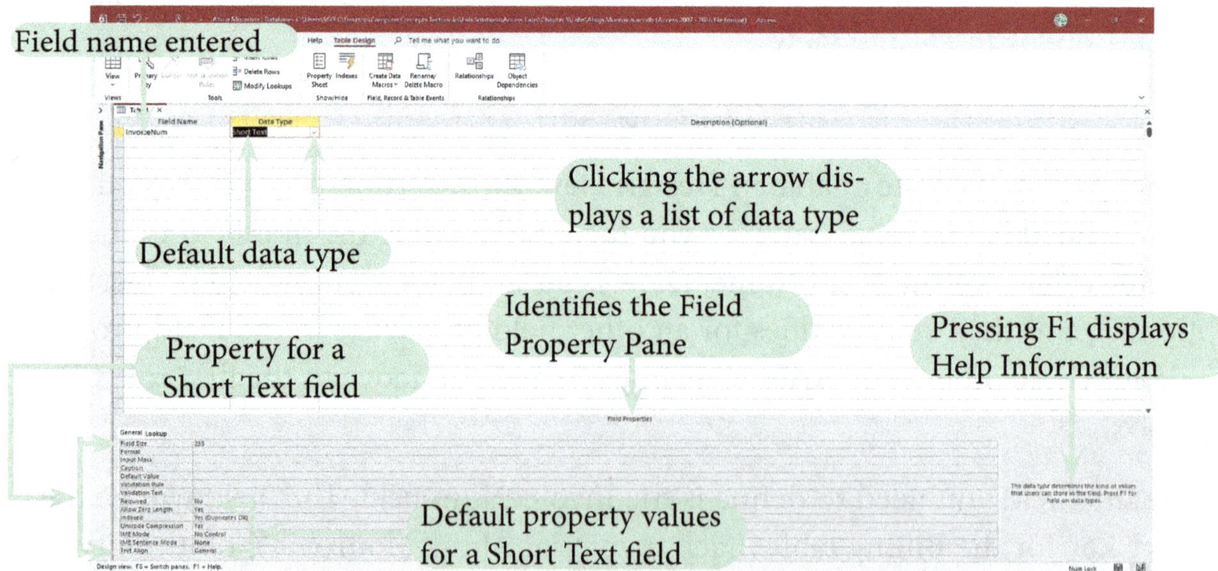

Figure 5-12 Table window after entering the first field name

The right side of the Field Properties pane now provides an explanation for the current property, Data Type.

Are You Having Trouble? If you make a typing error, you can correct it by clicking to position the insertion point, and then using either BACKSPACE to delete characters to the left of the insertion point or DELETE to delete characters to the right of the insertion point. Then type the correct text.

Because the InvoiceNum field values will not be used in calculations, you will accept the default Short Text data type for the field.

(2). **Press TAB** to accept Short Text as the data type and to advance to the **Description** (Optional) box.

Next you'll enter the Description property value as "Primary key." The value you enter for the Description property will appear on the status bar when you view the table datasheet. Note that specifying "Primary key" for the Description property does not establish the current field as the primary key; you use a button on the ribbon to specify the primary key in Design view, which you will do later in this session.

(3). **Type Primary key** in the **Description** (Optional) box and **Press ENTER.**

At this point, you have entered the first field (InvoiceNum) into the table and are ready to enter the remaining fields into the table. See **Figure 5-13**.

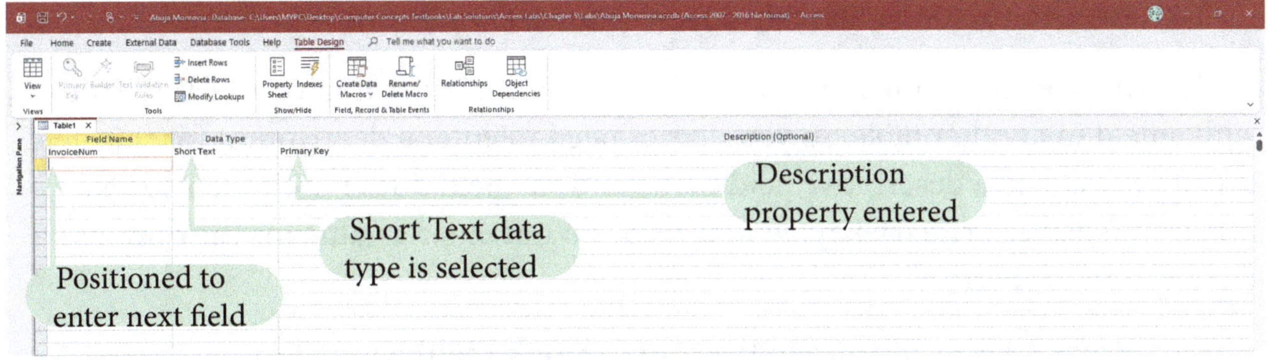

Figure 5-13 InvoiceNum field defined

Chioma's Billing table design (**Figure 1-11** shown earlier) shows VisitID as the second field. Because Chioma and other staff members need to relate information about invoices to the visit data in the Visit table, the Billing table must include the VisitID field, which is the Visit table's primary key. Recall that when you include the primary key from one table as a field in a second table to connect the two tables, the field is a foreign key in the second table.

To define the VisitID field:

(1). If the insertion point is not already positioned in the **second row's Field Name** box, **Click** the **second row's Field Name** box. Once properly positioned, **Type VisitID** in the box, and then **Press TAB** to advance to the **Data Type** box.

(2). **Press TAB** to accept **Short Text** as the field's data type. Because the VisitID field is a foreign key to the Visit table, you'll enter "**Foreign key**" in the **Description** (Optional) box to help users of the database understand the purpose of this field.

(3). **Type Foreign Key** in the **Description** (Optional) box and **press ENTER**.

The third field in the Billing table is the InvoiceAmt field, which will display the dollar amount of each invoice the clinic sends to the patients. The Currency data type is the appropriate choice for this field.

To define the InvoiceAmount field:

(1). In the **third row's Field Name** box, **Type InvoiceAmount** and then **Press TAB** to advance to the **Data Type** box.

(2). **Click** the **Data Type arrow**, **Click Currency** in the list, and then **Press TAB** to advance to the **Description** (Optional) box.

The InvoiceAmount field is not a primary key, nor does it have a relationship with a field in another table, so you do not need to enter a description for this field. If you've assigned a descriptive field name and the field does not fulfill a special function (such as primary key), you usually do not enter a value for the optional Description property.

(3). **Press TAB** to advance to the **fourth row's Field Name** box.

The fourth field in the Billing table is the InvoiceDate field. This field will contain the dates on which invoices are generated for the clinic's patients. You'll define the InvoiceDate field using the Date/Time data type.

To define the InvoiceDate field:

(1). In the **fourth row's Field Name** box, **Type InvoiceDate** and then **Press TAB** to advance to the **Data Type** box.

You can select a value from the Data Type list as you did for the InvoiceAmount field. Alternately, you can type the property value in the box or type just the first character of the property value.

(2). **Type d**. The value in the fourth row's Data Type box changes to "date/Time," with the letters "ate/Time" highlighted. See **Figure 5-14**.

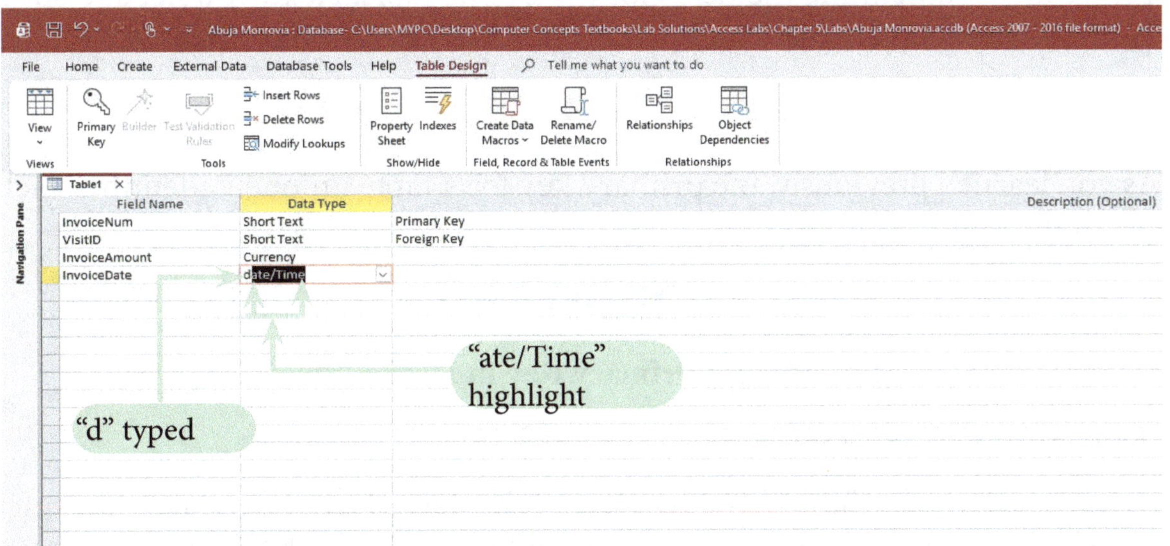

Figure 5-14 Selecting a value for the Data Type property

(3). **Press TAB** to advance to the Description (Optional) box. Note that Access changes the value for the **Data Type property** to **"Date/Time."**

(4). Because the InvoiceDate field does not need a special description, **press TAB**.

The fifth, and final, field to be defined in the Billing table is InvoicePaid. This field will be a Yes/No field to indicate the payment status of each invoice record stored in the Billing table. Recall that the Yes/No data type defines fields that store true/false, yes/no, and on/off field values. When you create a Yes/No field in a table, the default Format property is set to Yes/No.

To define the InvoicePaid field:

(1). In the fifth row's Field Name box, **Type InvoicePaid** and then **Press TAB** to advance to the **Data Type** box.

(2). **Type** y. Access completes the data type as "**Yes/No.**" **Press TAB** to **select** the **Yes/No** data type and move to the Description (Optional) box.

(3). Because the **InvoicePaid** field does not need a special description, **Press TAB.**

You've finished defining the fields for the Billing table. Next, you need to specify the primary key for the table.

Specifying the Primary Key

As you learned previously, the primary key for a table uniquely identifies each record in the table.

Smart Tips

Specifying a Primary Key in Design View

- Display the table in Design view.
- Click in the row for the field you've chosen to be the primary key to make it the active field. If the primary key will consist of two or more fields, click the row selector for the first field, press and hold down CTRL, and then click the row selector for each additional primary key field.
- On the Table Tools Design tab in the Tools group, click the Primary Key button.

According to Chioma's design, you need to specify InvoiceNum as the primary key for the Billing table. You can do so while the table is in Design view.

To specify InvoiceNum as the primary key:

(1). **Click** in the row for the InvoiceNum field to make it the current field.

(2). On the **Table Tools Design Tab** in the **Tools Group**, **Click** the **Primary Key** button. The Primary Key button is highlighted and a key symbol appears in the row selector for the first row, indicating that the InvoiceNum field is the table's primary key. See **Figure 5-15** on the next page.

In the table below, there are now five fields in the new Table1: InvoiceNum (Short Text, Primary Key); VisitID (Short Text, Foreign Key); InvoiceAmount (Currency); InvoiceDate (Date/Time); and InvoicePaid (Yes/No).

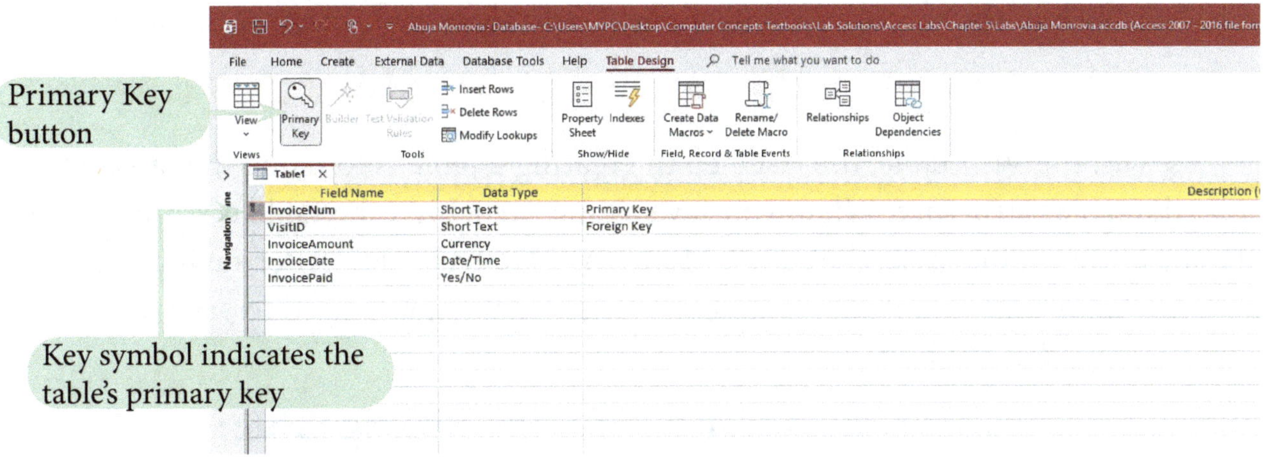

Figure 5-15 InvoiceNum field selected as the primary key

RENAMING FIELDS IN DESIGN VIEW

Chioma has decided to rename the InvoiceAmount field in the Billing table to InvoiceAmt. Since Amt is an appropriate abbreviation for Amount, this new name will be just as readable, yet a little shorter.

To rename a field in Design view:

(1). **Click** to position the insertion point to the right of the word "**InvoiceAmount**" in the third row's **Field Name** box, and then **Press BACKSPACE four times** to delete the letters "**ount**." The name of the fourth field is now **InvoiceAm**. Now add the final letter by pressing the letter **t**. The name of the new field is now **InvoiceAmt** as Chioma wants it to be. See **Figure 5-16**.

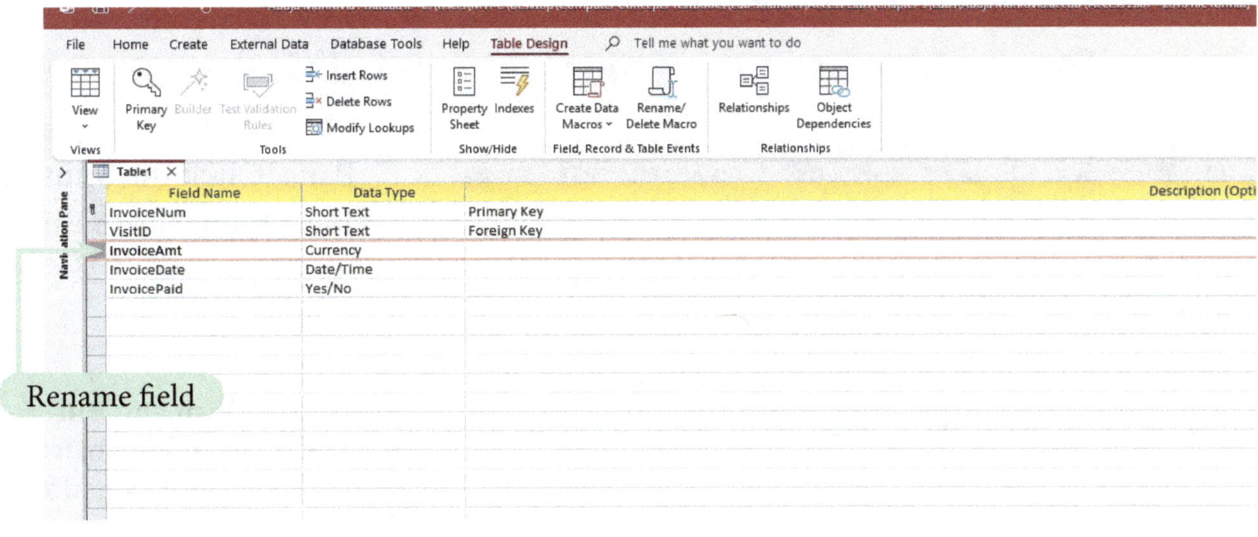

Figure 5-16 Billing table after renamed field

(2). Click in the row for the InvoiceAmt field to make it the current field.

SAVING THE BILLING TABLE STRUCTURE

As with the Visit table, the last step in creating a table is to name the table and save the table's structure. When you save a table structure, the table is stored in the database file (in this case, the Abuja Monrovia Database file). After saving the table, you can enter data into it. According to Whioma's plan, you need to save the table you've defined as "Billing."

To save, name, and close the Billing table:

(1). On the **Quick Access Toolbar**, **Click** the **Save** button . The Save As dialog box opens.

(2). With the default name **Table1 selected** in the **Table Name** box, **type Billing**, and then **Click OK**. The tab for the table now displays the name "**Billing**," and the Billing table design is saved in the **Abuja Monrovia** Database.

(3). **Click** the **Close** '**Billing**' button on the object tab. The Billing table closes, and the main portion of the Access window is now blank because no database object is currently open. The Abuja Monrovia database file is still open, as indicated by the filename in the Access window title bar.

You have now successfully created and saved the structures for the Visit and Billing tables; however, you have not yet added any data to these tables. You can view and work with these objects in the Navigation Pane.

To view objects in the Abuja Monrovia Database:

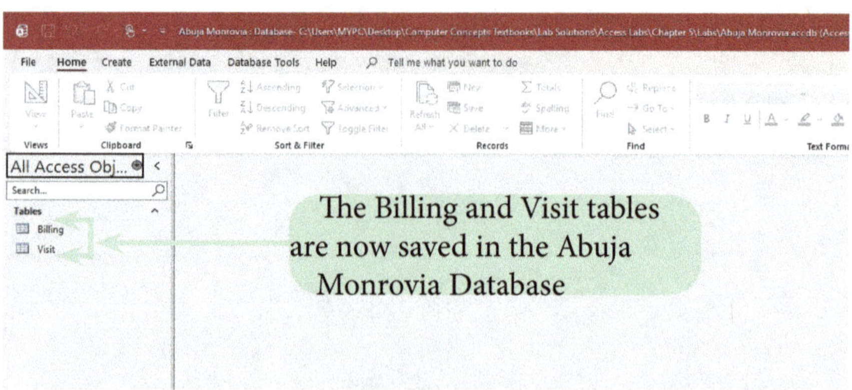

(1). On the Navigation Pane, **Click** the **Shutter Bar Open/Close** Button » to open it. See **Figure 5-17**.

Figure 5-17 Visit and Billing tables (database objects) displayed in the Navigation Pane

CLOSING A TABLE AND EXITING ACCESS

When you are finished working in an Access table, it's a good idea to close the table so that you do not make unintended changes to the table data. You can close a table by clicking its Close button on the object tab, as you did earlier. Or, if you want to close the Access program as well, you can click the program's Close button. When you do, any open tables are closed, the active database is closed, and you exit the Access program.

To close any opened tables and exit Access:

(1). **Click** the **Close** button ✕ on the program window title bar. Any opened tables would close, along with the Abuja Monrovia Database, and then the Access program closes.

Smart Tips

Saving a Database

Unlike the Save buttons in other Office programs, the Save button on the Quick Access Toolbar in Access does not save the active document (database). Instead, you use the Save button to save the design of an Access object, such as a table (as you saw earlier), or to save datasheet format changes, such as resizing columns. Access does not have or need a button or option you can use to save the active database.

Access saves changes to the active database automatically when you change or add a record or close the database. If your database is stored on a removable storage device, such as a USB drive, you should never remove the device while the database file is open. If you do, Access will encounter problems when it tries to save the database, which might damage the database. Make sure you close the database first before removing the storage device.

It is possible to save a database with a different name. To do so, you would click the File tab to open Backstage view, and then click the Save As option. You save the database in the default database format unless you select a different format, so click the Save As button to open the Save As dialog box. Enter the new name for the database, choose the location for saving the file, and then click Save. The database is saved with a new name and is stored in the specified location.

Now that you've become familiar with database concepts and Access, and created the Abuja Monrovia Database and the structures for the Visit and Billing tables, Chioma wants you to add records to the Visit table and work with the data stored in it to create database objects including a query, form, and report. You'll complete these tasks in the next session.

Quick Review

1. A(n) ____________________ is a single characteristic of a person, place, object, event, or idea.
2. You connect the records in two separate tables through a(n)____________________ that appears in both tables.
3. The ____________________, whose values uniquely identify each record in a table, is called a(n) ____________________ when it is placed in a second table to form a relationship between the two tables.
4. The ____________________ is the area of the Access window that lists all the objects in a database, and it is the main control center for opening and working with database objects.
5. What is the name of the field that Access creates, by default, as the primary key field for a new table in Datasheet view?
6. Which group on the Table Tools Fields tab contains the options you use to add new fields to a table?
7. What are the two views you can use to create a table in Access?
8. Explain how the saving process in Access is different from saving in other Office programs.

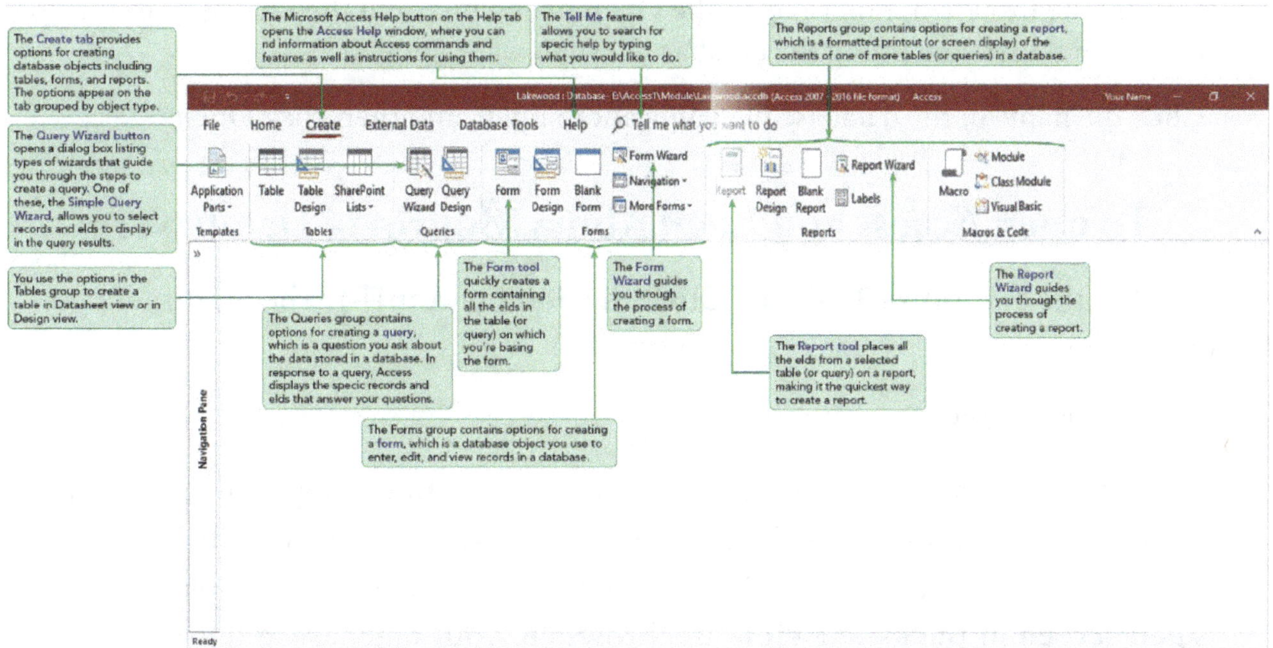

ENTERING DATA INTO TABLES

With the fields in place for the Visit table, you can now enter the field values for each record. However, if you closed Access, as instructed, after the previous session, you must first open Access and the Abuja Monrovia Database to be able to work with the Visit table. If you did not close Access in the previous session and the Abuja Monrovia Database is still open (see **Figure 5-17** in the previous session), you may skip the steps below that open Access and the Abuja Monrovia Database, and go directly to the steps to enter data into the Visit table.

Smart Tips

Opening a Database

- Start Access and display the Recent screen in Backstage view.
- Click the name of the database you want to open in the list of recently opened databases.

or

- Start Access and display the Recent screen in Backstage view.
- In the navigation bar, click Open Other Files to display the Open screen.

- Click the Browse button to open the Open dialog box, and then navigate to the drive and folder containing the database file you want to open.
- Click the name of the database file you want to open, and then click Open.

To open Access and Load Abuja Monrovia Database:

(1). On the **Windows Taskbar**, **Click** the **Start** button. The Start menu opens.

(2). **Click Access**.

(3). Access starts and displays the Recent screen in **Backstage View**. You may choose the **Abuja Monrovia Database** from the Recent list (with its location listed below the database name), or **Click Open Other Files** to display the **Open screen** in **Backstage view** and browse to your database and location. If you choose to open the Abuja Monrovia Database from the Recent list, skip **steps 4**, **5**, and **6**.

(4). If you choose to open other files from **step 3**, on the **Open screen**, **click Browse**. The Open dialog box opens, showing folder information for your computer.

Are You Having Trouble? If you are storing your files on OneDrive, click OneDrive, and then sign in if necessary, or else, click the location you use to store files.

(5). **Navigate** to the drive that contains your **Data Files**.

(6). **Navigate** to the **Access** > **Chapter 5 folder**, **Click** the database file named **Abuja Monrovia**, and then **Click Open**. The Abuja Monrovia Database opens in the Access program window.

Are You Having Trouble? If a security warning appears below the ribbon indicating that some active content has been disabled, click the Enable Content button. Access provides this warning because some databases might contain content that could harm your computer. Because the Abuja Monrovia database does not contain objects that could be harmful, you can open it safely. If you are accessing the file over a network, you might also see a dialog box asking if you want to make the file a trusted document; click Yes.

Note that the Abuja Monrovia Database contains two objects, the Billing and Visit tables you created at the end of the previous session of this chapter (kindly see **Figure 5-17**). The next step is for you to open the Visit table to begin adding records.

To open the Visit table:

(1). In the **Navigation Pane**, **Double-Click Visit** to open the **Visit table** in **Datasheet view**.

(2). On the **Navigation Pane**, **Click** the **Shutter Bar Open/Close Button** « to close the pane.

(3). **Click** the **first row** value for the **VisitID** field. See **Figure 5-18**.

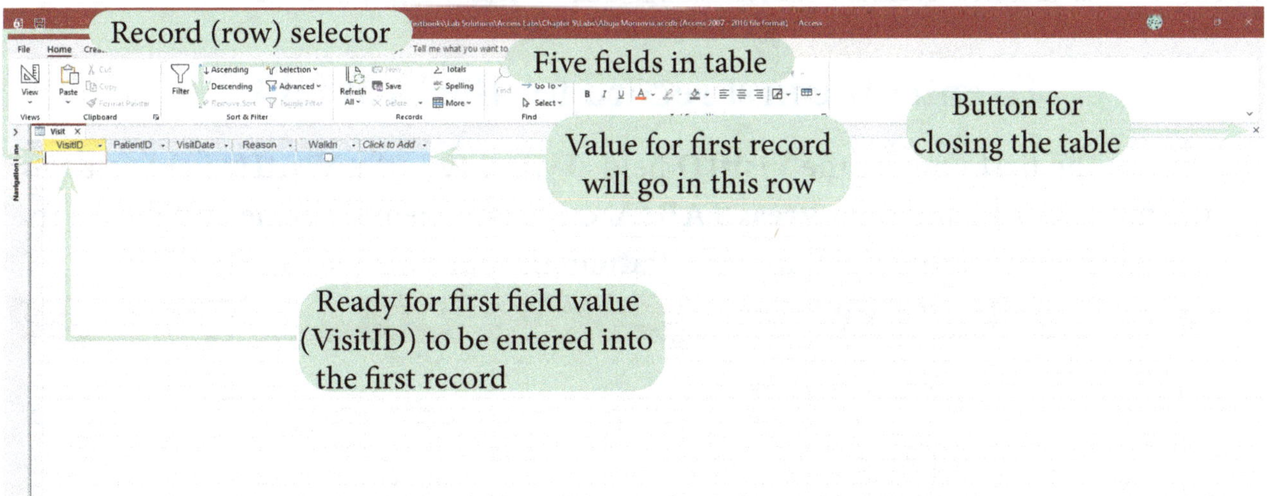

Figure 5-18 Visit table opened and ready to enter data

The Figure 5-18 above shows the Visit table open in Datasheet View. We see that there are five fields in the table. VisitID; PatientID; VisitDate; Reason; and WalkIn. The WalkIn field has a check box in it. The value of the first record will go in the first row in the table. The record row selector is a box with a star next to the first record. The Close button for closing the table is in the upper-right corner of the table.

You are now ready to begin adding records and are positioned in the first field (VisitID) of the first record. Chioma requests that you enter eight records into the Visit table, as show in **Figure 5-19** on the next page.

VisitID	PatientID	VisitDate	Reason	WalkIn
1495	13310	12/23/2020	Rhinitis	Yes
1450	13272	10/26/2020	Influenza	Yes
1461	13250	11/3/2020	Dermatitis	Yes
1615	13308	4/1/2021	COPD management visit	No
1596	13299	3/24/2021	Pneumonia	Yes
1567	13283	2/26/2021	Annual wellness visit	No
1499	13264	12/28/2020	Hypotension	No
1475	13261	11/19/2020	Annual wellness visit	No

Figure 5-19 Visit table records

To enter the first record for the Visit table:

(1). In the **first row** for the **VisitID** field, **Type 1495** (the **VisitID** field value for the first record), and then **Press TAB**. Access adds the field value and moves the insertion point to the right, into the **PatientID** column. See **Figure 5-20**.

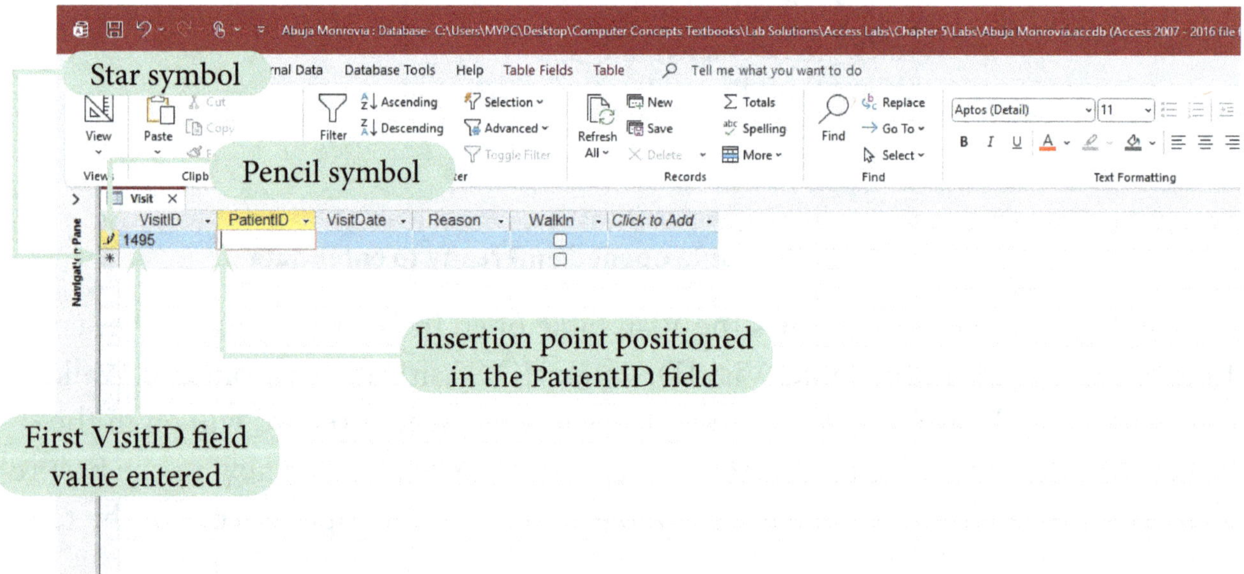

Figure 5-20 First field value entered

Note: be sure to type the numbers "0" and "1" and not "O" and "I" when entering numeric values, even though the field is of the Short Text data type.

Are You Having Trouble? If you make a mistake when typing a value, use BACKSPACE to delete characters to the left of the insertion point or use DELETE to delete characters to the right of the insertion point. Then type the correct value. To correct a value by replacing it entirely, press ESC, and then type the correct value.

Notice the pencil symbol that appears in the row selector for the new record. The **Pencil symbol** indicates that the record is being edited. Also notice the star symbol that appears in the row selector for the second row. The **Star symbol** identifies the second row as the next row available for a new record.

(2). **Type 13310** (the PatientID field value for the first record), and then **Press TAB**. Access enters the field value and moves the insertion point to the VisitDate column.

(3). **Type 12/23/20** (the VisitDate field value for the first record), and then **press TAB**. Access displays the year as "**2020**" even though you entered only the final two digits of the year. This is because the VisitDate field has the **Date/Time data type**, which automatically formats dates with four-digit years.

(4). **Type Rhinitis** (the Reason field value for the first record), and then **Press TAB** to move to the **WalkIn** column.

Recall that the WalkIn field is a Yes/No field. Notice the check box displayed in the WalkIn column. By default, the value for any Yes/No field is "No"; therefore, the check box is initially empty. For Yes/No fields with check boxes, you press TAB to leave the check box unchecked, or you press SPACEBAR to insert a checkmark in the check box. The record you are entering in the table is for a walk-in visit, so you need to insert a checkmark in the check box to indicate "Yes."

(5). **Press SPACEBAR** to insert a checkmark, and then **Press TAB**. The first record is entered into the table, and the insertion point is positioned in the VisitID field for the second record. The pencil symbol is removed from the first row because the record in that row is no longer being edited. The table is now ready for you to enter the second record. See **Figure 5-21** on the next page.

Figure 5-21 on the next page shows that the following values for the first record in the Visit table were entered: VisitID (1495), PatientID (13310), VisitDate (12/23/2020), Reason (Rhinitis), and WalkIn (check box selected).

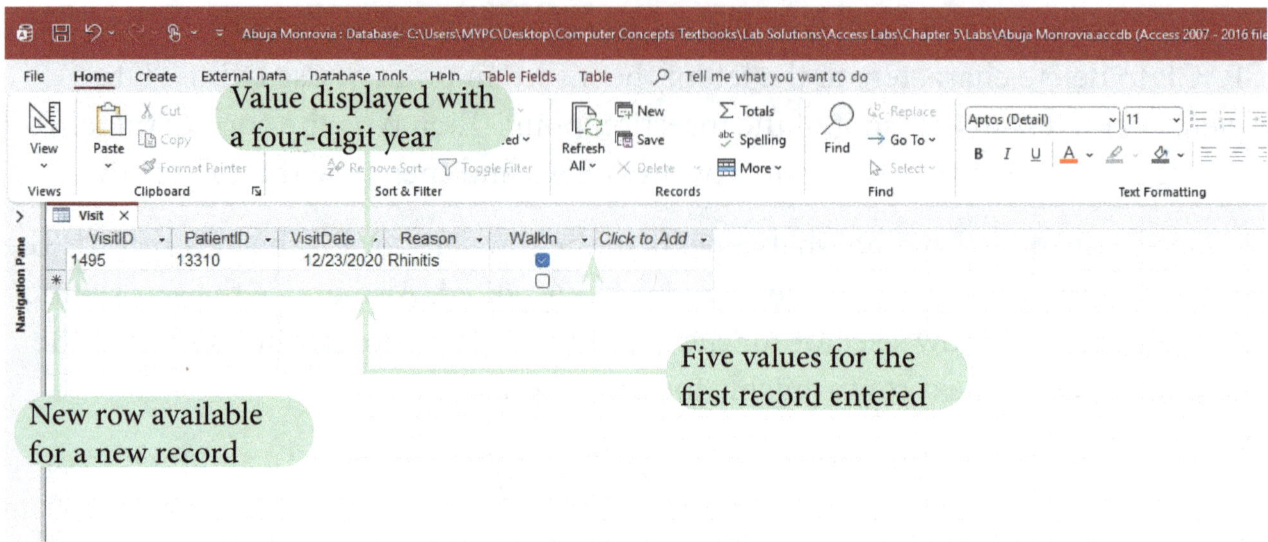

Figure 5-21 Datasheet with first record entered

To enter the remaining records in the Visit table:

(1). Referring to **Figure 5-19** (which contains a table), **enter** the **values** for **records 2** through **8**, **pressing TAB** to move from field to field and to the next row for a new record. Keep in mind that you **do not have to type all four digits of the year** in the **VisitDate** field values; you can enter only the final two digits, and Access will display all four. Also, for any **WalkIn** field values of "**No**," be sure to **Press TAB** to leave the check box empty.

You can also Press ENTER instead of TAB to move from one field to another and to the next row.

Are You Having Trouble? If you enter a value in the wrong field by mistake, such as entering a Reason field value in the VisitDate field, a menu might open with options for addressing the problem. If this happens, click the "Enter new value" option in the menu. You'll return to the field with the incorrect value selected, which you can then replace by typing the correct value.

Notice that not all of the Reason field values are fully displayed. To see more of the table datasheet and the full field values, you'll resize the Reason column.

(2). **Place** the **pointer** on the **vertical line** to the **right** of the **Reason** field name until the pointer changes to the column **resizing** pointer ✛, and then **double-click** the vertical line. All the Reason field values are now fully displayed. See **Figure 5-22** on the next page.

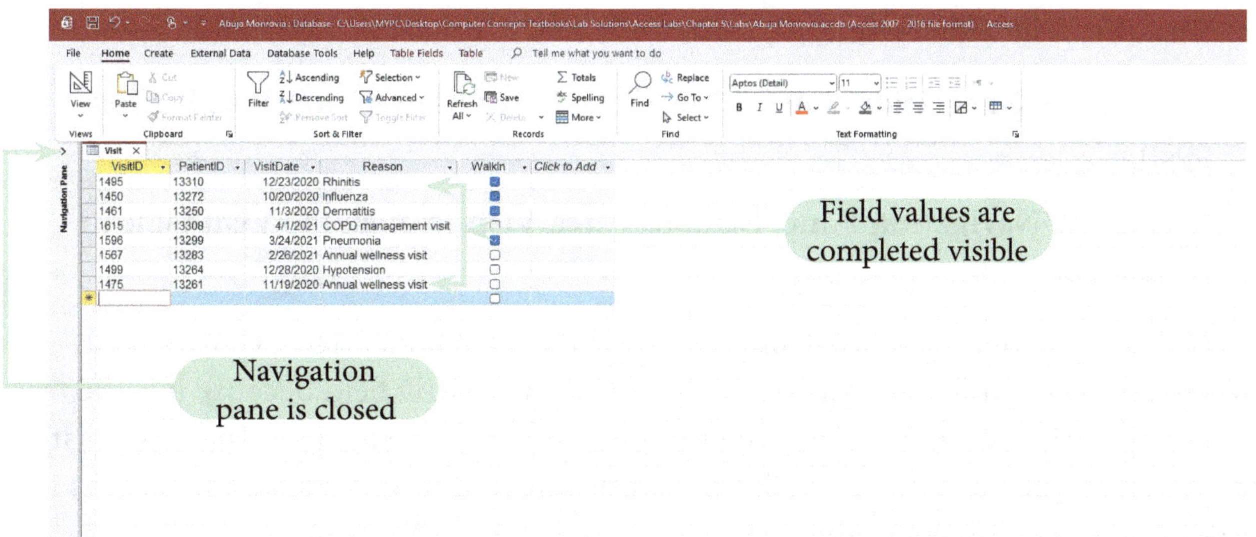

Figure 5-22 Datasheet with eight records entered

When you resize a datasheet column by double-clicking the column dividing line, you are sizing the column to its **Best Fit**—that is, so the column is just wide enough to display the longest visible value in the column, including the field name.

(3). Compare your table to the one in **Figure 5-22**. If any of the field values in your table do not match those shown in the figure, you can correct a field value by clicking to position the insertion point in the value, and then using BACK-SPACE or DELETE to delete incorrect text. Type the correct text and press EN-TER. To correct a value in the WalkIn field, click the check box to add or remove the checkmark as appropriate. Also, be sure the spelling and capitalization of field names in your table match those shown in the figure exactly and that there are no spaces between words. To correct a field name, double-click it to select it, and then type the correct name; or use the Rename Field option on the shortcut menu to rename a field with the correct name. Always perform a quality check.

Remember that Access automatically saves the data stored in a table; however, you must save any new or modified structure to a table. Even though you have not clicked the Save button, your data has already been saved. To ensure this is the case, you can close the table and then reopen it.

To close and reopen Visit table:

(1). **Click** the **Close 'Visit'** button ☒ on the object tab for the Visit table. When asked if you would like to save the changes to the layout of the Visit table, **Click**

Yes. The Visit table closes.

(2). On the **Navigation Pane**, **Click** the **Shutter Bar Open/Close Button** [»] to open it.

(3). In the **Navigation Pane**, **Double-Click Visit** to open the **Visit** table in **Datasheet view**.

Notice that after you closed and reopened the Visit table, Access sorted and displayed the records in order by the values in the VisitID field because it is the primary key. If you compare your screen to **Figure 5-22**, which shows the records in the order you entered them, you'll see that the current screen shows the records in order by the VisitID field values.

Chioma asks you to add two more records to the Visit table. When you add a record to an existing table, you must enter the new record in the next row available for a new record; you cannot insert a row between existing records for the new record. In a table with just a few records, such as the Visit table, the next available row is visible on the screen. However, in a table with hundreds of records, you would need to scroll the datasheet to see the next row available. The easiest way to add a new record to a table is to use the New button, which scrolls the datasheet to the next row available so you can enter the new record.

To enter additional records in the Visit table:

(1). If necessary, **Click** the first record's **VisitID** field value (**1450**) to make it the current record.

(2). In the **Records Group**, **Click** the **New** button. The insertion point is positioned in the next row available for a new record, which in this case is **row 9**. See **Figure 5-23** on the next page.

Figure 5-23 on the next page displays the Visit Datasheet with the Home Tab displayed on the Ribbon. The New button on the Home Tab was clicked, positioning the insertion point in the first field (VisitID) for the new record in row 9. We now see that the New button is dimmed because a new record is being currently created.

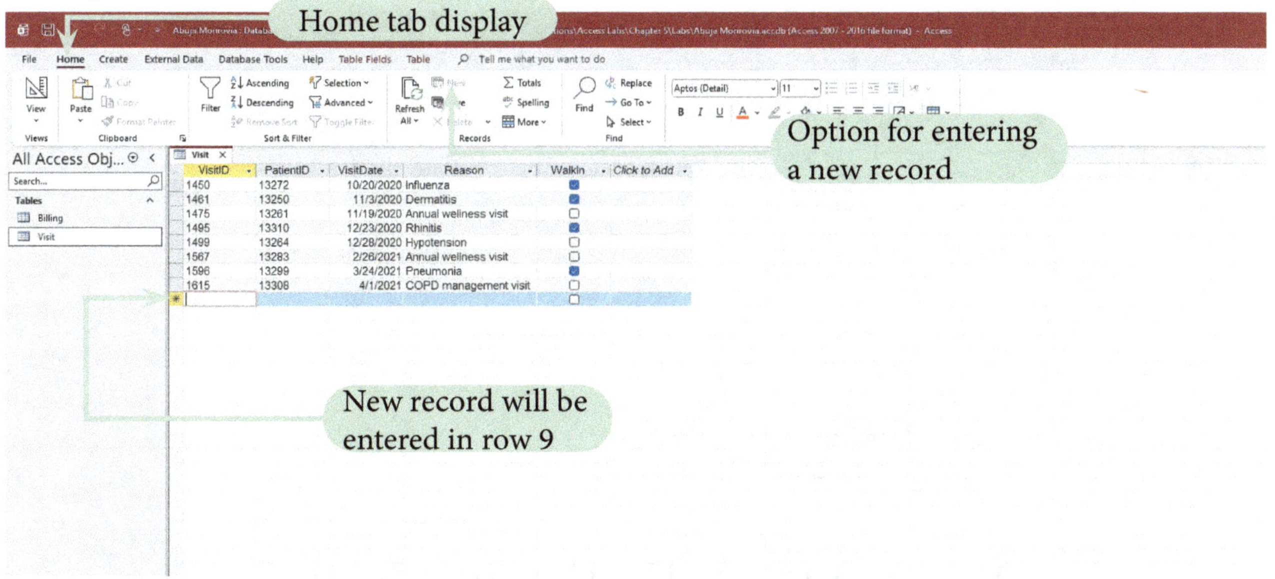

Figure 5-23 Datasheet with eight records entered

(3). With the insertion point in the **VisitID** field for the new record, **Type 1548** and then **Press TAB**.

(4). Complete the entry of this record by entering each value shown below, **Pressing TAB** to move from field to field:

PatientID = 13301

VisitDate = 2/10/2021

Reason = Hypothyroidism

WalkIn = No (unchecked)

(5). Enter the values for the next new record, as follows, and then **Press TAB** after entering the **WalkIn** field value:

VisitID = 1588

PatientID = 13268

VisitDate = 3/19/2021

Reason = Cyst removal

WalkIn = Yes (checked)

Your datasheet should look like the one shown in **Figure 5-24** on the next page.

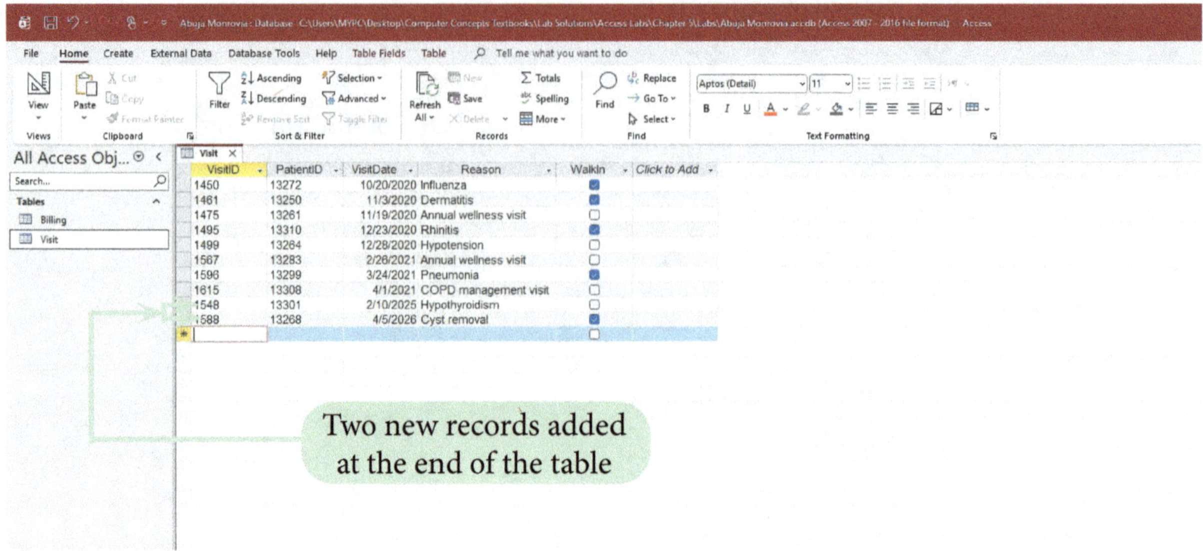

Figure 5-24 Datasheet with additional records entered

The new records you added appear at the end of the table, and are not sorted in order by the primary key field values. For example, VisitID 1548 should be the sixth record in the table, placed between VisitID 1499 and VisitID 1567. When you add records to a table datasheet, they appear at the end of the table. The records are not displayed in primary key order until you either close and reopen the table or switch views.

(6). **Click** the **Close 'Visit'** button ☒ on the object tab. The Visit table closes; however, it is still listed in the Navigation Pane.

(7). **Double-Click Visit** to open the table in **Datasheet view**. See **Figure 5-25**

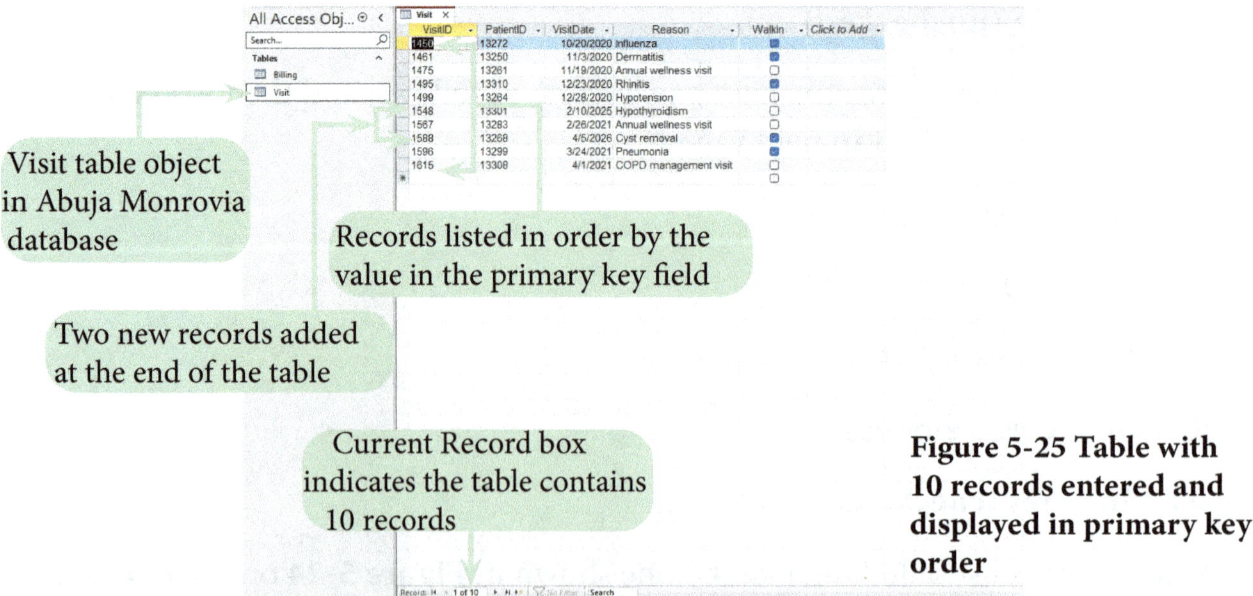

Figure 5-25 Table with 10 records entered and displayed in primary key order

The two records you added, with VisitID field values of 1548 and 1588, now appear in the correct primary key order. The table contains a total of 10 records, as indicated by the **Current Record box** at the bottom of the datasheet. The Current Record box displays the number of the current record as well as the total number of records in the table.

Each record contains a unique VisitID value because this field is the primary key. Other fields, however, can contain the same value in multiple records; for example, the Reason field has two values of "Annual wellness visit."

(8). **Click** the **Close** button on the program window title bar. The Visit table, along with the **Abuja Monrovia Database**, close, and then the Access program closes.

COPYING RECORDS FROM ANOTHER ACCESS DATABASE

When you created the Visit table, you entered records directly into the table datasheet. There are many other ways to enter records in a table, including copying and pasting records from a table into the same database or into a different database. To use this method, however, the two tables must have the same structure—that is, the tables must contain the same fields, with the same design, in the same order.

Chioma has already created a table named Appointment that contains additional records with visit data. The Appointment table is contained in a database named Chioma.accdb located in the Access > Chapter 5 folder included with your Data Files. The Appointment table has the same table structure as the Visit table you created.

Your next task is to copy the records from the Appointment table and paste them into your Visit table. To do so, you need to open the Chioma.accdb database.

To copy the records from the Appointment table:

(1). On the **Windows Taskbar**, **Click** the **Start** button. The Start menu opens.

(2). **Click Access.**

(3). **Click Open Other Files** to display the **Open screen** in Backstage view.

(4). On the **Open screen**, **Click Browse**. The Open dialog box opens, showing folder information for your computer.

Are You Having Trouble? If you are storing your files on OneDrive, click OneDrive, and then log in if necessary.

(5). Navigate to the drive that contains your Data Files.

(6). Navigate to the **Access > Chapter 5 folder**, **Click** the database file **Chioma.accdb**, and then **Click Open**. The **Chioma Database** opens in the Access program window. Note that the database contains only one object, the Appointment table.

Are You Having Trouble? If a security warning appears below the ribbon indicating that some active content has been disabled, click the Enable Content button. Access provides this warning because some databases might contain content that could harm your computer. Because the Chioma.accdb database does not contain objects that could be harmful, you can open it safely. If you are accessing the file over a network, you might also see a dialog box asking if you want to make the file a trusted document; click Yes.

(7). In the **Navigation Pane**, **Double-Click Appointment** to open the Appointment table in Datasheet view. The table contains **76** records and the same five fields, with the same characteristics, as the fields in the Visit table. See **Figure 5-26**.

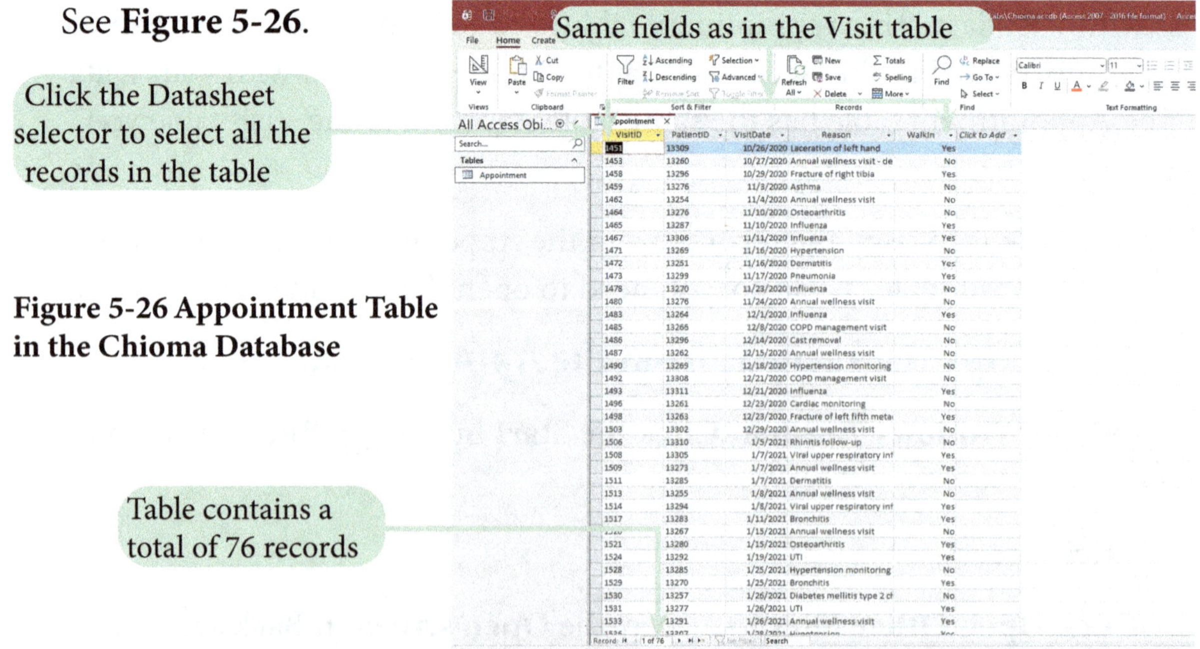

Figure 5-26 Appointment Table in the Chioma Database

Chioma wants you to copy all the records in the Appointment table. You can select all the records by clicking the datasheet selector, which is the box to the left of the first field name in the table datasheet, as shown in **Figure 1-26** on the previous page.

(8). **Click** the **Datasheet Selector** ☐ to the left of the **VisitID** field. All the records in the table are selected.

(9). In the **Clipboard group**, **Click** the **Copy** button to copy all the records to the Clipboard.

(10). **Click** the **Close** '**Appointment**' button ☒ on the object tab. A dialog box may open asking if you want to save the data you copied to the Clipboard. This dialog box opens only when you copy a large amount of data to the Clipboard. If asked, **Click Yes**. If opened, the dialog box closes, and then the **Appointment table closes**.

With the records copied to the Clipboard, you can now paste them into the Visit table. First you need to close the Chioma.accdb database while keeping the Access program open, and then open the Abuja Monrovia Database.

To close the Chioma.accdb database and then paste the records into the Visit table:

(1). **Click** the **File Tab** to open **Backstage view**, and then **Click Close** in the navigation bar to close the Chioma.accdb database. You return to a blank Access program window, and the **Home tab** is the active tab on the ribbon.

(2). **Click** the **File tab** to return to **Backstage view**, and then **Click Open** in the navigation bar. **Recent** is **selected** on the **Open screen**, and the recently opened database files are listed. This list should include the Abuja Monrovia Database.

(3). **Click Abuja Monrovia** to open the Abuja Monrovia Database file.

Are You Having Trouble? If the Abuja Monrovia Database file is not in the list of recent files, click Browse. In the Open dialog box, navigate to the drive and folder where you are storing your files, and then open the Abuja Monrovia Database file.

Are You Having Trouble? If the security warning appears below the ribbon, click the Enable Content button, and then, if necessary, click Yes to identify the file as a trusted document.

(4). In the **Navigation Pane**, **Double-Click Visit** to open the Visit table in Datasheet view.

(5). On the **Navigation Pane**, **Click** the **Shutter Bar Open/Close Button** « to close the pane.

(6). **Position** the **pointer** on the **star symbol** in the **row selector** for **row 11** (the next row available for a new record) until the pointer changes to a **right-pointing arrow** ➡, and then **click** to **select** the **row**.

(7). In the **Clipboard group**, **Click** the **Paste** button. The pasted records are added to the table, and a dialog box opens asking you to confirm that you want to paste all the records (76 total).

Are You Having Trouble? If the Paste button isn't active, click the row selection ➡ pointer on the row selector for row 11, making sure the entire row is selected, and then repeat **Step 7**.

(8). **Click Yes**. The dialog box closes, and the pasted records are selected. See **Figure 5-27**. Notice that the table now contains a total of 86 records—10 records that you entered previously and 76 records that you copied and pasted.

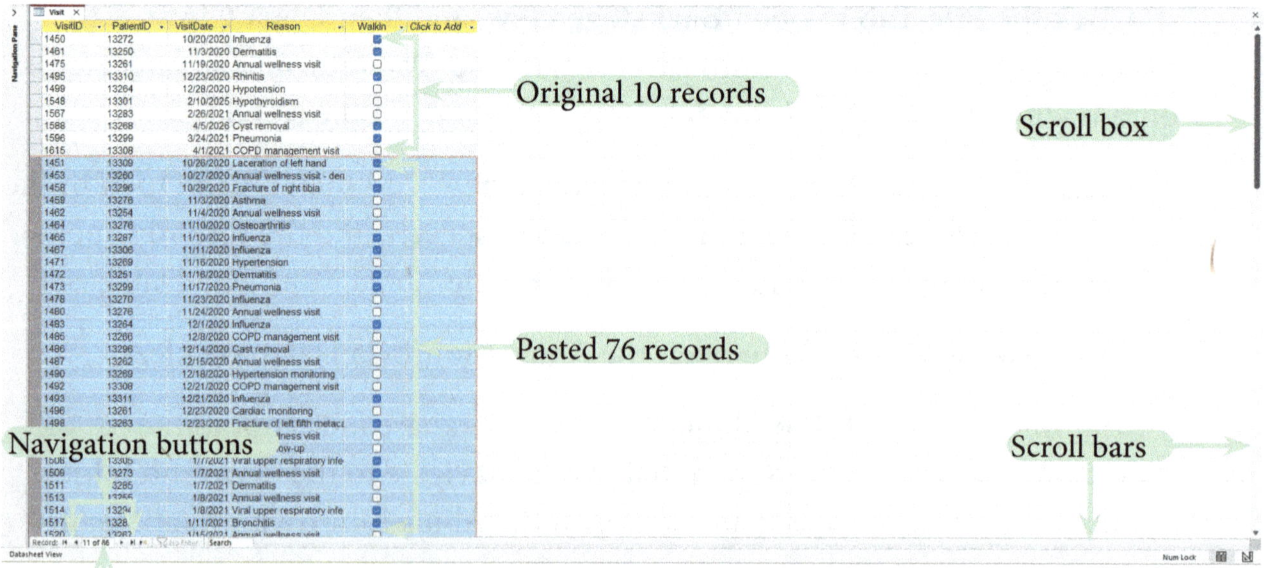

Figure 5-27 Visit table after copying and pasting records

Not all the Reason field values are completely visible, so you need to resize this column to its best fit.

(9). Place the pointer on the column dividing line to the right of the Reason field name until the pointer changes to the column resizing pointer ✛, and then double-click the column dividing line. The Reason field values are now fully displayed.

Navigating a Datasheet

The Visit table now contains 86 records, but only some of the records are visible on the screen. To view fields or records not currently visible on the screen, you can use the horizontal and vertical scroll bars to navigate the data. The **Navigation Buttons**, shown in **Figure 5-27** on the previous page and also described in **Figure 5-28**, provide another way to move vertically through the records. The Current Record box appears between the two sets of navigation buttons and displays the number of the current record as well as the total number of records in the table. **Figure 5-28** shows which record becomes the current record when you click each navigation button. The New (blank) record button works the same way as the New button on the Home tab, which you used earlier to enter a new record in the table.

Navigation Button	Record Selected
⏮	First record
◀	Previous record
▶	Next record
⏭	Last record
▶*	New (blank) record

Figure 5-28 Navigation buttons

Chioma suggests that you use the various navigation techniques to move through the Visit table and become familiar with its contents.

To navigate the Visit datasheet:

(1). **Click** the first record's **VisitID** field value (**1450**). The Current Record box shows that record **1** is the current record.

(2). **Click** the **Next record** button ▶. The second record is now highlighted, which identifies it as the current record. The second record's value for the **VisitID** field is selected, and the Current Record box displays "**2** of **86**" to indicate that the second record is the current record.

(3). **Click** the **Last record** button ▶|. The last record in the table, record 86, is now the current record.

(4). **Drag** the **scroll box** in the **vertical scroll bar up** to the top of the bar. **Record 86** is still the current record, as indicated in the Current Record box. Dragging the scroll box changes the display of the table datasheet, but does not change the current record.

(5). **Drag** the **scroll box** in the **vertical scroll bar back down** until you can see the end of the table and the current record (**record 86**).

(6). **Click** the **Previous record** button ◀. **Record 85** is now the current record.

(7). **Click** the **First record** button |◀. The first record is now the current record and is visible on the screen.

Earlier you resized the Reason column to its best fit, to ensure all the field values were visible. However, when you resize a column to its best fit, the column expands to fully display only the field values that are visible on the screen at that time. If you move through the complete datasheet and notice that not all of the field values are fully displayed after resizing the column, you need to resize the column again.

(8). **Scroll down** through the records and observe if the field values for the **Reason field** are fully displayed. The **Reason field** values for visit **1595** and visit **1606** are not fully displayed. With these records displayed, place the pointer on the column dividing line to the right of the Reason field name until the pointer changes to the column resizing pointer ✛, and then **Double-Click** the column dividing line. The field values are now fully displayed.

The Visit table now contains all the data about patient visits for Abuja Monrovia Community Health Services. To better understand how to work with this data, Chioma asks you to create simple objects for the other main types of database objects—queries, forms, and reports.

CREATING A SIMPLE QUERY

A **Query** is a question you ask about the data stored in a database. When you create a query, you tell Access which fields you need and what criteria it should use to select the records that will answer your question. Then Access displays only the information you want, so you don't have to navigate through the entire database for the information. In the Visit table, for example, Chioma might create a query to display only those records for visits that occurred in a specific month. Even though a query can display table information in a different way, the information still exists in the table as it was originally entered.

Chioma wants to see a list of all the visit dates and reasons for visits in the Visit table. She doesn't want the list to include all the fields in the table, such as PatientID and WalkIn. To produce this list for Chioma, you'll use the Simple Query Wizard to create a query based on the Visit table.

To start the Simple Query Wizard:

(1). On the **Ribbon**, **Click** the **Create Tab**.

(2). In the **Queries Group**, **Click** the **Query Wizard** button. The New Query dialog box opens.

(3). Make sure **Simple Query Wizard** is **selected**, and then **Click OK**. The first Simple Query Wizard dialog box opens. See **Figure 5-29**.

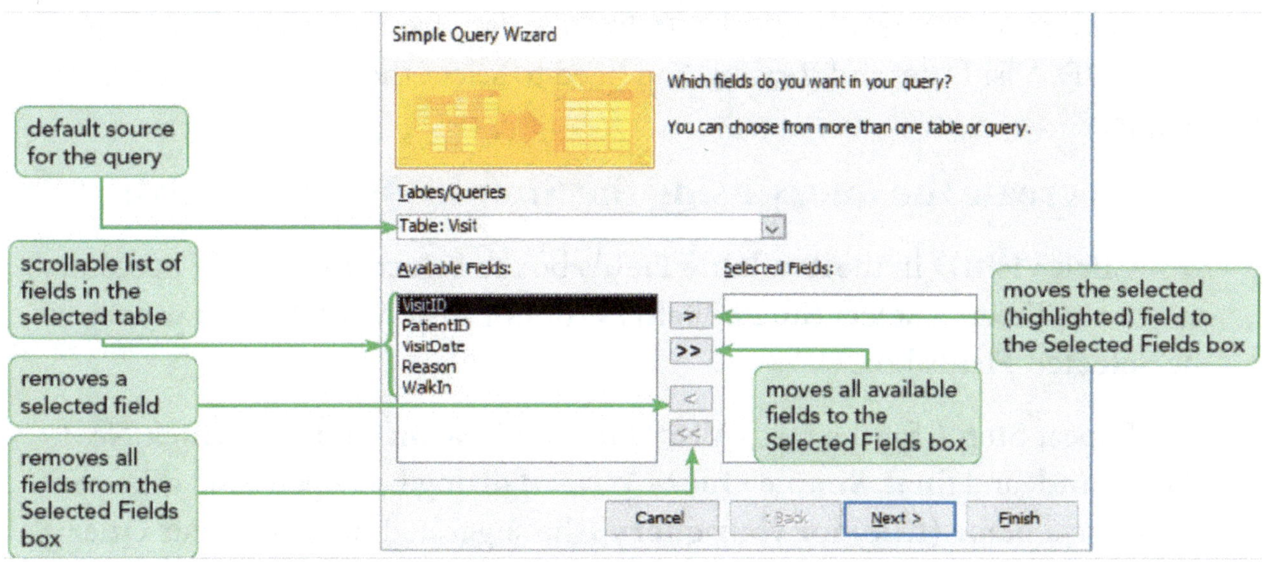

Figure 5-29 First Simple Query Wizard dialog box

Because the Visit table is open in the Abuja Monrovia Database, it is listed in the Tables/Queries box by default. If the database contained more objects, you could click the Tables/Queries arrow and choose another table or a query as the basis for the new query you are creating. In this case you could choose the Billing table; however, the Visit table contains the fields you need. The Available Fields box lists all the fields in the Visit table.

Are You Having Trouble? If the Visit table is not the default source for the query, click the Tables/Queries arrow to choose the Visit table (Table: Visit) from the list.

You need to select fields from the Available Fields box to include them in the query. To select fields one at a time, click a field and then click the Select Single Field [>] button. The selected field moves from the Available Fields box on the left to the Selected Fields box on the right. To select all the fields, click the Select All Fields [>>] button. If you change your mind or make a mistake, you can remove a field by clicking it in the Selected Fields box and then clicking the Remove Single Field [<] button. To remove all fields from the Selected Fields box, click the Remove All Fields [<<] button.

Each Simple Query Wizard dialog box contains buttons that allow you to move to the previous dialog box (Back button), move to the next dialog box (Next button), or cancel the creation process (Cancel button). You can also finish creating the object (Finish button) and accept the wizard's defaults for the remaining options.

Chioma wants her query results list to include data from only the following fields: VisitID, VisitDate, and Reason. You need to select these fields to include them in the query.

To create the query using the Simple Query Wizard:

(1). **Click VisitID** in the **Available Fields** box to **select the field** (if necessary), and then **Click** the **Select Single Field** [>] button. The **VisitID field moves** to the **Selected Fields** box.

(2). Repeat **Step 1** for the fields **VisitDate** and **Reason**, and then **Click Next**. The **second**, and **final**, **Simple Query Wizard** dialog box opens and asks you to choose a **name (title) for your query**. The suggested name is "**Visit Query**" because the query you are creating is based on the **Visit table**. You'll change the suggested name to "**VisitList**."

(3). **Click** at the end of the suggested name, use **BACKSPACE** to **delete** the word "**Query**" and the space, and then **Type List**. Now you can view the query results.

(4). **Click Finish** to complete the query. The query results are displayed in **Datasheet view**, on a new tab named "**VisitList**." A query datasheet is similar to a table datasheet, showing fields in columns and records in rows—but only for those fields and records you want to see, as determined by the query specifications you select.

(5). Place the **pointer** on the **column divider line** to the **right** of the **Reason field** name until the pointer changes to the column **resizing pointer** ✛, and then **Double-Click** the column divider line to resize the **Reason field**. See **Figure 5-30**.

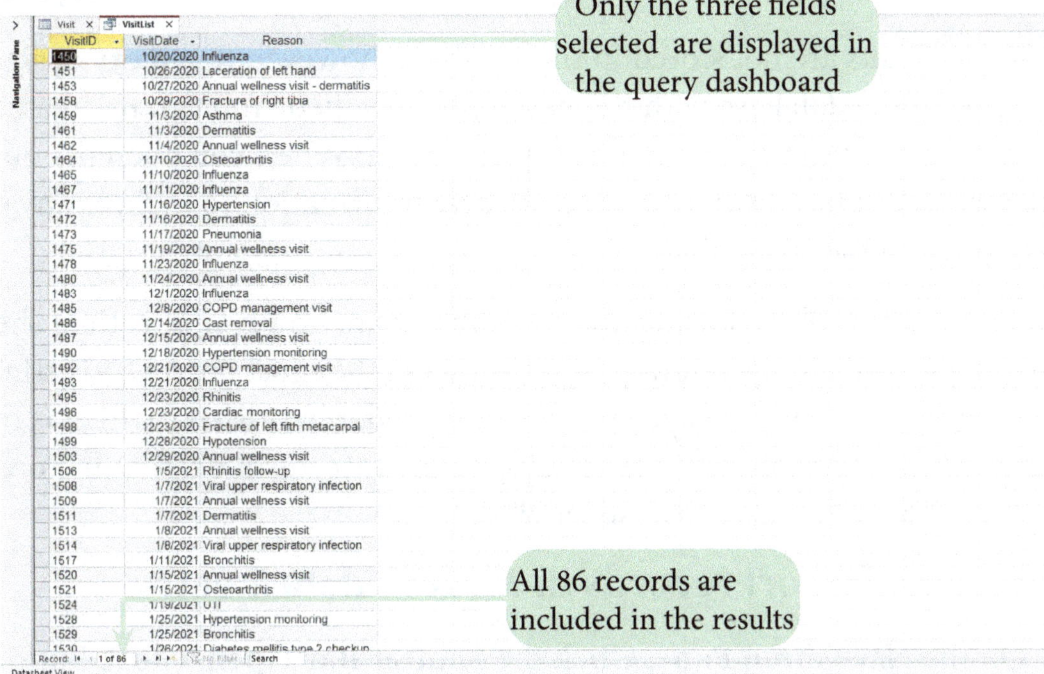

VisitID	VisitDate	Reason
1450	10/20/2020	Influenza
1451	10/26/2020	Laceration of left hand
1453	10/27/2020	Annual wellness visit - dermatitis
1458	10/29/2020	Fracture of right tibia
1459	11/3/2020	Asthma
1461	11/3/2020	Dermatitis
1462	11/4/2020	Annual wellness visit
1464	11/10/2020	Osteoarthritis
1465	11/10/2020	Influenza
1467	11/11/2020	Influenza
1471	11/16/2020	Hypertension
1472	11/16/2020	Dermatitis
1473	11/17/2020	Pneumonia
1475	11/19/2020	Annual wellness visit
1478	11/23/2020	Influenza
1480	11/24/2020	Annual wellness visit
1483	12/1/2020	Influenza
1485	12/8/2020	COPD management visit
1486	12/14/2020	Cast removal
1487	12/15/2020	Annual wellness visit
1490	12/18/2020	Hypertension monitoring
1492	12/21/2020	COPD management visit
1493	12/21/2020	Influenza
1495	12/23/2020	Rhinitis
1496	12/23/2020	Cardiac monitoring
1498	12/23/2020	Fracture of left fifth metacarpal
1499	12/28/2020	Hypotension
1503	12/29/2020	Annual wellness visit
1506	1/5/2021	Rhinitis follow-up
1508	1/7/2021	Viral upper respiratory infection
1509	1/7/2021	Annual wellness visit
1511	1/7/2021	Dermatitis
1513	1/8/2021	Annual wellness visit
1514	1/8/2021	Viral upper respiratory infection
1517	1/11/2021	Bronchitis
1520	1/15/2021	Annual wellness visit
1521	1/15/2021	Osteoarthritis
1524	1/19/2021	UTI
1528	1/25/2021	Hypertension monitoring
1529	1/25/2021	Bronchitis
1530	1/26/2021	Diabetes mellitis type 2 checkup

Figure 5-30 Query results

The VisitList query datasheet displays the three fields in the order you selected them in the Simple Query Wizard, from left to right. The records are listed in order by the primary key field, VisitID. Even though the query datasheet displays only the three fields you chose for the query, the Visit table still includes all the fields for all records.

Navigation buttons are located at the bottom of the window. You navigate a query datasheet in the same way that you navigate a table datasheet.

(6). **Click** the **Last record** button ▶|. The last record in the query datasheet is now the current record.

(7). **Click** the **Previous record** button ◀ . **Record 85** in the query datasheet is now the current record.

(8). **Click** the **First record** button |◀ . The first record is now the **current record**.

(9). **Click** the **Close 'VisitList'** button ✕ on the object tab. A dialog box opens asking if you want to save the changes to the layout of the query. This dialog box opens because you resized the Reason column.

(10). **Click Yes** to **Save** the query layout changes and **close** the **query**.

The query results are not stored in the database; however, the query design is stored as part of the database with the name you specified. You can re-create the query results at any time by opening the query again. When you open the query later, the results displayed will reflect up-to-date information to include any new records entered in the Visit table.

Chioma asks you to display the query results again; however, this time she would like to list the records in descending order showing the most current VisitID first. The records are currently displayed in ascending order by VisitID, which is the primary key for the Visit table. In order to display the records in descending order, you can sort the records in Query Datasheet view.

To sort records in a query datasheet:

(1). On the **Navigation Pane**, **Click** the **Shutter Bar Open/Close** Button » to open it.

(2). In the **Navigation Pane**, **Double-Click VisitList** to open the **VisitList Query** in **Datasheet view**.

(3). On the **Navigation Pane**, **Click** the **Shutter Bar Open/Close Button** « to **close** it.

(4). On the **Ribbon**, **Click** the **Home Tab**. The first record value in the **VisitID** field is highlighted; therefore, **VisitID** is the current field. Also note the data in the first record (VisitID: **1450**; **VisitDate**: **10/26/2020**; and **Reason**: **Influenza**).

(5). In the **Sort & Filter group**, **Click** the **Descending** button. The records are sorted in descending order by the current field (**VisitID**). Because the list of records is now sorted in descending order, the original first record (**VisitID 1450**) should now be the last record.

(6). **Scroll down** the list of records and see that the same data for **VisitID 1450** is now in the last record. Chioma has decided not to keep the data sorted in descending order and wants to return to ascending order.

(7). In the **Sort & Filter Group**, **Click** the **Remove Sort** button. The data returns to its original state in ascending order with **VisitID 1450** (and its corresponding data) listed in the first record.

(8). **Click** the **Close 'VisitList'** button ☒ on the object tab for the **VisitList query**. When asked if you would like to save the changes to the design of the **VisitList query**, **Click No**. The **VisitList query closes**.

Next, Chioma asks you to create a form for the Visit table so the staff at Abuja Monrovia Community Health Services can use the form to enter and work with data in the table easily.

Creating a Simple Form

As noted earlier, you use a **Form** to enter, edit, and view records in a database. Although you can perform these same functions with tables and queries, forms can present data in many customized and useful ways.

Chioma wants a form for the Visit table that shows all the fields for one record at a time, with fields listed one below another in a column. This type of form will make it easier for her staff to focus on all the data for a particular visit. You'll use the Form Wizard to create this form quickly and easily.

To create the form using the Form Wizard

(1). Make sure the **Visit Table** is still open in **Datasheet View**.

Are You Having Trouble? If the Visit table is not open, click the Shutter Bar Open/ Close Button » to open the Navigation Pane. Double-click Visit to open the Visit table in Datasheet view. Click the Shutter Bar Open/Close Button « to close the pane.

(2). On the **Ribbon**, **Click** the **Create Tab** if necessary.

(3). In the **Forms Group**, **Click** the **Form Wizard** button. The first **Form Wizard** dialog box opens. Make sure the **Visit Table** is the default data source for the form.

Are You Having Trouble? If the Visit table is not the default source for the form, click the Tables/Queries arrow to choose the Visit table (Table: Visit) from the list.

The first Form Wizard dialog box is very similar to the first Simple Query Wizard dialog box you used in creating a query.

(4). **Click** the **Select All Fields** button >> to move all the fields to the Selected Fields box.

(5). **Click Next** to display the **second Form Wizard** dialog box, in which you select a layout for the form. See **Figure 5-31**.

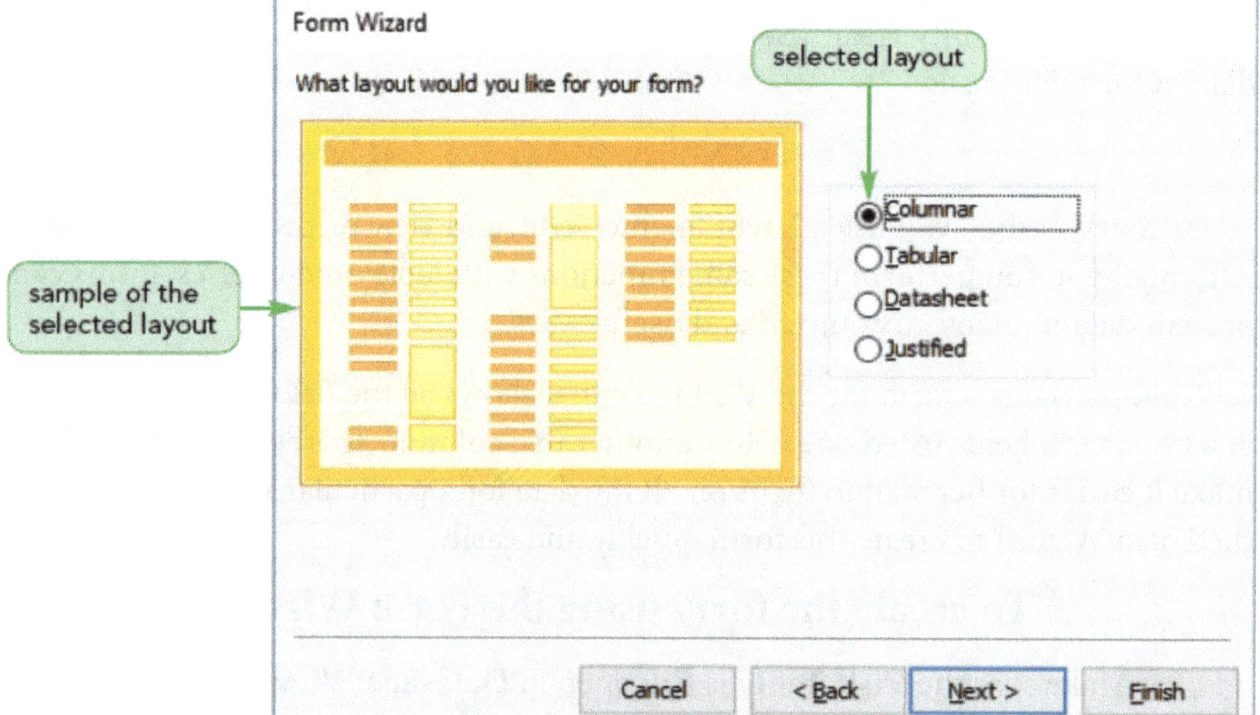

Figure 5-31 Choosing a layout for the form

The layout choices are Columnar, Tabular, Datasheet, and Justified. A sample of the selected layout appears on the left side of the dialog box.

(6). **Click** each **option** button and review the corresponding sample layout.

(7). Because Chioma wants to arrange the form data in a column with each field listed one below another, **Click** the **Columnar** option button (if necessary), and then **Click Next**.

The third and final Form Wizard dialog box shows the Visit table's name as the default name for the form name. "Visit" is also the default title that will appear on the tab for the form.

You'll use "VisitData" as the form name, and because you don't need to change the form's design at this point, you'll display the form.

(8). **Click** to position the insertion point to the right of '**Visit**' in the box, **Type Data**, and then **Click** the **Finish** button.

The completed form opens in Form view, displaying the values for the first record in the Visit table. The Columnar layout places the field captions in labels on the left and the corresponding field values in boxes to the right, which vary in width depending on the size of the field. See **Figure 5-32**.

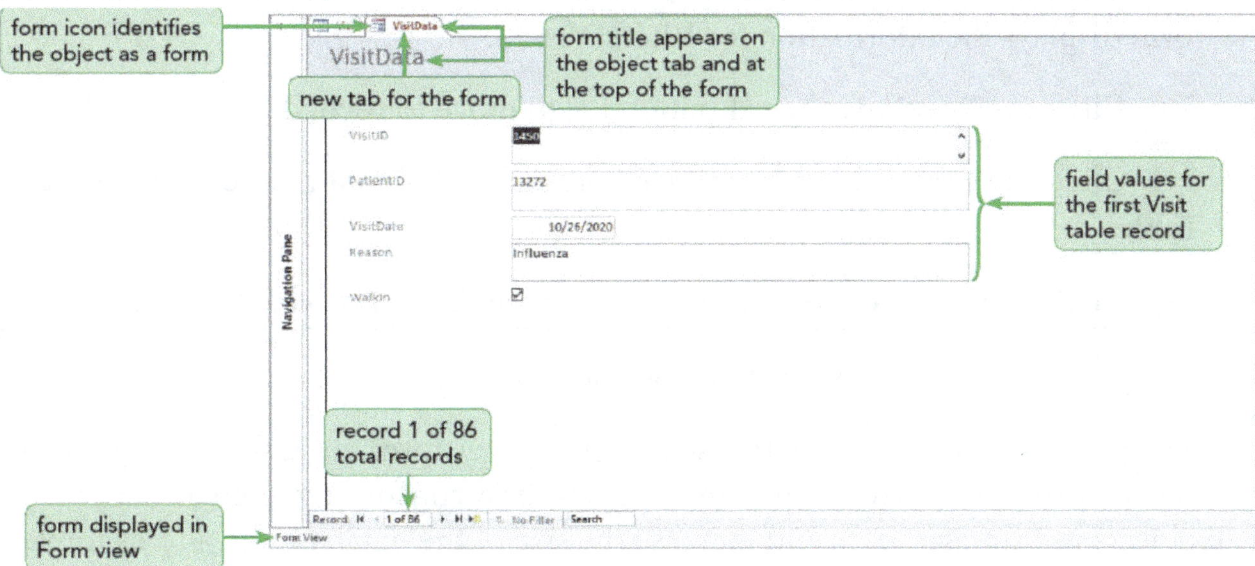

Figure 5-32 VisitData form in Form view

The form displays one record at a time in the Visit table, providing another view of the data that is stored in the table and allowing you to focus on the values for one record. Access displays the field values for the first record in the table and selects the first field value (VisitID), as indicated by the value being highlighted. Each field name appears on a separate line and on the same line as its field value, which appears in a box to the right. Depending on your computer's settings, the field value boxes in your form might be wider or narrower than those shown in the figure. As indicated in the status bar, the form is displayed in **Form View**. Later, you will work with a form in **Layout View**, where you can make design changes to the form while it is displaying data.

To view, enter, and maintain data using a form, you must know how to move from field to field and from record to record. Notice that the form contains navigation buttons, similar to those available in Datasheet view, which you can use to display different records in the form. You'll use these now to navigate the form; then you'll save and close the form.

To navigate, save, and close the form:

(1). **Click** the **Next record** button ▶. The form now displays the values for the second record in the **Visit table**.

(2). **Click** the **Last record** button ▶| to move to the last record in the table. The form displays the information for **VisitID 1623**.

(3). **Click** the **Previous record** button ◀ to move to **record 85**.

(4). **Click** the **First record** button |◀ to return to the first record in the **Visit Table**.

(5). **Click** the **Close 'VisitData'** button ✕ on the object tab to **close** the form.

SMARTSKILLS Saving Database Objects

In general, it is best to save a database object—query, form, or report—only if you anticipate using the object frequently or if it is time-consuming to create, because all objects use storage space and increase the size of the database file. For example, you most likely would not save a form you created with the Form tool because you can re-create it easily with one click. (However, for the purposes of this text, you usually need to save the objects you create.)

Chioma would like to see the information in the Visit table presented in a more readable and professional format. You'll help Chioma by creating a report.

CREATING A SIMPLE REPORT

A **Report** is a formatted printout (or screen display) of the contents of one or more tables or queries. You'll use the Report Wizard to guide you through producing a report based on the Visit table for Chioma. The Report Wizard creates a report based on the selected table or query.

To create the report using the Report Wizard:

(1). On the **Ribbon**, **Click** the **Create Tab**.

(2). In the **Reports Group**, **Click** the **Report Wizard** button. The first Report Wizard dialog box opens. Make sure the **Visit table** is the default data source for the report.

Are You Having Trouble? Trouble? If the Visit table is not the default source for the report, click the Tables/Queries arrow to choose the Visit table (Table: Visit) from the list.

The first Report Wizard dialog box is very similar to the first Simple Query Wizard dialog box you used in creating a query, and to the first Form Wizard dialog box you used in creating a form.

You select fields in the order you want them to appear on the report. Chioma wants to include only the VisitID, PatientID, and Reason fields (in that order) on the report.

(3). **Click VisitID** in the **Available Fields** box (if necessary), and then **Click** the **Select Single Field** button [>] to move the field to the **Selected Fields** box.

(4). **Repeat step 3** to add the **PatientID** and **Reason** fields to the **Selected Fields** box. The **VisitID**, **PatientID**, and **Reason** fields (in that order) are listed in the **Selected Fields** box to add to the report. See **Figure 5-33** on the next page.

(5). **Click Next** to open the second **Report Wizard** dialog box, which asks whether you want to add **Grouping Levels** to your report. This concept will be discussed later; Chioma's report does not have any **Grouping Levels**.

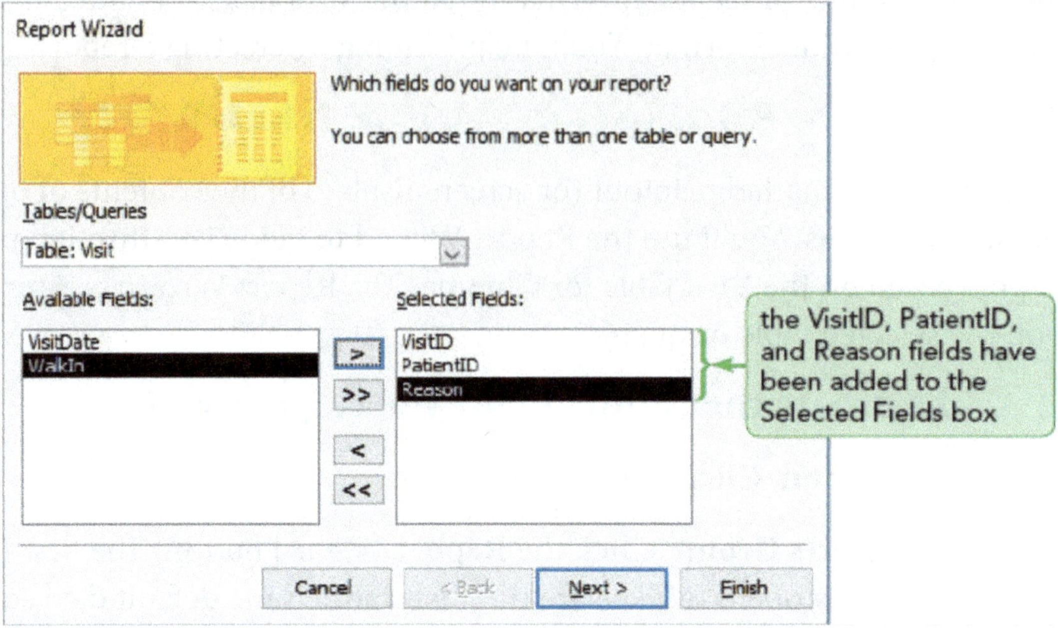

Figure 5-33 First Report Wizard dialog box

(6). **Click Next** to proceed to the **third Report Wizard dialog** box, which asks whether to sort records in a certain order by a particular field on the report. Chioma wants to **list** the **records** by the **VisitID** field in **ascending order**. Access allows up to **four levels** of **sorting**, although Chioma wants only one.

(7). **Click** the **arrow** in the **first sort option box**, and then **Click VisitID**. See **Figure 1-34**. The default option for sorting on the **VisitID field** is **ascending**.

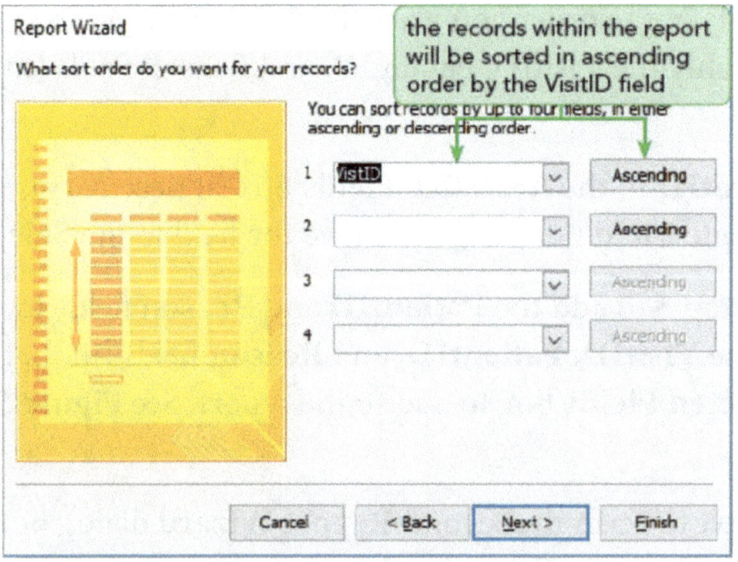

Figure 5-34 Third Report Wizard dialog box

(8). **Click Next** to proceed to the **fourth Report Wizard dialog** box, which asks you to select the layout for the report. You can **click** a **Layout option** to display an example of the layout.

(9). **Click** the **Tabular option** button (if necessary). Later you can select other options for a report; however, this report uses the current default options.

(10). **Click Next** to proceed to the **final Report Wizard dialog** box, in which you name the report. Chioma wants to name the report "**VisitDetails**."

(11). **Click** to **position** the **insertion point** to the **right** of '**Visit**' in the box, and then **Type Details**. **Click Finish** to preview the report. See **Figure 5-35**.

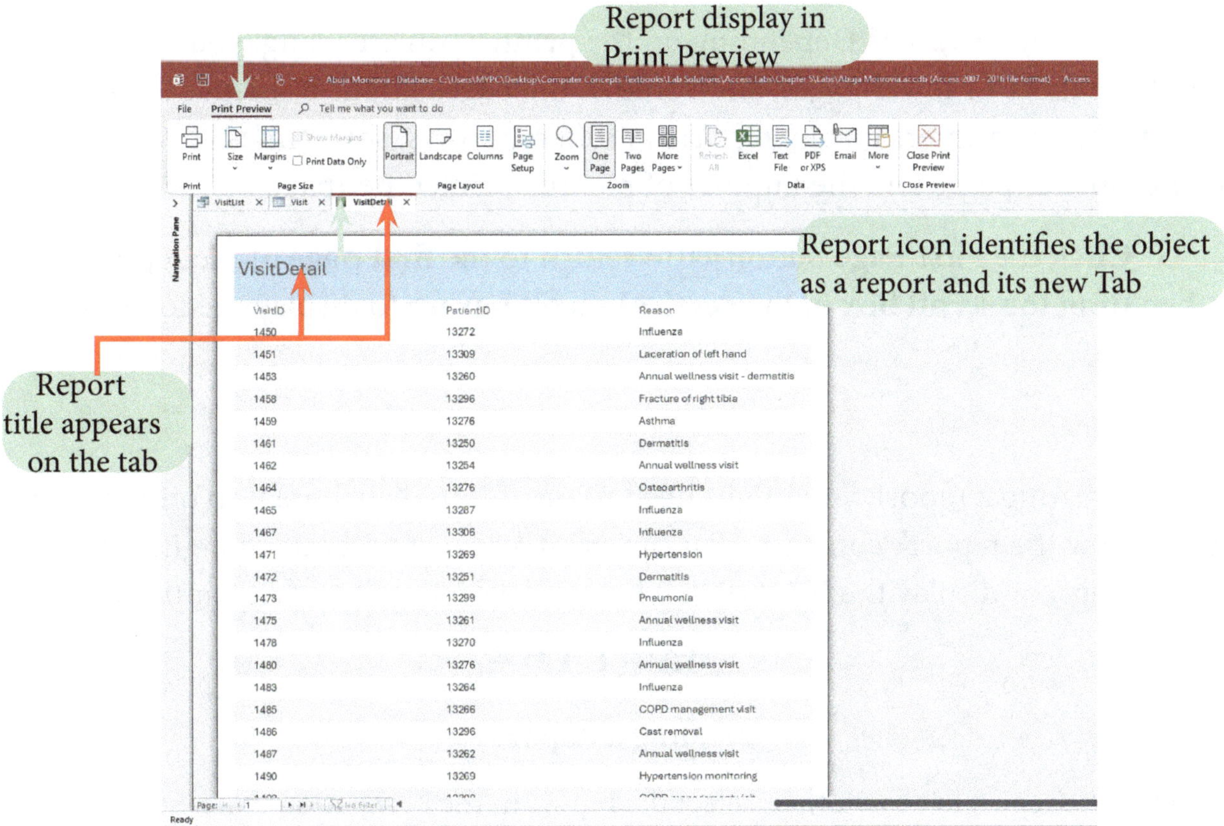

Figure 5-35 Report in Print Preview

The report shows each field in a column, with the field values for each record in a row, similar to a table or query datasheet. However, a report offers a more visually appealing format for the data. The report is currently shown in Print Preview. Print Preview shows exactly how the report will look when printed. Print Preview also provides page navigation buttons at the bottom of the window, similar to the navigation buttons you've used to move through records in a table, query, and form.

To navigate the report in Print Preview:

(1). **Click** the **Next Page** button ▶. The second page of the report is displayed in Print Preview.

(2). **Click** the **Last Page** button ▶| to move to the last page of the report.

(3). **Drag** the **scroll** box in the **vertical scroll bar down** until the bottom of the report page is displayed. The current date is displayed at the bottom left of the page. The notation "Page 3 of 3" appears at the bottom right of the page, indicating that you are on page 3 out of a total of 3 pages in the report.

Are You Having Trouble? Depending on the printer you are using, your report might have more or fewer pages, and some of the pages might be blank. If so, don't worry. Different printers format reports in different ways, sometimes affecting the total number of pages and the number of records printed per page.

(4). **Click** the First **Page** button |◀ to return to the **first page** of the report, and then **drag** the **scroll box** in the **vertical scroll bar up** to display the top of the report.

PRINTING A REPORT

After creating a report, you might need to print it to distribute it to others who need to view the report's contents. You can print a report without changing any print settings, or display the Print dialog box and select options for printing.

Smart Tips

Printing a Report

- Open the report in any view, or select the report in the Navigation Pane.
- Click the File tab to display Backstage view, click Print, and then click Quick Print to print the report with the default print settings.

or

- Open the report in any view, or select the report in the Navigation Pane.
- Click the File tab, click Print, and then click Print; or, if the report is displayed in Print Preview, click the Print button in the Print group on the Print Preview tab.

Chioma asks you to print the entire report with the default settings, so you'll use the Quick Print option in Backstage view.

Note: To complete the following steps, your computer must be connected to a printer. Check with your instructor first to see if you should print the report.

To print the report and then close it:

(1). On the **Ribbon**, **Click** the **File Tab** to open **Backstage view**.

(2). In the **Navigation Bar**, **Click Print** to display the Print screen, and then **Click Quick Print**. The report prints with the default print settings, and you return to the report in **Print Preview**.

Are You Having Trouble? If your report did not print, make sure that your computer is connected to a printer, and that the printer is turned on and ready to print. Then repeat **Steps 1** and **2**.

(3). **Click** the **Close 'VisitDetails'** button [×] on the object tab to **close** the **report**.

(4). **Click** the **Close 'Visit'** button [×] on the object tab to **close** the **Visit table**.

Are You Having Trouble? If you are asked to save changes to the layout of the table, click Yes.

You can also use the Print dialog box to print other database objects, such as table and query datasheets. Most often, these objects are used for viewing and entering data, and reports are used for printing the data in a database.

VIEWING OBJECTS IN THE NAVIGATION PANE

The Abuja Monrovia Database now contains five objects—the Billing table, the Visit table, the VisitList query, the VisitData form, and the VisitDetails report. When you work with the database file—such as closing it, opening it, or distributing it to others—the file includes all the objects you created and saved in the database. You can view and work with these objects in the Navigation Pane.

To view the objects in the Abuja Monrovia database:

(1). On the **Navigation Pane**, **Click** the **Shutter Bar Open/Close Button** [»] to open the pane. See **Figure 5-36**.

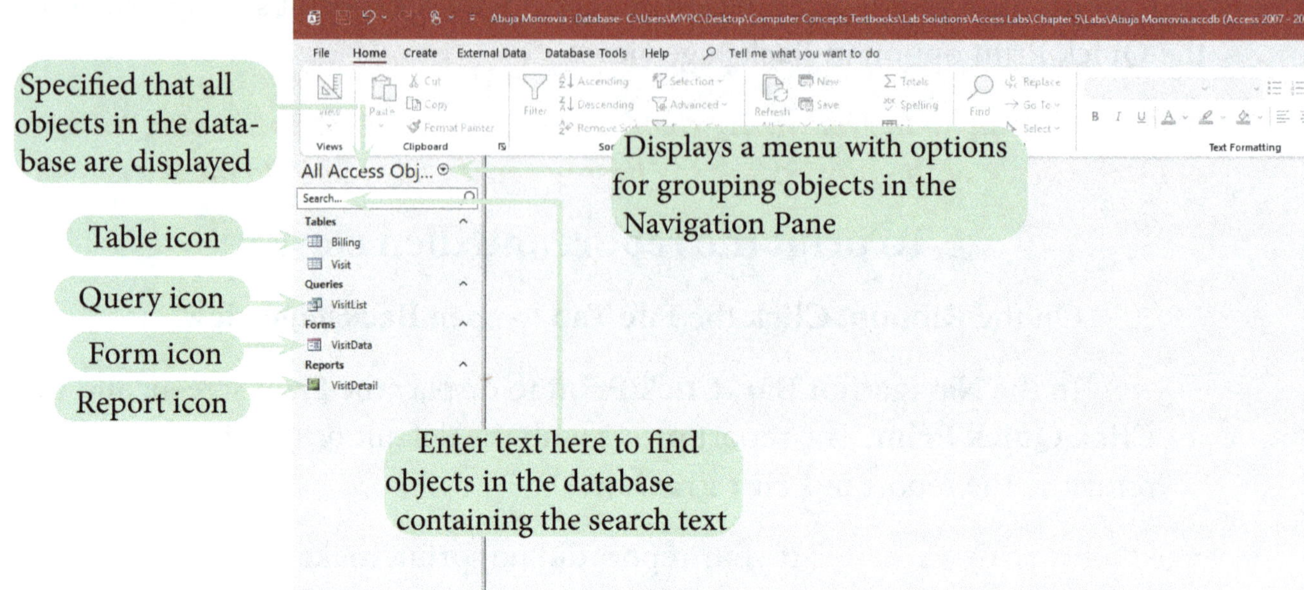

Figure 5-36 Abuja Monrovia database objects displayed in the Navigation Pane

Figure 5-36 above shows the Abuja Monrovia Database with no open objects. The Navigation Pane is open. "All Access Objects" at the top of the Navigation Pane, which specifies that all objects in the database are displayed. "All Access Objects" can be clicked to display a menu with options grouping objects in the Navig. Pane.

The Navigation Pane currently displays the default category, **All Access Objects**, which lists all the database objects in the pane. Each object type (Tables, Queries, Forms, and Reports) appears in its own group. Each database object (the Billing table, the Visit table, the VisitList query, the VisitData form, and the VisitDetails report) has a unique icon to its left to indicate the type of object. This makes it easy for you to identify the objects and choose which one you want to open and work with.

The arrow on the All Access Objects bar displays a menu with options for various ways to group and display objects in the Navigation Pane. The Search box enables you to enter text for Access to find; for example, you could search for all objects that contain the word "Visit" in their names. Note that Access searches for objects only in the categories and groups currently displayed in the Navigation Pane.

As you continue to build the Abuja Monrovia database and add more objects to it in later modules, you'll use the options in the Navigation Pane to manage those objects.

USING MICROSOFT ACCESS HELP

Access includes a **Help** system you can use to search for information about specific program features. You start Help by clicking the Microsoft Access Help tab on the ribbon, or by **Pressing F1**.

You'll use Help now to learn more about the Navigation Pane.

To search for information about the Navigation Pane in Help:

(1). On the **Ribbon, Click** the **Help Tab**. Multiple buttons are displayed, with the first being the Help button. **Click** the **Help** button. The **Access Help Window opens**.

(2). **Click** in the **Search** box (if necessary), **Type Navigation Pane,** and then **Press ENTER**. The Access Help window displays a list of topics related to the Navigation Pane.

(3). **Click** the topic **Show or hide the Navigation Pane in Access**. The Access Help window displays the article you selected. See **Figure 5-37**.

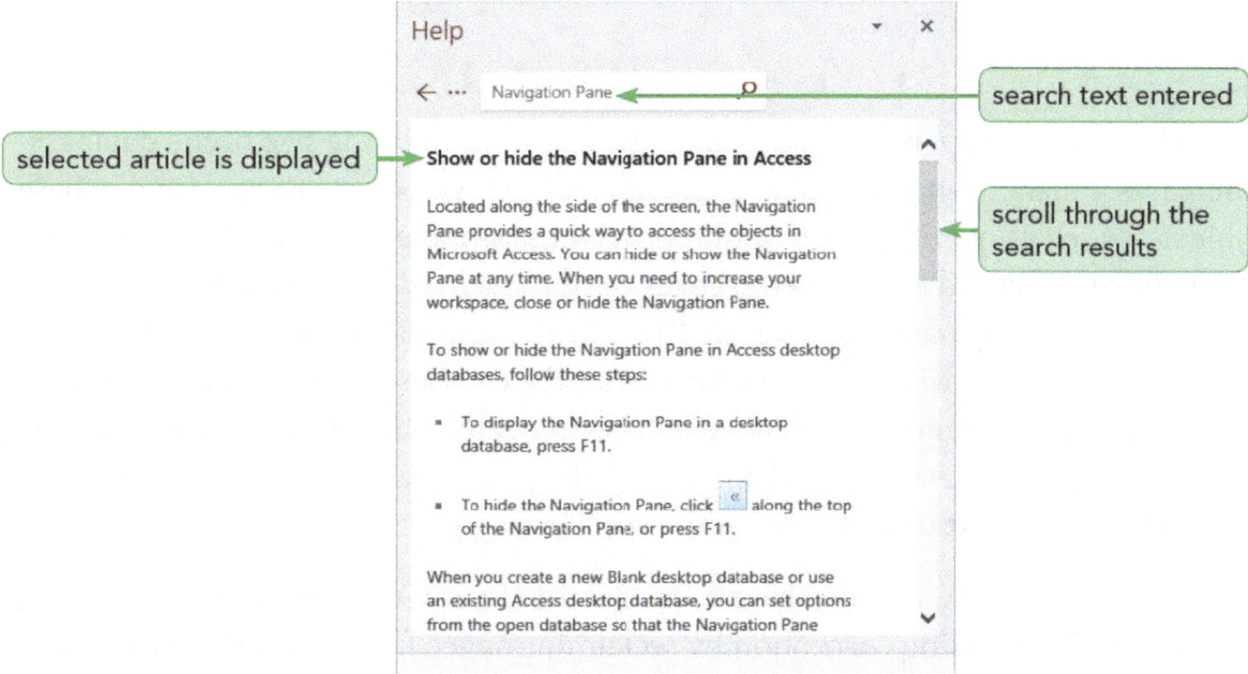

Figure 5-37 Article displayed in the Access Help window

Are You Having Trouble? If the article on managing database objects is not listed in your Help window, choose another article related to the Navigation Pane to read. Your Help window may also look different from the figure.

(4). **Scroll** through the article to read detailed information about working with the **Navigation Pane**.

(5). When finished, **Click** the **Close** button ☒ on the **Access Help window** to close it.

The Access Help system is an important reference tool for you to use if you need additional information about databases in general, details about specific Access features, or support with problems you might encounter.

MANAGING A DATABASE

One of the main tasks involved in working with database software is managing your databases and the data they contain. Some of the activities involved in database management include compacting and repairing a database and backing up and restoring a database. By managing your databases, you can ensure that they operate in the most efficient way, that the data they contain is secure, and that you can work with the data effectively.

COMPACTING AND REPAIRING A DATABASE

Whenever you open an Access database and work in it, the size of the database increases. Further, when you delete records or when you delete or replace database objects—such as queries, forms, and reports—the storage space that had been occupied by the deleted or replaced records or objects does not automatically become available for other records or objects. To make the space available, and to increase the speed of data retrieval, you must compact the database. Compacting a database rearranges the data and objects in a database to decrease its file size, thereby making more storage space available and enhancing the performance of the database. **Figure 5-38** on the next page illustrates the compacting process.

Compacting an Access database is crucial for improving performance, reducing file size, and preventing corruption by reclaiming space from deleted data, defragmenting the file, and fixing structural issues, making queries faster and the database more stable. It essentially rebuilds the database into a new, clean file, eliminating wasted space and optimizing its structure for efficient operation.

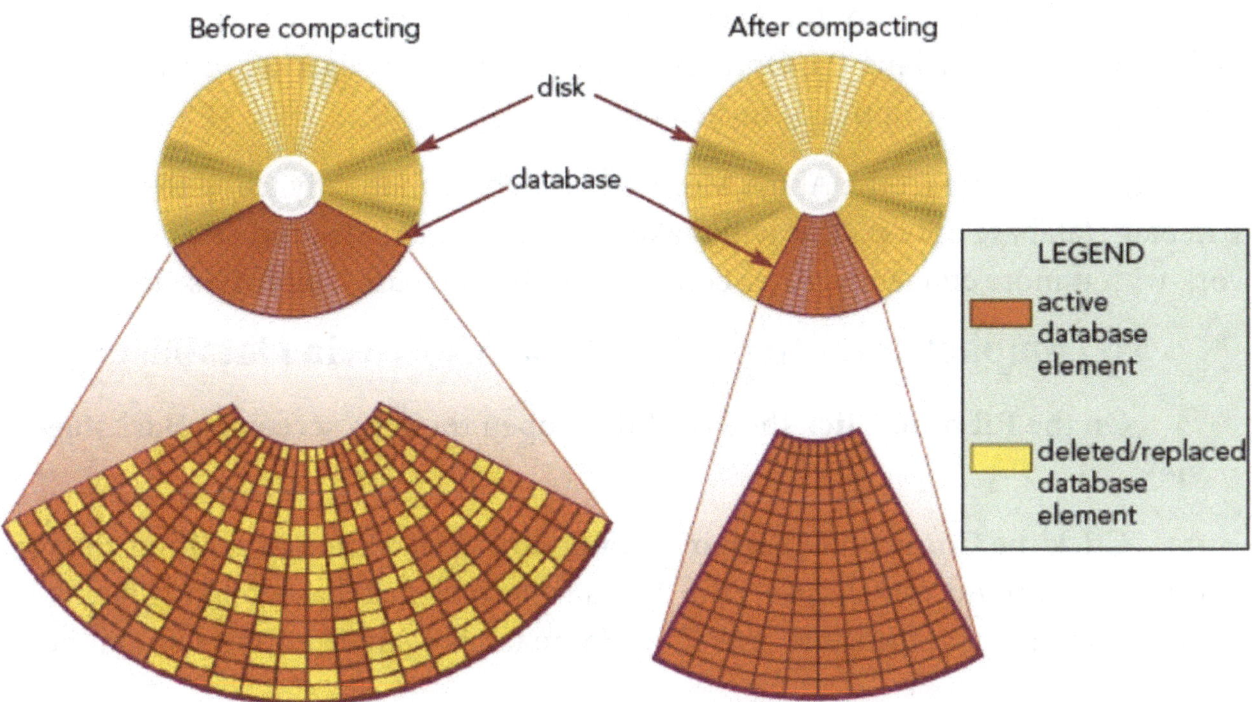

Figure 5-38 Compacting a database

When you compact a database, Access repairs the database at the same time, if necessary. In some cases, Access detects that a database is damaged when you try to open it and gives you the option to compact and repair it at that time. For example, the data in your database might become damaged, or corrupted, if you exit the Access program suddenly by turning off your computer. If you think your database might be damaged because it is behaving unpredictably, you can use the Compact & Repair Database option to fix it.

Smart Tips

Compacting and Repairing a Database

- Make sure the database file you want to compact and repair is open.
- Click the File tab to display the Info screen in Backstage view.
- Click the Compact & Repair Database button.

Access also allows you to set an option to compact and repair a database file automatically every time you close it. The Compact on Close option is available in the Current Database section of the Access Options dialog box, which you open from Backstage view by clicking the Options command in the navigation bar. By default, the Compact on Close option is turned off.

Next, you'll compact the Abuja Monrovia Database manually using the Compact & Repair Database option. This will make the database smaller and allow you to work with it more efficiently. After compacting the database, you'll close it.

To compact and repair the Abuja Monrovia Database:

(1). On the **Ribbon**, **Click** the **File Tab** to open the **Info screen** in **Backstage view**.

(2). **Click** the **Compact & Repair Database** button. Although nothing changes on the screen, Access compacts the Abuja Monrovia Database, making it smaller, and repairs it at the same time. The Home tab is again the active tab on the ribbon.

(3). **Click** the **File Tab** to return to **Backstage view**, and then **Click Close** in the navigation bar. The **Abuja Monrovia Database closes**.

BACKING UP AND RESTORING A DATABASE

Backing Up a database is the process of making a copy of the database file to protect your database against loss or damage. The Back Up Database command enables you to back up your database file from within the Access program while you are working on your database. To use this option, click the File tab to display the Info screen in Backstage view, click Save As in the navigation bar, click Back Up Database in the Advanced section of the Save Database As pane, and then click the Save As button. In the Save As dialog box that opens, a default filename is provided for the backup copy that consists of the same filename as the database you are backing up (for example, "Abuja Monrovia"), and an underscore character, plus the current date. This file naming system makes it easy for you to keep track of your database backups and when they were created. To restore a backup database file, you copy the backup from the location where it is stored to your hard drive, or whatever device you use to work in Access, and start working with the restored database file.

(You will not actually back up the Abuja Monrovia Database in this module unless directed by your instructor to do so.)

SMARTSKILLS Planning and Performing Database Backups

Experienced database users make it a habit to back up a database before they work with it for the first time, keeping the original data intact. They also make frequent backups while continuing to work with a database; these backups are generally on flash drives, recordable CDs or DVDs, external or network hard drives, or cloud-based storage (such as OneDrive) or Amazon Cloud. Also, it is recommended to store the backup copy in a different location from the original. For example, if the original database is stored on a flash drive, you should not store the backup copy on the same flash drive. If you lose the drive or the drive is damaged, you would lose both the original database and its backup copy.

If the original database file and the backup copy have the same name, restoring the backup copy might replace the original. If you want to save the original file, rename it before you restore the backup copy. To ensure that the restored database has the most current data, you should update the restored database with any changes made to the original between the time you created the backup copy and the time the original database became damaged or lost.

By properly planning for and performing backups, you can avoid losing data and prevent the time-consuming effort required to rebuild a lost or damaged database.

Smart Tips

Decision Making: When to Use Access vs. Excel

Using a spreadsheet application like Microsoft Excel to manage lists or tables of information works well when the data is simple, such as a list of contacts or tasks. As soon as the data becomes complex enough to separate into tables that need to be related, you see the limitations of using a spreadsheet application. The strength of a database application such as Access is in its ability to easily relate one table of information to another. Consider a table of contacts that includes home addresses, with a separate row for each person living at the same address. When an address changes, it's too easy to make a mistake and not update the home address for each person who lives there. To ensure you have the most accurate data at all times, it's important to have only one instance of each piece of data. By creating separate tables that are related and keeping only one instance of each piece of data, you ensure the integrity of the data. Trying to accomplish this in Excel is complex, whereas Access is specifically designed for this functionality.

Another limitation of using Excel instead of Access to manage data has to do with the volume of data. Although a spreadsheet can hold thousands of records, a database can hold millions. A spreadsheet containing thousands of pieces of information is cumbersome to use. Think of large-scale commercial applications such as enrollment at a college or tracking customers for a large company. It's hard to imagine managing such information in an Excel spreadsheet. Instead, you'd use a database. Finally, with an Access database, multiple users can access the information it contains at the same time. Although an Excel spreadsheet can be shared, there can be problems when users try to open and edit the same spreadsheet at the same time.

When you're trying to decide whether to use Excel or Access, ask yourself the following questions.

1. Do you need to store data in separate tables that are related to each other?
2. Do you have a very large amount of data to store?
3. Will more than one person need to access the data at the same time?

If you answer "yes" to any of these questions, an Access database is most likely the appropriate application to use.

Quick Review

1. To copy the records from a table in one database to another table in a different database, the two tables must have the same ____.
2. A(n) is a question you ask about the data stored in a database.
3. The quickest way to create a form is to use the ____.
4. Which view enables you to see the total number of pages in a report and navigate through the report pages?
5. In the Navigation Pane, each database object has a unique ____ to its left that identifies the object's type.
6. ____ a database rearranges the data and objects in a database to decrease its file size and enhance the speed and performance of the database.
7. a database is the process of making a copy of the database file to protect the database against loss or damage.

CHAPTER REVIEW

Lab Project 1

CHAPTER REVIEW LABS

This is the Data File needed for the Review Assignments: Company.accdb.

For Abuja Monrovia Community Health Services, Chioma asks you to create a new database to contain information about the vendors that the clinic works with to obtain medical supplies and equipment, and the vendors who service and maintain the equipment. Complete the following steps:

(1). **Create** a **new**, **blank database named Vendor** and **save** it in the folder where you are storing your files, as specified by your instructor.

(2). In **Datasheet view**, begin creating a table. **Rename** the **default ID primary key** field to **SupplierID**. Change the **data type** of the **SupplierID** field to **Short Text**.

(3). Add the following **10 fields** to the new table in the order shown; all of them are **Short Text** fields except **InitialContact**, which is a **Date/Time** field: **Company**, **Category**, **Address**, **City**, **Country**, **Mailing Code**, **Phone**, **ContactFirst**, **ContactLast**, and **InitialContact**. **Resize** the **columns** as necessary so that the complete field names are displayed.

(4). **Save** the **table** as **Supplier** and **close** the **table**.

(5). In **Design view**, begin creating a second table containing the following three **Short Text** fields in the order shown: **ProductID**, **SupplierID**, and **ProductName**.

(6). Make **ProductID** the **primary key** and use Primary key as its description. Use **Foreign key** as the description for the **SupplierID** field.

(7). Add a field called **Price** to the table, which is of the **Currency data type**.

(8). Add the following two fields to the table in the order shown: **TempControl** and **Sterile**. Both are of the **Yes/No data type**.

(9). **Add** the final field called **Units** to the table, which is of the **Number data**

CHAPTER REVIEW

Lab Project 1

CHAPTER REVIEW LABS

(10). **Save** the **table** as **Product** and **close** the **table.**

(11). Use the **Navigation Pane** to open the **Supplier table.**

(12). **Enter** the records shown in **Figure 5-39** into the Supplier table. For the first record, **enter** your **first name** in the **ContactFirst** field and your **last name** in the **ContactLast** field.

SupplierID	Company	Category	Address	City	Coun-try	Mailing Code	Phone	Contact First	Contact Last	Initial tact
ABC123	ABC Pharma-ceuticals	Supplies	234 Hope Street	La-gos	Nig	101001	840-555-8125	Student First	Student Last	9/22/2
HAR912	Harper Surgical, Supplies	Supplies	9 Clay Street	Mon	Lib	1000	555-737-3872	Betty	Zayzay	10/26/
DUR725	Ducor Medical Services	Equip-ment	16 Mc-Donald Street	Mon	Lib	1000	555-937-5088	Marie	Gaye	12/14/
TEN247	Tenneka Labs, LLC	Service	47 Allen Street	La-gos	Nig	101001	869-555-5392	Thomas	Tenneka	11/30/
BAZ412	Bazarrack Enterpris-es	Supplies	42 Bode T. Street	Abu-jah	Nig	900001	861-555-2201	Bakare	Moham-med	9/3/20

Figure 5-39 Supplier table records

Note: When entering field values that are shown on multiple lines in the figure, do not try to enter the values on multiple lines. The values are shown on multiple lines in the figure for page spacing purposes only.

(13). **Chioma** created a database named **Company.accdb** that contains a Business table with supplier data. The Supplier table you created has the same design as the Business table. **Copy all the records** from the **Business table** in the **Company.accdb** database (located in the **Access** > **Chapter 5 folder** provided with your Data Files) and then **paste** them at the end of the **Supplier Table** in

CHAPTER REVIEW

Lab Project 1

CHAPTER REVIEW LABS

the **Vendor Database**.

(14). **Resize** all datasheet columns to their best fit, and then **save** the **Supplier table**.

(15). **Close** the **Supplier table**, and then use the **Navigation Pane** to reopen it. Note that the records are displayed in primary key order by the values in the **SupplierID field**.

(16). Use the **Simple Query Wizard** to create a query that includes the **Company**, **Category**, **ContactFirst**, **ContactLast**, and **Phone** fields (in that order) from the **Supplier table**. Name the query **SupplierList**, and then **close** the **query**.

(17). Use the **Form Wizard** to create a form for the Supplier table. **Include all fields** from the **Supplier table** on the form and use the **Columnar layout**. **Save** the form as **SupplierInfo**, and then **close** it.

(18). Use the **Report Wizard** to create a report based on the **Supplier table**. Include the **SupplierID**, **Company**, and **Phone** fields on the report (in that order), and **sort** the **report** by the **SupplierID field** in **ascending order**. Use a **Tabular layout**, **save** the **report** as **SupplierDetails**, and then **close** it.

(19). **Close** the **Supplier table**, and then **compact** and **repair** the **Vendor Database**.

(20). **Close** the **Vendor database**.

CHAPTER REVIEW

Lab Project 2

CHAPTER REVIEW LABS

Self Challenge Lab

Data Files needed for this Self Driven Problem: Records.accdb

In this exercise we are going to use a made-up American technical career training school called Great Giraffe. Jeremiah Garver is the operations manager at Great Giraffe, which is located in Denver, Colorado. Great Giraffe offers part-time and full-time courses in areas of study that are in high demand by industries in the area, including cybersecurity, data science, digital marketing, and bookkeeping. Jeremiah wants to use Access to maintain information about the courses offered by Great Giraffe, the students who enroll at the school, and the payment information for students. He needs your help in creating this database. Complete the following steps:

(1). **Create** a **new**, **blank database** named **Career** and **save** it in the folder where you are storing your files, as specified by your instructor.

(2). In **Datasheet view**, begin creating a table. **Rename** the default ID primary key field to **StudentID**. Change the **data type** of the **StudentID** field to **Short Text**.

(3). **Add** the following **eight fields** to the new table in the order shown; all of them are **Short Text** fields: **FirstName**, **LastName**, **Address**, **City**, **State**, **Zip**, **Phone**, and **Email**. **Resize** the **columns** as necessary so that the complete field names are displayed.

(4). **Add** a field called **BirthDate** to the table, which is of the **Date/Time data type**.

(5). **Add** a final field called **Assessment** to the table, which is of the **Yes/No data type**.

(6). **Save** the table as **Student** and **close** the table.

(7). In **Design view**, begin creating a **second table** containing the following **three Short Text** fields in the following order: **SignupID**, **StudentID**, and

CHAPTER REVIEW

Lab Project 2

CHAPTER REVIEW LABS

InstanceID.

(8). **Make SignupID** the **Primary Key** and use **Primary key** as its **description**. **Use Foreign key** as the descriptions for the **StudentID** and **InstanceID** fields.

(9). **Add** the following two fields to the table in the following order: **TotalCost** and **BalanceDue**. Both are of the **Currency data type**.

(10). **Add** the final field called **PaymentPlan** to the table. It is of the **Yes/No data type**.

(11). **Save** the table as **Registration** and **close** the **table**.

(12). Use the **Navigation Pane** to **open** the **Student table**.

(13). **Enter** the records shown in **Figure 1-40** into the **Student table**.

StudentID	First-Name	Last-Name	Address	City	State	Zip	Phone	Email	BirthDate	Assess-ment
ALB7426	Student First	Student Last	378 North River Avenue	Denver	CO	80227	(303) 555-8364	student@ex-ample.com	3/28/1980	Yes
MAR4120	Jennifer	Marshall	185 St Clair Way	Denver	CO	80223	(303) 555-1434	j.marshall75@ example.com	4/7/1967	Yes
WAL5737	Michael	Walker	367 Lawler Avenue	Engle-wood	CO	80110	(303) 555-6369	m.walker61@ example.com	1/17/1971	No
PER4083	Richard	Perry	923 Charles Avenue	Denver	CO	80211	(303) 555-8773	r.perry15@ example.com	2/14/1987	Yes
DRE9559	Angeli-na	Abu	370 Dower Street	Denver	CO	80233	(303) 555-7491	a.abu80@ example.com	8/6/2000	No

Figure 5-40 Student table records

CHAPTER REVIEW

Lab Project 2

CHAPTER REVIEW LABS

(14). Jeremiah created a database named **Records.accdb** that contains a **MoreStudents table** with additional student data. The Student table you created has the same design as the **MoreStudents** table. Copy all the records from the **MoreStudents table** in the **Records.accdb** database (located in the **Access > Chapter folder** provided with your Data Files), and then **paste** them at the **end** of the **Student table** in the **Career database.**

(15). **Resize** all **datasheet columns** to their best fit, and then **save** the **Student table.**

(16). **Close** the **Student table**, and then use the **Navigation Pane** to reopen it. Note that the records are displayed in **primary key** order by the values in the **StudentID field.**

(17). Use the **Simple Query Wizard** to create a **query** that includes the **StudentID, FirstName, LastName,** and **Email** fields (in that order) from the **Student table. Save** the **query** as **StudentData**, and then **close** the **query.**

(18). Use the **Form Wizard** to create a form for the **Student table**. Include only the **StudentID, FirstName, LastName, Phone,** and **Email** fields (in that order) from the **Student table** on the form and use the **Columnar layout. Save** the form as **StudentInfo**, and then **close** it.

(19). Use the **Report Wizard** to create a report based on the **Student table**. Include the **StudentID, FirstName,** and **Email** fields on the report (in that order), and **sort** the **report** by the **StudentID field** in **ascending order**. Use a **Tabular layout, save** the report as **StudentList**, and then **close** it.

(20). **Close** the **Student table**, and then **compact** and **repair** the **Career database.**

(21). **Close** the **Career database.**

Cell Range (or simply a Range)	A group of cells in a rectangular block	125	3
Cell Reference	indicates the column and row in which the cell is located	122	3
Centered	text is positioned evenly between the left and right margins	82	2
Chart Sheet	contains other graphical elements like clip art, but it doesn't contain a grid for entering data values	119	3
Clearing	removes the data from the selected cells, leaving those cells blank but preserving the current worksheet structure	167	3
Clipboard	is a temporary Windows storage area that holds the selections you copy or cut so you can use them later	154, 217	3, 4
Column Selector	selects a field or a group of fields	297	5
Common Field	connects records in the separate tables	283	5
Content Placeholder	is a placeholder designed to contain text or graphic objects	202	4
Contextual Tab	appears only in context—that is, when a particular type of object is selected or is active	89, 197	2, 4
Copy	means you select it and place a duplicate of it on the Clipboard	217	4
COUNT Function	counts only numeric values	163	3
Crop	means to trim away part of a picture	235	5
Current Record box	displays the number of the current record as well as the total number of records in the table	319	5
Cut	remove the text or object from a file and place it on the Clipboard	217	4
Data	is a raw fact, such as text or numbers that has not been processed	5	1
Database or a Relational Database	a collection of related tables	282	5

Chapter 1

1. **Desktop Computer:** https://www.lifewire.com/how-fast-does-your-pc-need-to-be-832310
2. **The IBM PC:** https://www.itpro.com/hardware/361331/the-true-story-behind-the-ibm-personal-computer
3. **Microsoft Officer Applications:** https://www.stylus.co.za/customize-the-ribbon-in-a-microsoft-office-app/
4. **Microsoft Access**: https://www.malavida.com/en/soft/microsoft-access/
5. **Computer network:** iStock.com/adventtr; MilanTomazin/ Shutterstock.com
6. **Windows 11 Home Screen:** https://www.unocero.com/noticias/windows-11-pro-cuenta-microsoft/
7. **Windows 11 On-Screen Keyboard:** https://www.elevenforum.com/t/change-touch-keyboard-layout-in-windows-11.3188/
8. **Clarke Publishing Communications & BPO, Inc.:** Various Microsoft Office screenshots
9. **File Explorer Windows:** Clarke Publishing Communications & BPO, Inc.

Chapter 2

1. **Word Start Menu:** Clarke Publishing Communications & BPO, Inc.
2. **Various Microsoft Office Word Screenshots Figure 2-1 to Figure 2-5:** Clarke Publishing Communications & BPO, Inc.: www.clarkepublish.com
3. **Floppy Disk:** iStockPhoto.com: Iuliia Alekseeva
4. **Various Microsoft Office Word Screenshots Figure 2-7 to Figure 2-50:** Clarke Publishing Communications & BPO, Inc.: www.clarkepublish.com

Chapter 3

1. **Various Microsoft Office Excel Screenshots Figure 3-1 to Figure 3-40:** Clarke Publishing Communications & BPO, Inc.: www.clarkepublish.com

Chapter 4

1. **Dr. Deborah Dolopei:** https://www.instagram.com/p/C1Z0o-YtOdv/
2. **Dr. Shakinghands:** https://www.vecteezy.com/free-photos/doctor-hand-shake

3. **Woman of phone:** https://borgenproject.org/smartphones4good-startup-supplying-affordable-used-phones-to-female-entrepreneurs-in-africa/
4. **Various Microsoft Office PowerPoint Screenshots Figure 4-1 to Figure 4-50:** Clarke Publishing Communications & BPO, Inc.: www.clarkepublish.com

Chapter 5

1. **Various Microsoft Office Access Database Screenshots Figure 5-1 to Figure 5-40:** Clarke Publishing Communications & BPO, Inc.: www.clarkepublish.com

Computer Concepts & Applications

Since its introduction into the market on October 1, 1990, Microsoft Office has undergone significant transformations, evolving from a basic suite of productivity tools to a comprehensive, intelligent ecosystem integrated with advanced artificial intelligence (AI) capabilities like Copilot its next generation. This evolution has been driven by technological advancements, changing user needs, and the increasing integration of AI into everyday applications.

Early Years (1990s - Early 2000s)

Initially, Microsoft Office comprised core applications such as Word, Excel, PowerPoint, and Outlook. During this period, the focus was on improving user interface, compatibility, and functionality. The suite became a standard in both corporate and educational environments, facilitating document creation, data analysis, and communication. AI integration was minimal, primarily limited to basic spell check and grammar tools.

Mid-2000s to 2010s: The Rise of Cloud Computing and AI

With the advent of cloud computing, Microsoft introduced Office 365, enabling real-time collaboration, cloud storage, and seamless updates. AI began to play a more prominent role, with features like predictive text, intelligent grammar suggestions, and data insights powered by machine learning algorithms. Educational institutions adopted these tools for interactive learning, digital assignments, and collaborative projects, transforming traditional teaching methods.

2020s: AI-Driven Personalization and Automation

In this decade, AI became deeply embedded within Microsoft Office applications. Features such as AI-powered design suggestions in PowerPoint, automated data analysis in Excel, and contextual writing assistance in Word became standard. The integration of natural language processing (NLP) allowed users to interact with applications through voice commands and conversational interfaces. Education saw personalized learning experiences, adaptive assessments, and AI tutors supporting diverse student needs.

Predictions for 2030-2050: The Future of Microsoft Office and AI

Looking ahead, the evolution of Microsoft Office is expected to be profoundly influenced by advancements in AI, quantum computing, and augmented reality (AR). Key developments may include:

- **Ubiquitous AI Assistants:** Highly intelligent, context-aware virtual assistants capable of managing complex workflows, providing real-time insights, and offering personalized learning experiences.
- **Autonomous Document Creation:** AI systems that can generate, edit, and optimize documents autonomously based on user goals and preferences, reducing manual effort significantly.
- **Immersive Collaboration:** Integration with AR and virtual reality (VR) to facilitate immersive meetings, training sessions, and collaborative projects, especially in educational settings.
- **Enhanced Accessibility:** AI-driven tools to support users with disabilities, ensuring equitable access to educational and professional resources.
- **Data Privacy and Ethics:** As AI becomes more integrated, emphasis on ethical AI use, data privacy, and security will be paramount, influencing how Office applications handle user data and AI decision-making processes.

By 2050, Microsoft Office is likely to be a highly adaptive, intelligent environment that seamlessly integrates AI into every aspect of productivity and education, fostering innovation, efficiency, and inclusivity across all sectors.

www.ingramcontent.com/pod-product-compliance
Lightning Source LLC
LaVergne TN
LVHW081401110826
845149LV00010B/1629

9798995363910